Handbook of Systems Thinking Methods

Handbook of Systems Thinking Methods

Paul M. Salmon
Neville A. Stanton
Guy H. Walker
Adam Hulme
Natassia Goode
Jason Thompson
Gemma J. M. Read

CRC Press
Taylor & Francis Group
Boca Raton London New York

CRC Press is an imprint of the
Taylor & Francis Group, an informa business

First edition published 2023
by CRC Press
6000 Broken Sound Parkway NW, Suite 300, Boca Raton, FL 33487-2742

and by CRC Press
4 Park Square, Milton Park, Abingdon, Oxon, OX14 4RN

CRC Press is an imprint of Taylor & Francis Group, LLC

ISBN: 978-0-367-22017-4 (hbk)
ISBN: 978-1-032-27238-2 (pbk)
ISBN: 978-0-429-28162-4 (ebk)

DOI: 10.1201/9780429281624

Typeset in Times
by SPi Technologies India Pvt Ltd (Straive)

Contents

Preface

Society faces many current and forthcoming challenges that represent significant threats to human health and wellbeing and, in some cases, even existential threats to humanity. The scale and complexity of these issues is such that there are few scientific theories and methods available to support effective preparedness, management, and mitigation. As a result, the response is too often inadequate resulting in substantial personal, societal, environmental, and economic costs.

Road trauma is one example of a long-standing issue that continues to kill and injure on an unacceptable scale, annually causing over 1.3 million deaths, approximately 50 million injuries, and costing countries around 3% of their gross domestic product (World Health Organisation, 2021). On top of other long-standing issues such as workplace injury, inadequate disaster management, and public health issues such as disease, obesity, and mental health disorders there are an emerging set of global threats for which we are ill-prepared. The ongoing COVID-19 pandemic provides a sobering example; at the time of writing, the pandemic has killed over 6 million people, destroyed entire industries, disrupted social norms, and has cost the global economy in the region of $28 trillion (International Monetary Fund, 2020). Looking into the future, Artificial General Intelligence (AGI) is the forthcoming next generation of Artificial Intelligence (AI) which, if not adequately managed and controlled, could wipe out humanity (Bostrom, 2014; Salmon, 2021). Worryingly, there is little indication to suggest that these pressing global challenges can be resolved or managed with current approaches. As such, new ways of thinking about global challenges and the adoption of alternate theoretical and scientific perspectives is needed.

Systems thinking is a way of thinking about the world that helps us understand the many components and dynamic interactions that create behaviour, and problems, in different contexts. It provides "a way of seeing and talking about reality that helps us better understand and work with systems to influence the quality of our lives" (Kim, 1999). Throughout our research we have applied a set of analysis methods which can be used to apply a systems thinking lens to help understand complex problems. Whilst these methods have their origins in a diverse set of disciplines, and often have differences in their specific theoretical underpinnings, they share the common trait of enabling analysts to describe and understand complex systems in terms of their components, interactions, and emergent properties.

The central premise of this book is that complex issues, such as road trauma and the COVID-19 pandemic, are best understood and managed through the use of what we will call 'systems thinking methods'. Despite their utility, practical guidance on many of the methods can be hard to come by, and novice analysts are often deterred by their apparent complexity. Consequently, the methods are often underused, especially by practitioners working outside of academia. Accordingly, the aim of this book is to provide practical step-by-step guidance for a subset of systems thinking methods that the authors have used extensively in our own work. This guidance covers 12 methods categorised into four broad types: risk assessment, systems analysis and design, accident analysis, and computational modelling methods. The intention is to provide a 'how to' guide for each method, demonstrate how we have applied each method in practice, and encourage researchers, students, and practitioners to learn and apply them in their own work and studies.

Acknowledgements

Many colleagues and collaborators contributed to the case studies presented in this book. We would like to acknowledge specifically the contributions of Lisa Aiken, Melissa Bedinger, Tony Carden, Adrian Cherney, Clare Dallat, Amanda Clacy, Lauren Coventon, Dennis Desmond, Alison Gembarovski, Eryn Grant, David Lacey, Ben Lane, Jennifer Marks, Anjum Naweed, Rod McClure, Scott McLean, Sharon Newnam, Aaron Roberts, Erin Stevens, Nicholas Stevens and Patrick Waterson. Without these contributions it would not have been possible to write this book.

Author Biographies

Paul M. Salmon is a Professor in Human Factors and is the founding Director of the Centre for Human Factors and Sociotechnical Systems at the University of the Sunshine Coast. Paul has over 20 years' experience of applied Human Factors and systems thinking research in areas such as road and rail safety, aviation, defence, sport and outdoor recreation, healthcare, workplace safety, and cybersecurity. His research has focused on understanding and optimising human, team, organisational, and system performance through the application of Human Factors and systems theory and methods. He has co-authored 21 books, over 270 peer reviewed journal articles, and numerous book chapters and conference contributions. Paul's contribution has been recognised through various accolades, including the Chartered Institute for Ergonomics and Human Factor's 2019 William Floyd award and 2008 Presidents Medal, the Human Factors and Ergonomics Society Australia's 2017 Cumming memorial medal, and the International Ergonomics Association's 2018 research impacting practice award.

Neville A. Stanton is a Chartered Psychologist, Chartered Ergonomist and Chartered Engineer. He holds the Chair in Human Factors Engineering in the Faculty of Engineering and the Environment at the University of Southampton in the UK. He has degrees in Occupational Psychology, Applied Psychology and Human Factors Engineering and has worked at the Universities of Aston, Brunel, Cornell and MIT. His research interests include modelling, predicting, analysing and evaluating human performance in systems as well as designing the interfaces and interaction between humans and technology. Professor Stanton has worked on design of automobiles, aircraft, ships and control rooms over the past 30 years, on a variety of automation projects. He has published 45 books and over 350 journal papers on Ergonomics and Human Factors. In 1998 he was presented with the Institution of Electrical Engineers Divisional Premium Award for research into System Safety. The Institute of Ergonomics and Human Factors in the UK awarded him The Otto Edholm Medal in 2001, The President's Medal in 2008 and 2018, The Sir Frederic Bartlett Medal in 2012 and The William Floyd Medal in 2019 for his contributions to basic and applied ergonomics research. The Royal Aeronautical Society awarded him and his colleagues the Hodgson Prize in 2006 for research on design-induced, flight-deck, error published in The Aeronautical Journal. The University of Southampton has awarded him a Doctor of Science in 2014 for his sustained contribution to the development and validation of Human Factors methods.

Guy H. Walker is a Professor within the Centre for Sustainable Road Freight at Heriot-Watt University in Edinburgh. He lectures on transportation engineering and human factors. He is the author/co-author of over one hundred peer reviewed journal articles and 18 books. He has been awarded the Institute for Ergonomics and Human Factors (IEHF) President's Medal for the practical application of Ergonomics theory and Heriot-Watt's Graduate's Prize for inspirational teaching. Dr Walker has a BSc Honours degree in Psychology from the University of Southampton, a PhD in Human Factors from Brunel University, is a Fellow of the Higher Education Academy, a

Fellow of the Chartered Institute of Ergonomics and Human Factors, and a member of the Royal Society of Edinburgh's Young Academy of Scotland. His research has featured in the popular media, from national newspapers, TV and radio, through to an appearance on the Discovery Channel.

Adam Hulme is a Research Fellow at the Centre for Human Factors and Sociotechnical Systems at the University of the Sunshine Coast. He has a bachelor's degree in sport and exercise science, an honours degree in Psychological Metacognition, a master's degree in Public Health, and a PhD in Sports Injury Epidemiology and Systems Human Factors and Ergonomics. Adam's current research is focused on the development and testing of a new integrated safety management toolkit that includes both proactive systems-based risk assessment and retrospective incident analysis methods. He has 47 publications including one edited book, five book chapters, 36 peer reviewed journal articles, and various international conference papers on safety. Adam is becoming increasingly passionate about closing the research-practice gap in human factors and safety science, ensuring that methods to enhance the safety of complex systems are usable and effective in real settings.

Natassia Goode is a Program Manager at WorkSafe Victoria, and Adjunct Fellow within the Centre for Human Factors and Sociotechnical Systems at the University of the Sunshine Coast. Natassia's PhD and honours research in Psychology investigated how people learn about complex systems. Over the last nine years, her work has focused on applying systems thinking to fundamentally change the way that organisations manage safety, and reduce incidents. She has worked with several industries to design more effective incident reporting and investigation systems, including healthcare, construction, and outdoor education. She has written book on designing incident reporting systems, and co-authored over 60 peer reviewed journal articles. Her roles at WorkSafe Victoria have involved providing expert advice on applying systems thinking to improve health and safety enforcement and compliance activities. She now coordinates WorkSafe Victoria's programs in the Healthcare and Social Assistance Industry.

Jason Thompson holds a PhD in Medicine, masters in Clinical Psychology, and a Bachelor of Science with Honours. Dr Thompson's work is focused on the translation of research into practice across the areas of transportation safety, public health, post-injury rehabilitation, and health system design. Since 2014, Dr Thompson has published over 40 articles related to the investigation of injury and transport systems alongside methods machine learning and computational social science. His work has pioneered the use of agent-based models and convolutional neural networks in areas of traditional health system, urban design, and transportation safety research involving vulnerable road users.

Dr Thompson is an Australian Research Council Discovery Early Career Research Fellow. His DECRA project is focused on challenges associated with the introduction of autonomous vehicles on the operation and sustainability of Australia's $5 billion personal injury insurance market. In particular, how this transition will change injury rates and types, the operation of compensation and rehabilitation systems, and how these systems' responses will shape the autonomous vehicle market, itself.

Gemma J. M. Read is a Senior Research Fellow at the Centre for Human Factors and Sociotechnical Systems at the University of the Sunshine Coast. Gemma holds a PhD in Human Factors from Monash University and has undergraduate qualifications in psychology and law. She has worked in the field of transportation human factors since 2006 in both academic and government roles. She has co-authored two books and over 40 journal articles applying Human Factors and systems thinking approaches across areas associated with transport safety, workplace safety, and sport and outdoor recreation.

Section I

Introduction to Systems Thinking

1 Introduction to Systems Thinking in Human Factors and Ergonomics and Safety Science

THE CALL TO ACTION

A convergence of ideas and themes have led to this book. The main proposition is this: research and application are becoming disconnected. The science-base is increasing its focus on complex systems problems, yet the application of systems thinking methods is lagging behind (Holman et al., 2020; Leveson, 2017; Salmon et al., 2017; Woods and Dekker, 2000; Walker et al., 2010, 2017). This book is an attempt to correct this situation. It is about putting cutting-edge systems thinking methods into the hands of those who need to apply them in their practice.

Systems thinking methods are different from deterministic methods. Deterministic methods – broadly speaking – focus on reducing systems down to their component parts. The assumption is that if individual components can be understood, then so can the overall system (Walker et al., 2010). Deterministic methods have driven disciplines such as Human Factors and Ergonomics (HFE) and safety science forward for over half a century. They remain entirely appropriate for deterministic problems. They are not, however, appropriate for *all* classes of problem (e.g. Ackoff, 1973; Sterman, 2000; Lee, 2001, Sharma, 2006). Such problems include near-term, and profound, changes in global work and societal systems, artificial intelligence, new developments in information and communication technology, the fourth industrial revolution, the internet of things, ecological sustainability, climate change adaptation, food and farming systems, and many other global risks and existential threats. Problems such as these are increasingly recognised as 'systems problems' (Dul et al., 2012; Wilson, 2014; Salmon et al., 2017). A growing body of recent research highlights the limitations of applying reductionist methods to tackle them (Walker et al., 2010, 2016; Woods and Dekker, 2000; Salmon et al., 2017).

HFE and safety scientists are today confronted with an uncomfortable and disconcerting truth (Holman et al., 2021). The continued use of reductionist methods to tackle systems problems of great complexity are likely to fail. Worse, their use could

DOI: 10.1201/9780429281624-1

be detrimental to the health and sustainability of those systems they seek to support. An exclusive reliance on deterministic methods could see HFE and safety science fail at a time when they are needed most. Something needs to be done.

Earlier work by the authors in Walker et al. (2010) provides a telling indication of the methodological issues we face. The term 'complexity' is a bellwether word that first occured in the title of a paper in the journal *Ergonomics* in 1958 (Chiles, 1958). Since then, well over 80 papers have featured either the word 'complex' or 'complexity' in their titles, with more than 90% having been published since 1990 (Walker et al., 2010). The key issue is not whether the traditional HFE and safety science domains have become more, or less, complex. It is that the problems 'left over' from the successful application of deterministic approaches are systemic and growing in number. In many domains the limits of deterministic approaches have exposed new classes of problem featuring a prominent, and complex, human dimension (e.g. EASA, 2010; Coury et al., 2010; CAA, 2011). Not just that, but in many sectors safety trends are plateauing as deterministic approaches reach the limits of their effectiveness (Salmon et al., 2017). In some sectors, safety performance is even worsening (Salmon et al., 2019, 2020). The methods that have yielded prior improvements in performance are becoming less effective when applied to the problems that remain: the complex, systemic ones. This, in turn, creates a corresponding pressure to find new methods and ensure they enter the practitioner mainstream (Salmon et al., 2017, 2022). This book is a direct attempt to address this issue.

The aim of this book is to provide the reader with practical guidance on how to apply a sub-set of methods that have their theoretical roots in systems theory, complexity science, and systems thinking. 'Systems thinking methods' are better matched to many extant HFE and safety science challenges and their uptake needs to be accelerated. In this book we will draw from an eclectic body of work to show what can be done, and how to do it. Before then, it is important to introduce the core elements of the systems thinking philosophy on which these methods are based.

A SHORT INTRODUCTION TO SYSTEMS THINKING

Systems thinking represents "a way of seeing and talking about reality that helps us better understand and work with systems to influence the quality of our lives" (Kim, 1999). In this book the term 'systems thinking' is used to describe a philosophy which is applied to help understand and optimise behaviour in work and societal systems. An axiom of systems thinking is that behaviour, safety, and accidents are emergent properties arising from non-linear interactions between multiple components across sociotechnical systems (Rasmussen, 1997; Leveson, 2004). From this point of view, responsibility for behaviour and safety spans all levels of work systems, up to and including regulatory bodies, government, and international organisations. The overall 'system', therefore, becomes the unit of analysis when attempting to understand behaviour.

A system can be viewed as "any group of interacting, interrelated, or interdependent parts that form a complex and unified whole that has a specific purpose"

(Kim, 1999). Meadows (2008) points out that a system must comprise three core components:

- Elements (or parts),
- Interconnections (or interactions), and
- A function or purpose (or goal).

We encounter and engage with many systems as part of our everyday lives. For example, this chapter was written at the University of the Sunshine Coast, in Queensland, Australia. Universities can be conceptualized as systems (see Perrow, 1984; Rouse, 2016). They comprise elements such as academic staff, administrators, students, lecture theatres, tutorial rooms, research laboratories, teaching and research materials, policies, and procedures, to name just a few. The system has purposes relating to the creation of new knowledge, educating society, and generating wealth. Interconnections between these university elements are required to achieve the purposes. There are interactions between academic staff and administrators, between academic staff and students, between research staff, participants, data, and between all staff and the materials required for their work. Whenever there are groups of interacting, interrelated, and interdependent parts that work together towards a specific goal, they can be considered a system. Analysing the components in isolation would not fully describe what a university is or how positive or negative outcomes emerge from the interactions between the components.

Systems are all pervasive. While the University of the Sunshine Coast can be considered a system in its own right, it is also a sub-component, or sub-system, of an even larger national education and research system. Australia's other 43 Universities are also large systems, but at the level of the national 'super-system' they are a sub-system. Likewise, if the boundary is widened still further to include the global educational system, then the other co-author's host institutions (the University of Melbourne, the University of Southampton, Heriot-Watt University, etc.) become sub-systems of their national educational systems, and the national educational systems are sub-systems of the global system. Working in the other direction the system boundary can be re-drawn yet again, this time around the lead author's research centre (the Centre for Human Factors and Sociotechnical Systems). This sub-system is also a system with its own components, interconnections, and core functions. A key task in analysing systems is to draw the boundaries around them and recognise that those boundaries are permeable, relying on a steady flow of resources, information, and other means of interconnection.

Systems thinking is interdisciplinary and has its roots in fields as diverse as biology, communications theory, systems theory, and complexity science. It does not reside in any one discipline. Within HFE and safety science, systems thinking is a core philosophy that informs various models, including sociotechnical systems theory (Trist and Bamforth, 1951), Rasmussen's risk management framework (Rasmussen, 1997), Leveson's Systems Theoretic Accident Model and Processes (STAMP; Leveson, 2004); Dekker's Drift into Failure (DIF) model (Dekker, 2011), and Grant et al.'s systems thinking safety tenets (Grant et al., 2018). From these diverse traditions emerge some core features we can expand upon here.

Open Systems

Systems theory first emerged within the biological and physical sciences. Von Bertalanffy (1950) outlined a set of open-systems principles with application to living organisms. Skyttner (2001) subsequently outlined the properties of general systems theory for open systems, based on the work of Von Bertalanffy, as well as numerous other systems theorists. According to Skyttner, open systems display the following characteristics:

- Interrelationship and interdependence of components and their attributes: the parts of the system are interconnected rather than disparate.
- Holism: the system exhibits emergent properties that cannot be identified from analysing the components; the whole is more than the sum of its parts.
- Goal seeking: the system has a goal or end state.
- Transformation processes: the system transforms inputs into outputs to attain its goal(s).
- Inputs and outputs: inputs are taken from the environment and transformed; outputs are returned to the environment.
- Entropy: systems tend towards disorder or randomness without intervention.
- Regulation: the interrelated components constituting the system must be regulated for goals to be obtained. Regulation can be achieved through control and feedback loops.
- Hierarchy: systems comprise sub-systems nested within one another in a hierarchical structure.
- Differentiation: specialised units performed specialised functions within a system.
- Equifinality: from the same initial conditions, systems have different alternative ways of achieving the same goal.
- Multifinality: from the same initial conditions, systems can obtain different goals and objectives.

These properties provide the basis for *adaptive capacity*. Systems that are able to achieve goals and avoid entropy are also those that can adapt to external environmental conditions (i.e. to changes to inputs) by using alternative means (i.e. having the property of equifinality). Matters such as how regulatory mechanisms within the system operate, and the amount of differentiation available within the system, can affect its capacity to adapt and achieve its goals.

Complexity Theory

The defining feature of systems thinking is, arguably, complexity. Complexity science is concerned with attempting to understand and respond to problems that are dynamic and unpredictable, multi-dimensional, and comprise various interrelated components. Fundamental to complexity science, and thus systems thinking, is a focus on the interactions among elements within the 'complex system', rather than on the role and contribution of those elements in isolation (Ottino,

2003; Batty and Torrens 2005; Senge 2006). Complexity is difficult to define (Cilliers, 1998). Various authors have outlined characteristics exhibited by complex systems (e.g. Von Bertalanffy, 1968; Cilliers, 1998; Skyttner, 2005; Holland, 2014). According to Luke and Stamatakis (2012), complex systems exhibit the following properties:

> …they are made up of a large number of heterogenous elements; these elements interact with each other; the interactions produce an emergent effect that is different from the effects of the individual elements alone; and, this effect persists over time and adapts to changing circumstances.
>
> (p. 2)

Cilliers (1998) outlined the following characteristics of complex systems:

- *Complex systems comprise multiple components.* Complex systems comprise many components that interact dynamically with one another. The existence of many components is necessary, although this is not sufficient to denote complexity – dynamic interactions between components are also required (Cilliers, 1998).
- *Interactions between components are multiple, rich, and non-linear.* Interactions between components are abundant and can be non-linear in nature, meaning that there is asymmetry between input and output, and that small events can produce large outcomes and vice versa (Dekker, 2011). Emergent properties arising from interactions mean that "the action of the whole is more than the sum of its parts" (Holland, 2014, p. 2). Holland (2014) explains emergence by discussing the 'wetness' of water. Wetness is not something that can be assigned to individual water molecules, rather it is an emergent property arising from the interaction of water molecules.
- *Interactions are short-range in nature.* Information received by components mainly comes from neighbouring components and long-range interactions are limited. However, as there are many interactions between components, it is possible to influence non-neighbouring components through just a few interactions. For example, most workers are not likely to interact with the CEO of their employing organisation; however, they interact with their supervisors who in turn interact with managers and senior managers who in turn interact with the CEO.
- *There are recurrent loops in the interactions.* The effect of an activity can feed back onto itself either directly or through other components. These feedback loops can be positive (reinforcing) or negative (balancing), and both are necessary.
- *Complex systems are open systems.* Complex systems are open systems meaning that it is difficult to define their boundaries and that they interact with their environment. As a result of these interactions complex systems have an influence on their environment and are influenced by their environment in return (Cilliers, 1998).

- *Components are ignorant of the system and its behaviour.* Components within the system often only have access to local information and hence do not fully comprehend the behaviour of the overall system, nor the effects of their actions on the behaviour of the overall system.
- *Complex systems are dynamic and do not operate in a state of equilibrium.* Systems must have constant inputs to maintain functioning. Without constant inputs, complex systems are unable to function (Cilliers, 1998).
- *Complex systems have a history and path dependence.* Within complex systems there is a dependence on initial conditions whereby past behaviour is co-responsible for present behaviour. This means that decisions and actions made previously (even many years previously) influence the here and now (Cilliers, 1998).

Evolution of Systems Thinking

The field of complexity science is extensive, involving multiple traditions, disciplines, methods, techniques, and analytical tools. For instance, Castellani's (2018) 'Map of the complexity sciences' (Figure 1.1) spans several decades and visualises the historical progression of five major intellectual traditions:

- Dynamical systems theory,
- Systems science,
- Complex systems theory,
- Cybernetics, and
- Artificial intelligence.

These traditions share several philosophical and theoretical similarities, as well as practical commonalities in terms of the approaches used to study and examine complex phenomena. Figure 1.1 indicates that there is no single, unified understanding of what exactly complexity is from an operational standpoint. When invoking the term systems thinking it can refer to distinct, as well as shared features embedded in Figure 1.1. Researchers and practitioners have a multitude of highly capable scientific approaches and modelling techniques at their disposal to understand complexity and complex problems so long as a suitable justification for their selection is offered. This is key. This book aims to help researchers and practitioners make those conceptual groundings, and analytical decisions, transparent.

THE SYSTEMS THINKING TENETS

The work of Grant et al. (2018) brings some much-needed order to the current state of systems thinking in HFE and safety science. It does so by synthesising a set of key tenets from a range of systems thinking models (Rasmussen, 1997; Leveson, 2004; Dekker, 2011). This has led to the specification of a set of work system features which interact to create both safe and unsafe performance. Labelled the systems thinking tenets, they encapsulate much of the current state of thinking in this area and include 15 characteristics of work systems that are believed to create both safe and unsafe performance (Grant et al., 2018, see Table 1.1). These tenets graphically illustrate the uncomfortable truth we are now living with in the HFE

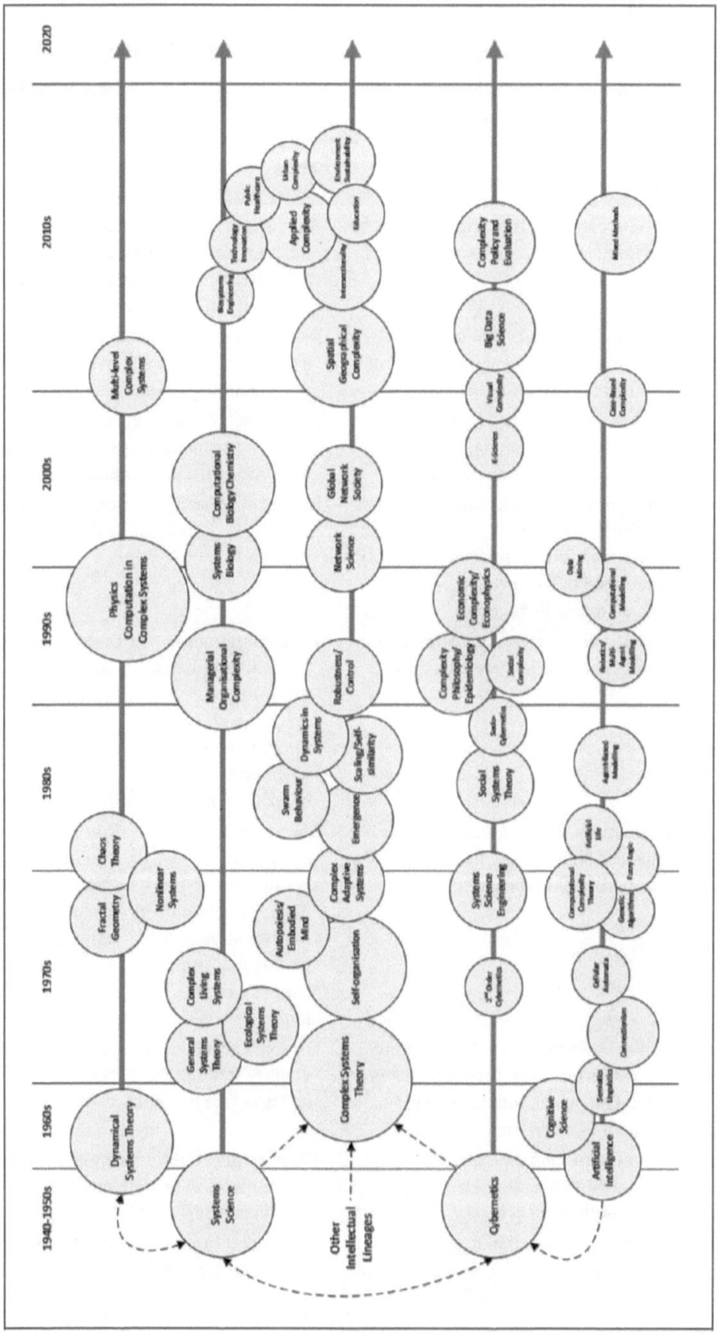

FIGURE 1.1 Historical evolution of complexity science. (Adapted from Castellani, 2018).

TABLE 1.1
Grant et al. (2018) Systems Thinking Tenets

Tenet	Definition	Unsafe System Description
Vertical integration	Interactions between elements within and across levels of the system hierarchy.	Decisions and actions do not filter through the system and impact behaviour. Information regarding the current status of the system across levels is not used when making decisions.
Constraints	System elements that impose limits on, or influence, other system elements.	A constraint that has failed to perform its function and/or restrict an appropriate response, behaviour, or the desired variability in performance.
Normal performance	Routine behaviours that are typically performed within a system, regardless of formal rules and procedures.	Routine and expected behaviours that played a contributory role (i.e. were a 'normal' part of the aetiological mechanism).
Performance variability	System elements vary their behaviour in response to changing conditions in the system and its environment.	Behaviours are adjusted to cope with changing circumstances; however, the outcome of the adjustment is not desirable.
Emergence	Outcomes that result from interactions between elements in the system that cannot be fully explained or reliably predicted in advance by examining the elements in isolation.	Emergent behaviours or outcomes that are unsafe or undermine the goals of the system.
Tight and loose coupling	The degree of interdependence that exists between system elements.	Tight coupling: Cascading failures that propagate quickly and widely through the system when one element breaks down. Loose coupling: Loss of control regulating behaviours. Too much independence between elements.
Linear and non-linear interactions	Interactions that produce effects that are either proportional or not proportional to the causes.	Linear: Interactions are readily predictable, predefined, and fixed with no allowances for adaptation when alternatives are required. Non-linear: Seemingly inconsequential events have small or large effects which cannot necessarily be predicted.
Feedback loops	Self-reinforcing and self-correcting forms of feedback between system elements which influence the system's behaviour.	Feedback mechanisms are not controlled and amplify through the system, increasing risk and accident potential.
Modularity	Sub-systems and elements that are designed to operate independently of one another.	The system is tightly integrated and complex, substitutions cannot be made (e.g. no contingencies).
Sensitive dependence on initial conditions	Characteristics of the originally designed system that influence system behaviour at a later point in time.	Initial conditions and their influence on the system create unsafe behaviours.

(*Continued*)

TABLE 1.1 (Continued)

Tenet	Definition	Unsafe System Description
Decrementalism	Minor and accepted modifications to system elements that gradually create a significant change in system behaviour.	Constant small changes eventually create unsafe behaviours and practice through migration and drift.
Unruly technologies	Unforeseen and unpredictable behaviours of technologies that are introduced into the system.	Technologies that behave in unanticipated ways.
Contribution of the protective structure	The contribution of the formal and organised structure that is intended to protect and optimise system safety.	Protective structure competes for resources with negative effects on behaviour and safety.

and safety science arenas. It is no longer safe to assume that our targets of analysis are simple, predictable, stable, and characterisable. Decomposing problems comprised of these tenets, analysing the component parts, and reassembling the whole does not guarantee the insight required. Indeed, by severing the interconnections which bind systems together it becomes just as likely that an erroneous analysis outcome will emerge. Instead, we need to think in systems.

HOW TO BLOW UP THE DEATH STAR

This book is all about systems thinking methods. A common criticism is that systems thinking methods are hard to pin down, a little ephemeral, and can seem uncertain in their outcomes. This makes communicating systems thinking methods a challenge. It explains a large part of why the uptake of systems thinking methods has been slow. To convey their utility, in this book we present numerous examples drawn from our own experience of applying them to a multitude of domains all around the world. From submarines, aeroplanes, automated vehicles, and darknet marketplaces, to hospitals, railway level crossings, led outdoor activities, and even entire road transport systems: all feature in this book. To aid understanding of what systems methods can do no singular example has surpassed our analysis of the Death Star from the film franchise Star Wars (Walker et al., 2016). Originally developed as a teaching aid, out of all our previous work it was this inter-galactic canter through systems thinking methods which was featured in National Geographic. When all other approaches to communicating the benefit of systems thinking methods fail, we need to be radical. An entirely sober and scientific treatment of systems thinking methods is covered in the following chapters, but if you, as a reader, are still a little unsure what a systems thinking method actually is, or what it can do, a Star Wars-based case study may help (Figure 1.2).

We are not alone in using the occasional Star Wars metaphor in the service of science communication. Researchers have used Star Wars to teach complex topics in psychiatry because it is "well known to students, registrars, and consultants alike" (Freidman and Hall, 2015, p. 432). It has facilitated an examination of how people interact with political philosophy (Geraci and Recine, 2014), it has

NATIONAL
GEOGRAPHIC

The real science inspired by 'Star Wars'

From Darth Vader's breathing to the dual sunsets of Tatooine, we take a look back at the real studies inspired by the "Star Wars" universe.

BY MICHAEL GRESHKO

PUBLISHED DECEMBER 1, 2019 • 13 MIN READ

Walker applied real-world techniques to the notoriously explosion-prone Death Star as an example for his civil engineering students of analyzing flaws in big engineering projects. After obtaining its plans from an official technical manual, he and his colleagues gave themselves the equivalent of four days—the amount of time the Rebel Alliance had with the plans—to test two different flaw-finding techniques.

FIGURE 1.2 Originally designed as a learning and teaching aid, Walker et al.'s (2016) application of systems methods to detect flaws in the Death Star from the Star Wars film franchise led to unexpected impacts in science communication. This article appeared in National Geographic, https://www.nationalgeographic.com/science/article/151209-star-wars-science-movie-film.

contributed to a better understanding of the behavioural processes underlying immersion in virtual worlds (Guitton, 2012), and a surgical assessment of Darth Vader's respiratory difficulties informs learning and teaching in pulmonology (Berg et al., 2014). If other disciplines are happy to provide a clinical diagnosis of lead character Anakin Skywalker's borderline personality disorder (e.g. Tobia et al., 2015), and to use Jabba the Hutt as a visual metaphor for nuclear migration in cellular biology (e.g. Morris, 2000), then a quick case study describing how Star Wars can be used to communicate complex ideas about systems thinking methods begins to seem quite sensible. Indeed, it flushes out several important key issues which do not necessarily emerge from 'real' case studies.

Our case study example focused on pro-active risk assessment methods. Risk assessment is a critical and mandated component of safety management (Kirwan and Ainsworth, 1992; Reason, 1997; Hollnagel, 2004; 2008; ISO, 2018; Dallat et al., 2019; Hulme et al., 2019; 2021). The intention is to proactively identify and manage risks by developing controls which either prevent them from occurring or mitigate their effects once they do occur. Formal risk assessment involves the use of structured methods to proactively identify potential hazards and vulnerabilities that may create adverse outcomes during work tasks (Chemweno et al., 2018).

For those readers who have not seen Star Wars, imagine having to identify a critical flaw in a highly complex planetoid sized orbital battle station, under extreme time pressure, and with no clear idea at the outset where the vulnerability will lie. This was the challenge faced by the Rebel Alliance, the film's protagonists. Risk assessment methods could be used to identify critical vulnerabilities, only this time the purpose was not to develop risk controls, it was to develop strategies to exploit the vulnerabilities.

In Walker et al. (2016) we compared how a traditional deterministic risk assessment method would compare against a systems thinking method in discovering the Death Star's vulnerabilities. The first option was to employ a traditional error identification method of the sort contemporaneous with the film's release in 1977 and still in widespread use today (Hazard and Operability Study, HAZOP). Methods like HAZOP tend to focus on the work of human operators at the sharp end. They seek to identify the different kinds of errors they could make and the risks that might arise, usually at an atomistic level of task decomposition. The second option was to use a systems thinking method (the Work Domain Analysis phase of Cognitive Work Analysis). This approach is different. Rather than focus on the sharp end, this method, like most systems thinking methods, attempts to describe the overall work system, the interactions that occur, and the emergent properties arising from these interactions. Despite the filmic overtones and hints of levity, this case study enabled us to put forward a powerful demonstration of how to match systems thinking methods to complex problems at a deeper, more fundamental level. In doing so, it provides a powerful demonstration of what systems thinking methods can do, and challenges equally powerful beliefs about how difficult or time consuming such methods are to apply.

The Death Star

Back to Star Wars and how to destroy a Death Star. As orbital battle-stations go, the Death Star certainly had impressive civil engineering credentials. It was a 160km diameter spheroid constructed from Quadanium steel, a high strength material allegedly mined from asteroids (Windham et al., 2013). The internal superstructure was devoted to a large Sienar Fleet Systems SFS-CR27200 Hypermatter Reactor and its ancillary systems. This supplied all the power needs for propulsion, life support, and defensive and offensive weapons systems. Principle among the last item here was the Super Laser, possessing enough power to destroy entire planets. The core of the Death Star is devoted largely to energy and offensive purposes. This means accommodation and living space resides on the surface, protected by turbo-laser towers and a magnetic shield system. In the film, the critical vulnerability that was discovered and exploited by the Rebel Alliance was a small thermal exhaust port located in the equatorial trench (Windham et al., 2013). This provided a route from the surface of the station directly to the main reactor core.

The exhaust port had clearly been risk-assessed by the Imperial Empire – the antagonists and Death Star creators – as it was protected by a ray shield. The port

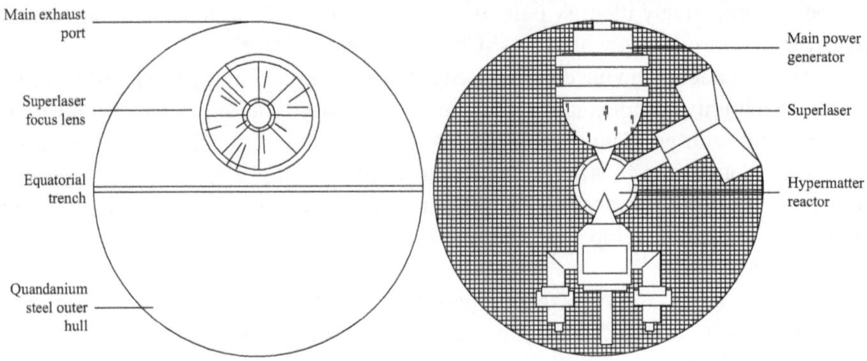

FIGURE 1.3 High-level schematic of the DS1 Orbital Battle station.

was small, being only two metres in diameter, meaning the weapon system required to breach the ray shield defences could not be launched remotely with sufficient accuracy. It would instead have to be launched from a small, agile weapons platform (such as a one-person fighter) operating in extremely close proximity. This was something the evil Imperial Empire considered near impossible in view of the other layers of defence in operation. These included squadrons of close fire support TIE (Twin Ion Engine) Fighters, the Taim and Bak xx-9 heavy turbo-laser towers, and the Borstel SB-920 maned laser cannons, to name just three (Figure 1.3).

The Star Wars plot hinges on the Imperial Empire's hubris and the perceived impossibility of exploiting this weakness, a perception that proved (explosively) false. In the closing scenes of the film, the hero and budding Jedi Knight, Luke Skywalker (no relation to author Prof Guy Walker), managed to launch a proton torpedo into the thermal exhaust port resulting in an explosive chain reaction which destroyed the entire battle station. The friendly Rebel forces achieved this feat by escorting a small X-Wing fighter craft to the surface of the Death Star. Manoeuvring a small weapons platform in such proximity neutralised the effects of the Death Star's main weapon, and also its surface weapons, which could not fire for risk of damaging the Death Star's hull. This left just the enemy TIE Fighter squadrons, reducing the asymmetry of the conflict to 'one-on-one' combat in which the Rebel Alliance emerged victorious. Prior to the attack the risk assessment activity needed to be run on a set of stolen plans, and in very short order. The Death Star was en route to the planet Yavin 4 to destroy the base from which any attack would be launched from. In a race against time and system complexity, which type of analysis method will win through? 1. A deterministic method, or 2. A systems thinking method?

> We'll be safe enough once we make the jump to hyperspace.

The first issue this fictitious case study exposes is that one size does not fit all. Systems thinking methods do not suit every type of problem, just as deterministic methods do not. All methods need to be matched to the extant problem they are targeting. This is a bigger issue than disciplines such as HFE or safety science tend to admit.

In the case of HFE, which is used to understand and optimise human health and wellbeing, there is an implicit guarantee in the use of HFE methods that, provided they are used properly, they will produce certain types of useful products. All methods should provide structure and the potential for repeatability. In addition, they should provide a route into making the discipline more accessible to all (Stanton and Young, 2002; Wilson, 2014). Given there are hundreds of HFE methods to choose from (Stanton et al., 2013), and little explicit guidance on how to choose between alternatives, it is easy to understand why practitioners and organisations often develop their own or else find their favourites and continue to use them (Stanton and Young, 1998).

The benefits of this pragmatism are clear. Familiarity with a particular methodology can cut costs, decrease training times within organisations, and make knowledge transfer much smoother (Crawford et al., 2016). Over time, the continued application of particular methods creates a legacy. Within high hazard industries, for example, regulations begin to specify them and further solidify their usage. As a result, it is not uncommon for an organisation to have just two or three commonly used methods, often developed decades previously, despite the discipline having 200 or more, many of which are more up to date.

There are currently three broad ways in which HFE methods are selected:

1. Legacy and preference;
2. Methods selection processes; and
3. Method integration/triangulation.

In the first case, developers of a particular method will establish its effectiveness and diffuse those benefits more widely, either through the academic literature, academic or industry training and/or the provision of consultancy services. Many significant methods in the discipline follow this path. Many practitioners continue to apply the methods they know well, which is understandable given constraints on their time to complete analyses and learn new approaches (Shorrock and Williams, 2016; Salmon et al., 2020). It is also common for methods to arise as a response to specific regulatory prerequisites. The advantage of this approach is that methods are made to leave the laboratory and be subject to the rigours of practical application. The disadvantage is that by inventing a particular approach, every problem begins to look like a suitable candidate. The further disadvantage is that some regulatory prerequisites may themselves not be fit for purpose in a rapidly changing world.

The second, and much less well-developed path, is the decision support meta-method approach. This describes semi-formal approaches used to match a wide(r) range of HFE methods to a client or sponsor's needs (e.g. Wilson and Sharples, 2015; Williams and Salmon, 2016). Stanton and Baber (1996) attempted to formalise this by putting forward a meta-method process based on establishing the reliability, validity, and utility of methods, which was latterly extended by Stanton and Young (1999a, 1999b, 2002) in the development of a decision-support flowchart shown in Figure 1.4. This presents method selection as a closed loop process with three feedback loops. The first feedback loop validates the selection of the methods against the selection criteria. The second feedback loop validates the methods against the adequacy of the intended intervention. The third feedback loop validates the initial criteria against the

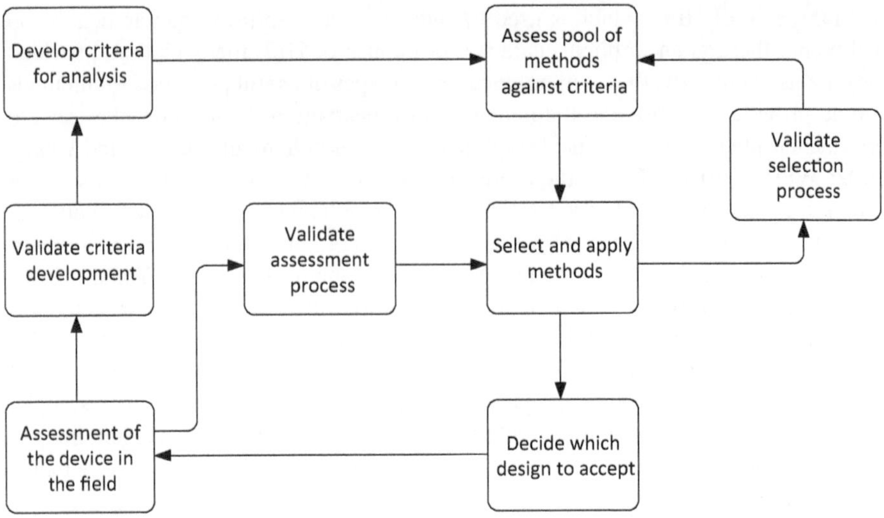

FIGURE 1.4 Stanton and Young's method selection flow chart. (Adapted from Stanton and Young, 1999b).

adequacy of the intervention. The process is logically sound, but each step is open to interpretation, especially when selecting methods under time pressure.

The final approach is method triangulation and integration. Through triangulation, weaknesses in one method can be mitigated by the strengths of another (Corlett et al., 1995) and vice versa. Methods can be combined like this in any way which would suit the work system under analysis, so long as theoretical compatibility between the methods is maintained (Stanton et al., 2013; Salmon and Read, 2019). Method triangulation reinforces the point that methods do not, nor do they have to, exist in isolation. This tends to be a feature of several systems HFE methods and is reflected in this book.

> An analysis of the plans provided by Princess Leia show…

The three approaches to method selection illustrate the difficulties faced by practitioners in a new world of increasing systems problems. Method selection is contingent, and the primary contingency factor – in our view – are the systems attributes of the target problem. Grant et al.'s systems thinking tenets (2018) makes these systems attributes explicit. The more that a target problem exhibits these features – and destroying a Death Star is a good example – the more that methods covered in this book become well suited.

A more formal way of expressing this contingent approach to method selection is via Relative Predictive Efficiency (RPE; Crutchfield, 1994). The formula is:

$$RPE = E/C$$

E is 'excess entropy' or the extent to which a system can be adequately modelled using any HFE method. An approximation of E can be derived from a comparison between the system behaviours predicted by an HFE method compared to those behaviours actually observed. Any disparity between 'expected' and 'observed', and in what quantity, represents an approximation to 'excess entropy'. For example, in the Star Wars case study, the simple organisation (the Rebel Alliance) was able to perform complex unexpected actions not predicted by any normative analysis of how attacks on space-borne battle-stations 'should' be performed. It would therefore measure low on the parameter E. The Rebel Alliance's agility makes its behaviours unpredictable and difficult to model.

C is 'statistical complexity'. It is a measure of the size and complexity of the system's model (or method output) at any given scale of observation. This can be measured in several ways using metrics from complexity theory (e.g. Hornby, 2007). It is possible to consider the number of 'build symbols' in the system model, the sophistication of the model, or the model's connectivity (the maximum number of links present in both), and so on.

A highly complex model – either deterministic or systems-based – that does not predict the behaviours of the system under analysis has poor RPE. It represents either wasted time and effort for sub-optimal outcomes, or an analysis of a system that is so large it becomes unwieldy. The sought-after outcome is the reverse. What we hope this book can help readers achieve is simple, parsimonious, efficient models of low complexity that nonetheless generate deep insight. In other words, high RPE.

> Any attack made by the Rebels against this station would be a useless gesture, no matter what technical data they've obtained

Returning to the Death Star, the engineering plans were obtained (read 'stolen') and a critical vulnerability was discovered and exploited. Estimates vary, but according to 'Wookipedia' the time available for the Rebel Alliance to undertake this analysis was four days. In Walker et al. (2016) we pitted the two competing approaches against each other. The two methods represent opposites in terms of their high-level approach. The former is deterministic and reductionist while the latter is systemic and holistic. Application of the RPE algorithm would suggest that the method which more closely matches the extant nature of the 'problem' will exhibit higher predictive efficiency. It would be the one that allows the Rebel Alliance to perform the analysis correctly and within the short amount of time inferred by the film.

The HAZOP and CWA analyses were driven from technical data contained in the Haynes Workshop Manual for the DS-1 Orbital Battle station (Windham et al., 2013). The time it took to complete each analysis was logged, as was the ability of each method to detect the vulnerability exploited in the film. An expert workshop was also held to construct the HE-HAZOP outputs, review the Work Domain Analysis phase of the CWA method, and facilitate narratives on how the Death Star could be destroyed. It is important to note that the participants in the study were aware of the actual vulnerability exploited in the film. After all, everyone (or nearly everyone) has seen it. The goal was to provide the best opportunity possible for each method to detect it. To that end, participants were guided in detail on how to perform the analyses 'strictly by the book' (in this case Stanton et al., 2013) and to pay close attention to whether the 'real' vulnerability was emerging. After the two analysis methods were complete, an RPE score was calculated to gauge the fit of the two methodological approaches to the host problem. What was found?

The Empire Strikes Back

HAZOP and its derivatives (including Human Error-HAZOP) are a well-established approach originally applied to engineering diagrams (Kletz, 1991; Kirwan, 1992; Kirwan and Ainsworth, 1992; Swann and Preston, 1995). It involves the analyst applying guidewords to each step in a task analysis to identify potential problems. It is the very definition of deterministic. It takes a complete entity (the system used to operate and defend the Death Star) and reduces it down to its component tasks. These tasks are individually analysed in order to construct detailed narratives about how the system could fail, and how to prevent it from doing so. Collecting these narratives together – in effect, recombining the system in its entirety – enables clusters of failure mechanisms and design solutions to be identified.

The most important part of any HAZOP analysis is assembling a HAZOP team with the right combination of skills and experience (Swann and Preston, 1995). Two Death Star Subject Matter Experts (SMEs) with extensive knowledge of the 'Star Wars Universe' were recruited via social media, Edinburgh Comicon, and a news item in the Edinburgh Evening News. An example of the detailed HE-HAZOP analysis outputs they produced is shown in Tables 1.2 and 1.3.

> Don't be too proud of this technological terror you've constructed.

CWA (Vicente, 1999) is a systems analysis and design framework that is typically used in applications that aim to understand and/or optimise complex systems (Bisantz and Burns, 2008; Stanton et al., 2017). The framework comprises five phases and provides a series of modelling approaches that focus on identifying the constraints imposed on behaviour within the system under analysis (Vicente, 1999). The first phase, Work Domain Analysis (WDA), is used to develop an event- and actor-independent model of the system under analysis (Vicente, 1999; Naikar, 2013). This involves using the abstraction hierarchy method to describe the system across five levels of abstraction in order to identify the primary goals of the system, the processes and functions that are undertaken within the system, the values and criteria used to determine whether the system is achieving its goals, and the objects used when undertaking processes and functions. This constraints-based approach can be used to explore all the possibilities for action contained in a system, and indeed, the possible ways in which it could fail.

Extracts of the completed WDA analysis are shown in Figure 1.5 (Walker et al., 2016). It comprises 115 functional nodes and 354 means-ends links. The functional purposes of the Death Star, as defined by the source documents, are as follows:

- Present the galaxy with a powerful symbol;
- Subjugate worlds;
- Enable the galaxy to be ruled unchallenged; and
- Enact the doctrine of 'rule through fear'.

TABLE 1.2
HE-HAZOP Analysis of Task Step 2.1.1.5: Expel Excess Energy

Task Step	Guide-Word	Error	Consequence	Cause	Recovery	Design Improvements
Expel excess energy	Less than	Amazing it works in the first place. Giant amount of energy.	Kills everyone on board. Too hot/cold in a giant metal ball.	**Exhaust port too small**.	Better sensing capabilities and control.	More sensors on exhaust port and source of radiation. Blow up more planets to test function. More test firing. Learn lessons/data from smaller guns on star destroyer. **Make port bigger to aid flow**.
	More than	Optimum amount of energy balance is required, so becomes out of balance.	Functions out of balance. General wear on essential part of station. Compounding problem of fixing problem might cause more exhaust emissions.	Power source.		
	As well as	Channel for something unwanted to enter like vermin or rubbish. Material properties of port constrains design. Bits of port fall off. Bits melt off and re-harden somewhere else.		Floating space debris – but there is a force field. Poor design. Substandard materials. Cost cutting. Planning for maintenance in the future.	Do a cognitive work analysis (Melissa). Roberts to maintain everything.	Control of thermal exhaust. Cooling/ extractor fans. **Put a grate over the port**. Re-route flows to other exhaust ports.
	Other than	n/a	n/a	n/a	n/a	n/a
	Repeated	Access gate to exhaust port activating too much.	Kill's everyone on board. Too hot/cold in a giant metal ball. Out of balance. General wear on essential part of station. Compounding problem of fixing problem might cause more exhaust emissions. Destroys main power generator. Puts life support at risk.	Overzealous with the Super-Laser. Overuse of the weapon creates too much thermal energy/ radiation – needs a cooldown period.	Stop overusing it. Full shutdown. Reboot Death Star.	Designated cool down period (procedural change). Back up for life support.

(Continued)

TABLE 1.2 (Continued)

Task Step	Guide-Word	Error	Consequence	Cause	Recovery	Design Improvements
	Sooner than	Poor control.	Wasting energy and output.	Lack of sensing capability.	Human (alien) intervention of some sort.	Better monitoring system. Control room improvements. Improve SCADA[1] system. Improve operator vigilance. Diagnostic capability via remote desktop or similar technology.
	Later than		Don't want too much energy in the core. Might slow things down, effect system performance. Strains systems if too much being demanded.			
	Mis-ordered	Energy for laser being expelled out of exhaust port and wasted.	Don't want energy for laser to shoot out of exhaust port and be wasted. Diminishes ability of the Death Star to fire main weapon system.	Faulty mechanics. Something not opening. Faulty sensors.	Shut it all down and reboot.	Mechanical/system interlock. Automatic fail safes.

TABLE 1.3
HE-HAZOP Analysis of Task Step 2.1.1.6: Control and Monitor Thermal Exhaust Port Ray Shield Generators

Task Step	Guide-Word	Error	Consequence	Cause	Recovery	Design Improvements
Control and monitor thermal exhaust port ray shield generators	Less than	Malfunction.	Not strong enough. Allows unwanted matter into exhaust port.		Reset strength of forcefield – quick disablement of port. Isolate problem.	Control dial. Touchscreen interface.
	More than		Heat and radiation prevented from getting out – ray shield too strong.	Fried equipment due to heat.		
	As well as	Controls set incorrectly. Malfunction.	Make problem worse. Make port vulnerable. Or make port less effective in dissipating heat.	Sabotage. Incompetence.		Threshold warning system. Condition monitoring. System feedback.
	Sooner than	Push dial too far due to inattention or slip.	Kills everyone aboard. Too hot/cold in a giant metal ball. Out of balance. General wear on essential part of station.	Delays in monitoring system – not in real time.		Safe zone – system optimisation. Engineered level.
	Later than		Compounding problem of fixing problem might cause more exhaust emissions. Destroys generator. Life support at risk.	Other warnings and control room ergonomics. Other demands and priorities. Workload and teamworking. Understaffing.		
	Mis-ordered	Distraction, concurrent demands, operator forgets what they are doing, error of commission.				Liaison between forcefield and exhaust port monitoring teams. Get the droids to do it…but don't trust them completely.
	Part of					

The values and priority measures, which can be used to determine the extent to which the system is meeting those purposes, include the ability of the Death Star to exhibit 'operational autonomy and self-containment', the 'relative firepower' it can muster, and the overall 'levels of fear inspired in the masses'. Purpose-related functions include: 'provide offensive and defensive capabilities', 'provide energy and propulsion' and 'provide command capability'. The object related processes which underlie these more general functions include: 'generate power', 'destroy enemy ships', 'capture, shift or redirect space-born objects' and 'accommodate enormous reactor core and super-laser'. All of these functions and priority measures could potentially be achieved with a wide variety of technical means. Currently (or rather previously), the Death Star combines a wide range of specific physical objects such as Phylon Q7 tractor beam projectors, Taim and Bak xx-9 heavy turbo-laser towers, the main power generator, the Quadanium steel armoured hull and, of course, the thermal exhaust ports. All of these, and more, service the higher-level functions.

A WDA validation and insight extraction process was undertaken. The first phase involved the SMEs scrutinising the WDA and identifying any missing elements or links. Both SMEs were satisfied that the WDA contained all relevant components. The next phase invited the SMEs to trace through the WDA, examining functions, processes, and their interconnections, and using these to drive further narratives for how the system could fail. The WDA had a significant effect on the ability of the SMEs to do this. They judged the WDA as being inferior compared to HE-HAZOP in terms of "fine detail" but felt it was better at "providing a good overview", "better for strategising", and perhaps most importantly, "a lot quicker".

A common misconception is that systems thinking methods take a long time. This belief alone is often enough to prevent systems methods being used. In this case study (as well as numerous other real-world case studies) the WDA took a similar amount of time to develop as the Human Error-HAZOP. It required approximately 10 hours' input by the lead investigators, and a further two hours for the SMEs to validate. This is a total of 12 hours, or 1.5 working days, compared to the 10.5 days required to run the complete Human Error-HAZOP analysis with the same resources. In fact, the WDA is a representation of the entire system, whereas the HE-HAZOP analyses just *two* out of 160+ tasks.

> I used to bullseye womp rats in my T-16.

Not only was the systems thinking method significantly quicker, it also succeeded in detecting the actual vulnerability exploited in the film (i.e. the exhaust port), along with many others. These included seizing control of the navicomputer and steering the Death Star into a planet or star; poisoning its atmosphere; or uploading a destructive computer virus. This last vulnerability informed the narrative of the 1994 film 'Independence Day', but of course, the first such virus was released into the wild in 1983 and was not available for use in the original 1977 Star Wars plot. Despite computer viruses not existing in 1977, the 'possibility for action' was very much present in the WDA and available for scrutiny.

The outputs of the two analyses, combined with the time it took for both to be completed, suggest that RPE for the WDA was higher than the comparable

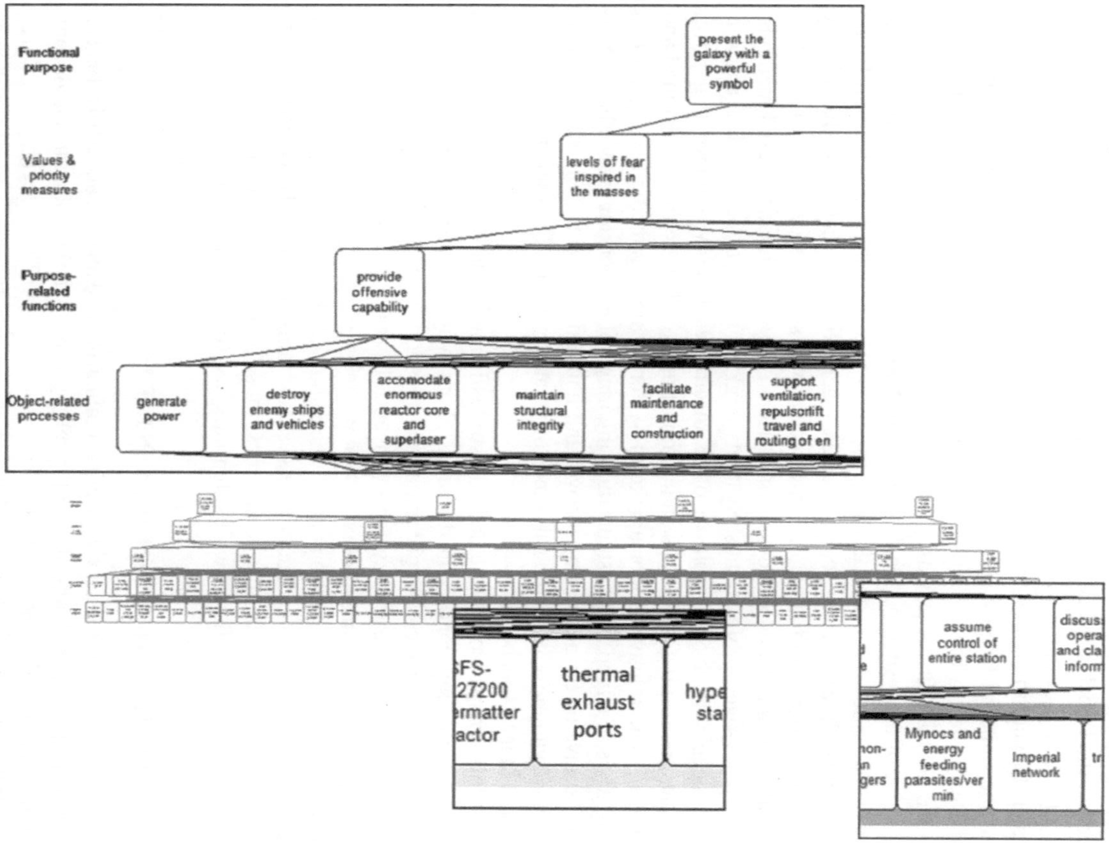

FIGURE 1.5 Extract of the Work Domain Analysis (WDA) of the DS-1 Orbital Battle Station. (Adapted from Walker et al., 2016).

HE-HAZOP analysis in this case (Walker et al., 2016). Further, when applied to the problem of destroying a Death Star with a proton torpedo fired into a thermal exhaust port, the WDA had a higher predictive efficiency than the deterministic Human Error-HAZOP (Walker et al., 2016). Not just that, but the WDA could also be completed well before the Death Star hovered into view over the horizon of Yavin 4, ready to destroy the Rebel base with its Super-Laser. It is thus the method recommended to the Rebel Alliance should the need arise to destroy another Death Star… which of course they did in 1983 (Return of the Jedi) and 2015 (The Force Awakens). Ever alert to a commercial opportunity, HFE and safety science can be of equal service to the Imperial Empire should they wish for a more resilient and jointly optimised Death Star in future. This may or may not involve a recommendation to fit grills over all the thermal exhaust ports.

The aim of analysing the Death Star from the film franchise Star Wars is not to trivialise the issue of systems thinking methods and their application in HFE and safety science. It arises from science outreach activities in which cross-disciplinary insights need to be communicated in effective ways to a wide range of different stakeholders, from students and first-time users of system thinking-based methods, through to seasoned professionals. Systems thinking is a core part of HFE and it is likewise becoming a core topic in other disciplines too. We believe that novel approaches to bridging these gaps and demonstrating value are to be welcomed and encouraged. The relatively poor uptake of systems thinking methods suggests that existing ways of advocating systems thinking may not be working, so we hope readers will forgive us for attempting a novel and unusual approach. With these caveats in place, we can move on to the substance of what we are trying to say with this case study, and indeed this book.

What this comparison of methods illustrates is the fundamental role of systems properties in making contingent decisions about which methods to apply to what problems. This is an increasingly important question because a) the paradigm has shifted towards greater use of systems concepts, b) many research grand challenges occur at the non-linear intersection of people and technology, and c) every time a method is used tacit assumptions about the nature of the problem to be solved are made. This is why our introduction to systems thinking methods has travelled to outer space to demonstrate that, sometimes, those assumptions can be at odds with what we are trying to achieve, with potentially disastrous consequences.

A REAL-LIFE THERMAL EXHAUST PORT TO FIND

Back on Earth we undertook a very similar comparison (Walker et al., 2014). Instead of a fictional space-born battle station the setting was a UK Magnox nuclear decommissioning site (Walker, et al., 2014). Instead of a Rebel Force seeking to destroy it, the challenge was to risk-assess decommissioning activities that had never been carried out on this site before. The risk was rather extreme. The activity being risk-assessed was the removal of two High Dose Rate Items (HDRIs) from a disused cartridge cooling pond. Both HDRIs were giving radiation dose readings of 14 Sieverts (Sv), which is the equivalent of a 5.25 second unshielded exposure time before the annual legal allowable dose uptake of 20mSv is exceeded. This reinforces the need to make *absolutely sure* that all risks were assessed. Instead of Human Error-HAZOP compared with WDA, the full CWA approach was compared with the Structured What If Technique (SWIFT). The outcomes of this earth-bound and very 'real' study are remarkably similar (Walker et al., 2014). CWA succeeded in detecting an additional key risk not spotted by SWIFT. Not just that, it did so for comparable total time, resources, and effort. Based on findings like these the call to action is clear. systems thinking methods have much to offer practitioners and will play an important part in the future of HFE and safety science. The practical realities, however, are often more challenging. They can perhaps be summed up in the words of the senior safety engineer involved in the decommissioning study. They agreed that the systems thinking method

> with its multiplicity could assist in these uncertain tasks [but] the risk adverse nature of our industry and relative lack of knowledge of [systems thinking-based methods] is not conclusive to its use on site. [the deterministic method] is a clear, concise, well understood and a recognised methodology. [the systems thinking method] is less well known, rather abstract in its application, difficult to follow, and not very certain in terms of its methodology and output.
>
> (Walker et al., 2014, p. 1632)

These are common beliefs and concerns but the evidence, as we will show in this book, is quite different.

We hope this book contributes to a process of change. We hope many more people can determine the right time to apply systems thinking methods and realise the benefits of doing so. Making them more accessible to more people is our key motivation. Only then will systems thinking methods become as mainstream as the current, deterministic, portfolio of HFE methods. The analytical weapon of a Jedi Knight, perhaps? An elegant weapon for a more civilised time. With that imagery in mind, along with some key systems thinking concepts, it is time to turn to an entirely serious and sober treatment of systems thinking methods.

NOTE

1 Supervisory Control and Data Acquisition.

REFERENCES

Ackoff, R. L. (1973). Science in the systems age: Beyond IE OR and MS. *Operations Research*, 21(3), 661–671.

Batty, M. & Torrens, P. M. (2005). Modelling and prediction in a complex world. *Futures*, 37, 745–766.

Berg, R. M., Ronit, A., Toksvang, L. N. & Plovsing, R. R. (2014). Using Darth Vader as a case study on pulmonary pathophysiology. *Proc Physiol Soc* 31, C76.

Bertalanffy, L. V. (1950). The theory of open systems in physics and biology. *Science*, 111, 23–29.

Bisantz, A. M. & Burns, C. M. (2008). *Applications of Cognitive Work Analysis*. Taylor & Francis Group, Boca Raton, FL.

CAA (2011). *Safety Plan 2011 – 2013*. London: Civil Aviation Authority.

Castellani, B. (2018). Map of the complexity sciences. *Art & Science Factory*, accessed 31 January. https://www.art-sciencefactory.com/complexity-map_feb09.html

Chemweno, P., Pintelon, L., Muchiri, P. N., Van Horenbeek, A. (2018). Risk assessment methodologies in maintenance decision making: A review of dependability modelling approaches. *Reliability Engineering & System Safety*, 173, 64–77.

Chiles, W. D. (1958). Effects of elevated temperatures on performance of a complex mental task. *Ergonomics*, 2(1), 89–96.

Cilliers, P. (1998). *Complexity and Postmodernism*. Boca Raton, FL: Routledge.

Corlett, E. N., Wilson, J. R., & Corlett, N. (Eds.). (1995). *Evaluation of Human Work*. CRC Press.

Coury, B. G., Ellingstad, V. S., & Kolly, J. M. (2010). Transport accident investigation: The development of human factors research and practice. *Reviews of Human Factors and Ergonomics*, 6(1), 1–33.

Crawford, J. O., Davis A., Walker, G. H., Cowie H., & Ritchie P. (2016). Evaluation of knowledge transfer for occupational safety and health in an organisational context: Development of an evaluation framework. *Policy and Practice in Health and Safety*, 14(1), 7–21.

Crutchfield, J. P. (1994). The calculi of emergence: Computation, dynamics and induction. *Journal of Physics D*, 75, 11–54.

Dallat, C., Salmon, P. M., & Goode, N. (2019). Risky systems versus risky people: To what extent do risk assessment methods consider the systems approach to accident causation? A review of the literature. *Safety Science*, 119, 266–279.

Dekker, S. (2011). *Drift into Failure: From Hunting Broken Components to Understanding Complex Systems*. CRC Press.

Dul, J., R. Bruder, P. Buckle, P. Carayon, P. Falzon, W.S. Marras, J.R. Wilson, & B. van der Doelen. (2012). A strategy for human factors/ergonomics: Developing the discipline and profession. *Ergonomics*, 55:4, 377–395.

European Aviation Safety Authority (2010). *Annual Safety Review 2010*. Cologne, Germany: EASA.

Freidman, S. H. & Hall, R. C. W. (2015). Using star wars' supporting characters to teach about psychopathology. *Australasian Psychiatry*, 23(4), 432–434.

Geraci, R. M. & Recine, N. (2014). Enlightening the galaxy: How players experience political philosophy in star wars: the old republic. *Games and Culture*, 9(4), 255–276.

Grant, E., Salmon, P. M., Stevens, N. J., Goode, N., & Read, G. J. M. (2018). Back to the future: What do accident causation models tell us about accident prediction? *Safety Science*, 104, 99–109.

Guitton, M. J. (2012). Living in the Hutt Space: Immersive process in the Star Wars role-play community of second life. *Computers and Human Behaviour*, 28(5), 1681–1691.

Holland, J. H. (2014). *Complexity: A Very Short Introduction*. Oxford: OUP.

Hollnagel, E. (2004). Task analysis. In: Bainbridge, W. S. (Eds.), *Berkshire Encyclopedia of Human-Computer Interaction* (pp. 707–711). Great Barrington, MA, USA: Berkshire Publishing Group.

Hollnagel, E. (2008). The changing nature of risk. *Ergonomics Australia*, 22(1–2), 33–46.

Holman, M., Walker, G. H. & Lansdown, T. (2020). Radical systems thinking and the future role of computational modelling in ergonomics. *Ergonomics*, 63(8), 1057–1074,

Holman, M., Walker, G. H., Lansdown, T., Stanton, N. A., Salmon, P. M., & Read, G. M. (2021). The binary based model (BBM) for improved human factors method selection. *Human Factors*, 63(8), 1408–1436.

Hornby, G. S. (2007). Modularity, reuse, and hierarchy: measuring complexity by measuring structure and organization. *Complexity*, 13(2), 50–61.

Hulme, A., Stanton, N., Walker, G. H., Waterson, P., & Salmon, P. (2019). What do applications of systems thinking accident analysis methods tell us about accident causation? A systematic review of applications between 1990 and 2018. *Safety Science*, 117, 164–183.

Hulme, A., Stanton, N. A., Walker, G. H., Waterson, P., & Salmon, P. M. (2021). Complexity theory in accident causation: Using AcciMap to identify the systems thinking tenets in 11 catastrophes. *Ergonomics*, 64(7), 821–838.

International Organization for Standardization (2018). ISO 31000 – Risk management – Guidelines.

Kim, D. H. (1999). *Introduction to Systems Thinking* (Vol. 16). Waltham, MA: Pegasus Communications.

Kirwan, B. (1992). Human error identification in human reliability assessment. Part 1: Overview of approaches. *Applied Ergonomics*, 23(5), 299–318.

Kirwan, B., & Ainsworth, L. K. (Eds.). (1992). *A Guide to Task Analysis*. Boca Raton, FL: Taylor & Francis Group.

Kletz, T. (1991). *An Engineer's View of Human Error*, Second Edition, Rugby, Warwick-Shire: Institution of Chemical Engineers.

Lee, J. D., (2001). Emerging challenges in cognitive Ergonomics: Managing swarms of self-organising agent based automation. *Theoretical Issues in Ergonomics Science*, 2 (3), 238–250.

Leveson, N. (2004). A new accident model for engineering safer systems. *Safety Science*, 42(4), 237–270.

Leveson, N. G. (2017). Rasmussen's legacy: A paradigm change in engineering for safety. *Applied Ergonomics*, *59*, 581–591.

Luke, D. A. & Stamatakis, K. A. (2012). Systems science methods in public health: Dynamics, networks, and agents. *Annual Review of Public Health*, 33, 357–376.

Meadows, D. H. (2008). *Thinking in Systems: A Primer*. London: Earthscan.

Morris, N. R. (2000). Nuclear migration: From fungi to the mammalian brain. *Journal of Cell Biology*, 148(6), 1097–1101.

Naikar, N. (2013). *Work Domain Analysis: Concepts, Guidelines and Cases*. Boca Raton, FL: CRC Press.

Ottino, J. M. (2003). Complex systems: American Institute of Chemical Engineers. *AIChE Journal*, 49(2), 292.

Perrow, C. (1984). *Normal Accidents: Living with High-Risk Technologies*. New York: Basic Books.

Rasmussen, J. (1997). Risk management in a dynamic society: A modelling problem. *Safety Science*, 27(2/3), 183–213.

Reason, J. (1997) *Managing the Risks of Organizational Accidents*. London: Ashgate.

Rouse, W. B. (2016). *Universities as Complex Enterprises: How Academia Works, Why it Works these Ways and Where the University Enterprise is Heading*. Hoboken, NJ: John Wiley & Sons.

Salmon, P. M., Read, G. J., Walker, G. H., Stevens, N. J., Hulme, A., McLean, S., & Stanton, N. A. (2020). Methodological issues in systems Human Factors and Ergonomics: Perspectives on the research–practice gap, reliability and validity, and prediction. *Human Factors and Ergonomics in Manufacturing & Service Industries*.

Salmon, P. M., Read, G. J., Walker, G. H., Stevens, N. J., Hulme, A., McLean, S., & Stanton, N. A. (2022). Methodological issues in systems Human Factors and Ergonomics: Perspectives on the research–practice gap, reliability and validity, and prediction. *Human Factors and Ergonomics in Manufacturing & Service Industries*, 32(1), 6–19.

Salmon, P. M., & Read, G. J. M. (2019). Many-model thinking in systems ergonomics: A case study in road safety. *Ergonomics*, 62(5), 612–628.

Salmon, P. M., Walker, G. H., Read, G., Goode, N., & Stanton, N. (2017). Fitting methods to paradigms: Are ergonomics methods fit for systems thinking? *Ergonomics*, 60(2), 194–205.

Salmon, P. M., Read, G. J. M., Beanland, V., Thompson, J., Filtness, A., Hulme, A., McClure, R., Johnston, I. (2019). Bad behaviour or societal failure? Perceptions of the factors contributing to drivers' engagement in the fatal five driving behaviours. *Applied Ergonomics*, 74, 162–171.

Senge, P.M. (2006) *The Fifth Discipline: The Art and Practice of the Learning Organization.* New York: Doubleday.
Sharma, S. (2006). An exploratory study of chaos in human machine system dynamics. *IEEE Transactions on Systems, Man, and Cybernetics-Part A: Systems and Humans*, 36(3), 319–326.
Shorrock, S. T., & Williams, C. A. (2016). Human factors and ergonomics methods in practice: Three fundamental constraints. *Theoretical Issues in Ergonomics Science*, *17*(5–6), 468–482.
Skyttner, L. (2001). *General Systems Theory: Ideas and Applications*. Singapore: World Scientific.
Skyttner, L. (2005). *General Systems Theory: Problems, Perspectives, Practice*. Singapore: World Scientific.
Stanton, N., & Young, M. (1998). Is utility in the mind of the beholder? A study of ergonomics methods. *Applied Ergonomics*, *29*(1), 41–54.
Stanton, N. A., & Young, M. S. (1999a). What price ergonomics? *Nature*, 399(6733), 197–198.
Stanton, N. A., & Young, M. S. (2002). *Guide to Methodology in Ergonomics: Designing for Human Use*. CRC Press.
Stanton, N. A., & Baber, C. (1996). A systems approach to human error identification. *Safety Science*, 22 (1–3), 215–228.
Stanton, N. A., Salmon, P. M., Rafferty, L., Walker, G. H., Baber, C., & Jenkins, D. P. (2013). *Human Factors Methods: A Practical Guide for Engineering and Design*. (2nd Edition). Farnham, UK: Ashgate.
Stanton, N.A., & Young, M.S. (1999b). *A Guide to Methodology in Ergonomics: Designing for Human Use*. London, Stanton & Young: Taylor & Francis Group, 2003.
Sterman, J. (2000). *Business Dynamics*. Boston: McGraw-Hill, Inc.
Swann, C. & Preston, M. (1995). Twenty-five years of HAZOPs. *Journal of Loss Prevention in The Process Industries*, 8(6), 349–353.
Tobia, A., Draschil, T., Zimmerman, A., Brief, J., Peters, S., Rudge, D., & Velez, D. (2015). Darth Vulcan? In support of Anakin Skywalker suffering from borderline personality disorder. *Psychiatry Research*, 229, 625–626.
Trist, E., & Bamforth, K. (1951). Some social and psychological consequences of the longwall method of coal getting. *Human Relations*, 4, 3–38.
Vicente, K.J. (1999). *Cognitive Work Analysis: Towards Safe, Productive, and Healthy Computer-Based Work.* Mahwah, NJ: Lawrence Erlbaum Associates.
Von Bertalanffy, L. V. (1968). *General Systems Theory*. New York: George Braziller Inc.
Walker, G.H., Cooper, M., Thompson, P., & Jenkins, D. (2014). Practitioner versus analyst methods: A nuclear decommissioning case study. *Applied Ergonomics*, 45, 1622–1633.
Walker, G. H., Salmon, P. M., Bedinger, M. & Stanton, N. A. (2016). What the death star can tell us about ergonomics methods. *Theoretical Issues in Ergonomics Science*, 17(4), 402–422.
Walker, G. H., Salmon, P. M., Bedinger, M., & Stanton, N. A. (2017). Quantum ergonomics: Shifting the paradigm of the systems agenda. *Ergonomics*, 60(2), 157–166.
Walker, G. H., Stanton, N. A., Salmon, P. M., Jenkins, D. P., & Rafferty, L. (2010). Translating concepts of complexity to the field of ergonomics. *Ergonomics*, 53(10), 1175–1186.
Williams, C., & Salmon, P. M. (2016). 10 The challenges of practice-oriented research. *Human Factors and Ergonomics in Practice: Improving System Performance and Human Well-Being in the Real World*, 143.
Wilson, J. R. (2014). Fundamentals of systems ergonomics/human factors. *Applied Ergonomics*, 45(1), 5–13.
Wilson, J. R., & Sharples, J. R. (2015). *Evaluation of Human Work*. CRC: Boca Raton, FL.
Windham, R., Reiff, C., & Trevas, C. (2013). *Imperial Death Star DS-1 Orbital Battle Station: Owner's Workshop Manual*. Sparkford, UK: Haynes.
Woods, D. D. and Dekker, S. (2000). Anticipating the effects of technological change: A new era of dynamics for human factors. *Theoretical Issues in Ergonomics Science*, 1 (3), 272–282.

Section II

Risk Assessment Methods

2 The Networked Hazard Analysis and Risk Management System (Net-HARMS)

BACKGROUND

Net-HARMS (Dallat et al., 2018) is a systems theory-based risk assessment method that combines Hierarchical Task Analysis (HTA; Annett et al., 1971; Stanton, 2006) with principles of the Event Analysis of Systemic Teamwork framework (EAST; Stanton et al., 2018) and the Systematic Human Error Reduction and Prediction Approach (SHERPA; Embrey, 1986). The method supports the proactive identification of risks within complex sociotechnical systems by providing a description of the system under analysis on which a taxonomy is applied to identify task and emergent risks.

The method was developed based on a review of existing risk assessment approaches which found that a majority of existing methods are not aligned with contemporary models of accident causation and do not support the identification of risks across work systems (Dallat et al., 2018). Net-HARMS thus provides two key advances over existing risk assessment methods: first, it enables analysts to identify risks across the overall system, as opposed to 'sharp-end' risks only, and second, it enables analysts to identify 'emergent risks' that arise when different risks interact.

Net-HARMS was designed to be easy to learn and use (Dallat et al., 2018; see Figure 2.1 for Net-HARMS procedure). Applying Net-HARMS involves first developing a HTA that describes the overall system under analysis. The HTA is then converted into a task network which shows the high-level tasks required for safe delivery of the work in question along with the relationships between tasks.

Once the task network is finalised, task risks are identified by applying the Net-HARMS risk mode taxonomy (see Figure 2.3) to each node within the task network. For risk modes that are deemed credible (i.e. could conceivably occur), the analyst provides a description of the risks and their consequences, ratings of their probability

DOI: 10.1201/9780429281624-2

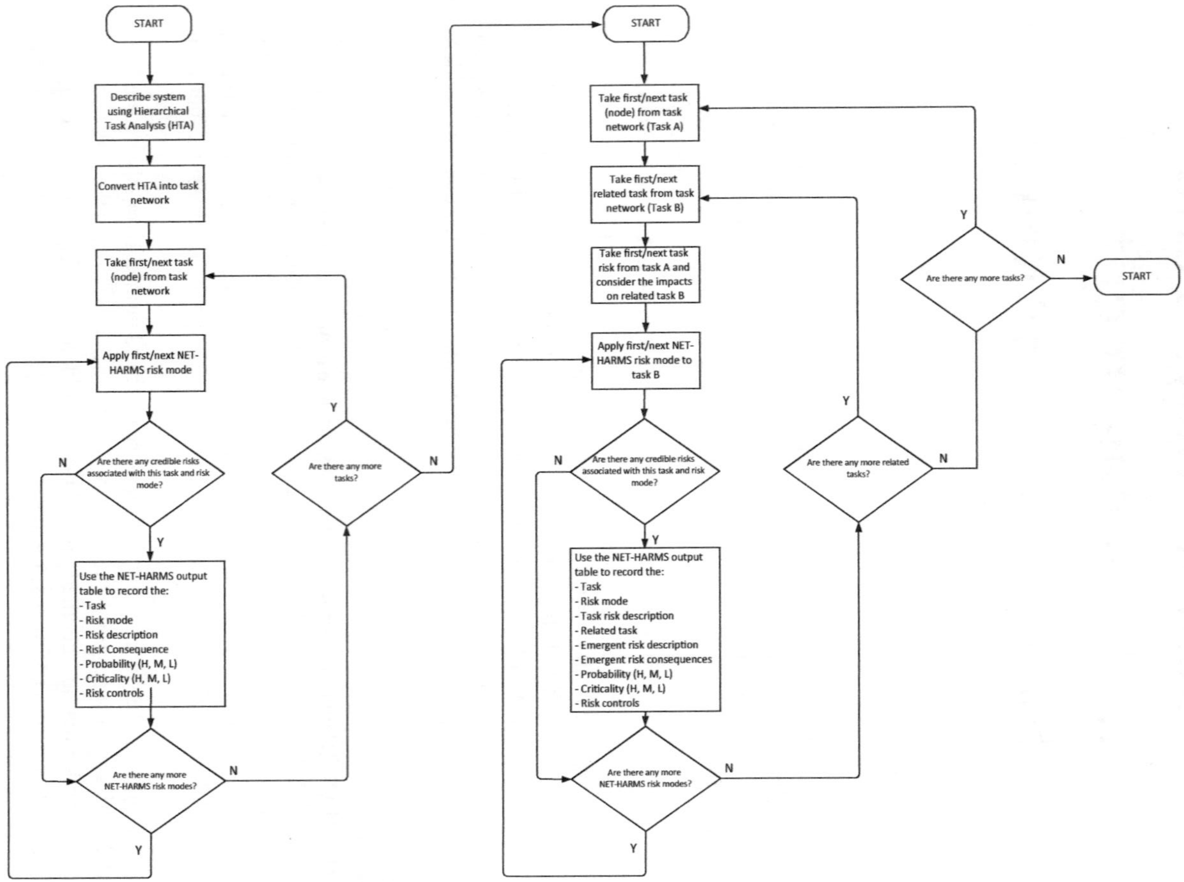

FIGURE 2.1 Net-HARMS procedure. The task risk phase of Net-HARMS is visualised on the left, and the emergent risk phase on the right.

and criticality (low, medium or high), and suggested risk controls. Following the task risk identification phase, the risk mode taxonomy is applied once more to identify emergent risks that are likely to arise during instances when the identified task risks influence the conduct of other tasks. The emergent risk identification phase of Net-HARMS is a critical feature that attempts to support the identification and management of new and unexpected risks that are created when performance elsewhere in the work system is sub-optimal. The emergent risk component of the analysis is based on the relationships between tasks described in the task network and involves assessing performance when risks from a related task influence the performance of the task in question. Again, for each credible emergent risk the analyst provides a description of the identified emergent risks and consequences, ratings of probability and criticality (low, medium or high), and suggested risk controls.

DOMAIN OF APPLICATION

Net-HARMS was originally developed for the led outdoor activity sector (Dallat et al., 2019); however, the approach is generic in nature and can be applied in any domain for risk assessment purposes. For example, Net-HARMS was recently used to identify task and emergent risks across the railway level crossing system design lifecycle and has been used in the elite sports context to identify the potential risks threatening cycling team performance (Hulme et al., 2021b). At the time of writing, Net-HARMS is the newest systems-based risk assessment method in HFE and further applications of the method are encouraged.

PROCEDURE AND ADVICE

A flowchart depicting the Net-HARMS procedure is presented in Figure 2.1. Step-by-step guidance is presented below.

As shown in Figure 2.1 HTA and task networks are used as the initial methods for modelling the system, on which the Net-HARMS risk modes taxonomy is applied to each task to identify task risks (left flow). When all task risks have been identified, the analyst progresses to the emergent risk phase and examines the relationships between tasks to identify emergent risks. The Net-HARMS risk modes taxonomy is again applied and is used to identify emergent risks.

STEP 1: DEFINE AIMS OF THE ANALYSIS AND THE TASK OR SYSTEM UNDER ANALYSIS

The first step in applying Net-HARMS involves clearly defining the aims and boundaries of the analysis. One of the key strengths of the Net-HARMS method is that it has the capacity to identify risks across overall organisations (e.g. an airline company) and even entire systems (e.g. the overall aviation system), as opposed to risks only at the sharp end of system operation (e.g. flying the plane). It is therefore recommended that analyses focus on overall systems rather than front-line tasks or operations only. This can include levels beyond the organisational level up to and including regulatory and government levels. By 'overall systems' here we mean that the tasks within a Net-HARMS analysis should ideally cover all levels beyond the organisational level of a system, from the government, policy, and including regulatory and government levels through to the equipment, technology, and working environment. This is not to say that Net-HARMS cannot be applied to sub-systems within the broader system of interest. Indeed, if the analysis aim is to identify risks in micro or meso systems embedded within the broader macro system of interest, then it is important to understand that the risk controls identified following the analysis may be limited in scope and effectiveness. Hence, a critical aspect of the first step involves clearly defining the levels of the system that will be considered during the analysis. For example, based on the aims, resources, access to data, and time available, it may be only necessary to consider risks within a particular organisation. Alternatively, on larger scale projects with more resources it may be appropriate to consider risks also at the regulatory and government levels of the system in question.

A useful step at this stage is to develop a clear description of the system under analysis in terms of the system hierarchy and which actors work together to achieve them. The Actormap method (Svedung and Rasmussen, 2002) is useful for this purpose (see Chapter 9).

STEP 2: CONSTRUCT A HIERARCHICAL TASK ANALYSIS FOR THE TASK/SCENARIO/SYSTEM UNDER ANALYSIS

Once the task/system under analysis and aims of the analysis are clearly defined, a HTA for the system under analysis is created. HTA (see Chapter 5) works by decomposing systems and behaviour into a hierarchy of goals, sub-ordinate goals, operations, and plans. In doing so, it focuses on "what an operator is required to do, in terms of actions and/or cognitive processes to achieve a system goal" (Kirwan and Ainsworth, 1992, p. 1). HTA outputs specify the overall goal of a system, the sub-goals to be undertaken to achieve this goal, the operations required to achieve each of the sub-goals specified and the plans that trigger engagement and achievement of the different goals and operations. The plans component of HTA specifies the sequence, and under what conditions, different sub-goals must be achieved in order to satisfy the requirements of a super-ordinate goal.

Creating the HTA initially involves collecting specific data regarding the system under analysis in terms of its goals and sub-goals and the operations that are undertaken to achieve them. Data should be collected regarding the goals and tasks involved, the human and non-human agents involved, the interactions between humans and non-human agents, the ordering of tasks and information on the factors that influence behaviour, task performance, and task outcome. A number of different approaches can be used to collect this data, including observations, concurrent verbal protocols, structured or semi-structured interviews (e.g. the Critical Decision Method), questionnaires and surveys, walkthrough analysis and documentation review (e.g. incident reports, standard operating procedures). The data collection approach selected is dependent upon project constraints, such as time, resources, access, and the number of analysts available.

Once the initial draft HTA is complete it is useful to have suitable SMEs review it. The HTA should subsequently be refined based on the SMEs feedback. It is normal practice for the HTA to go through many iterations before it is finalised.

STEP 3: CREATE TASK NETWORK

Once the HTA is finalised, the next step involves constructing a task network to support the identification of task and emergent risks. Task networks (see Figures 2.2, 2.4 and 2.5) are used to represent HTA outputs in the form of a network which shows key tasks and the relationships between them (Stanton et al., 2013, 2018). This enables analysts to understand the interactions and coupling that exists between tasks across the work system. Within Figure 2.2 the circular nodes represent tasks and the arrows linking the tasks represent relationships between tasks. Relationships between tasks are included in the task network when:

1. Tasks are undertaken sequentially (e.g. task three is undertaken after completion of task two);
2. Task are undertaken together (e.g. tasks one and two are undertaken together);
3. The outcomes of one task influence the conduct of another (e.g. the outcomes of task four influence how task five is undertaken); or
4. The conduct of one task is dependent on completion of the other (e.g. task five cannot be undertaken until task four is completed).

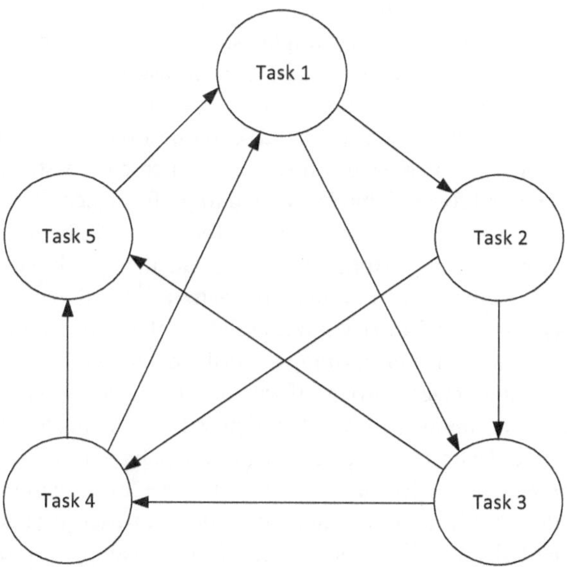

FIGURE 2.2 Task network example showing the relationships between five tasks.

Task networks are constructed by taking the first layer of sub-goals from the HTA and identifying which of them are related with one another based on the relationship types described above. For example, in the railway level crossing design and operation example presented in the case study section of this chapter (see Figure 2.5), the tasks 'Select site for upgrade' and 'Announce upgrade program' are linked as the upgrade program announcement cannot be made until the appropriate sites are selected. The task of 'Risk management' is linked to 'Monitor performance' as risks cannot be managed without performance monitoring to identify the risks that require management.

Once the initial draft task network is complete, it is useful to have SMEs review it. The task network should then be refined based on the SMEs feedback. Changes at this stage typically involve the addition and/or removal of relationships between tasks or the modification of task descriptions. Like HTA, it is normal practice for the task network to go through various iterations before it is finalised.

It should also be noted that whilst we recommend that the development of a task network should be based on HTA, it is possible to develop a task network based directly on raw data collection activities, SME input and verification, and document reviews (Stanton, 2014). This is often the case where project constraints do not enable development of an initial HTA. If the HTA step is indeed not utilised, then we recommend that analysts use SME review to ensure that the core tasks of interest are included in the task network.

STEP 4: IDENTIFY TASK RISKS

As shown in Figure 2.1, the task network and risk mode taxonomy are first used to identify 'task risks' that might emerge during the conduct of each task. Task risks are defined as those risks that might occur if tasks are not completed optimally. Identifying task risks involves taking each task step from the task network and using the Net-HARMS risk mode taxonomy (see Figure 2.3) and domain expertise to identify any credible risks for the task in question. As shown in Figure 2.3, this requires the analyst to consider 10 ways in which each task can be performed sub-optimally. The first five risk modes related to performance of the task itself, and include the task being mistimed, not being performed at all, being performed inadequately in some way, being performed with an inadequate task object (e.g. a poorly designed piece of equipment), or an inappropriate task being performed in place of the required task. The next four risk modes relate to the information that is communicated between agents when performing the task, and include the required information not being communicated, wrong information being communicated, inadequate information being communicated, and the required communication being mistimed. The final risk mode relates to the environment in which the task is undertaken being adverse in some way (e.g. adverse weather, high noise levels, extremely busy).

All risk modes should be systematically applied to each task, with the analyst considering what risks could emerge as a result. For example, for the railway level crossing design and operation example, the analyst may determine that the risk mode 'Task mistimed' is credible for the task 'Select site for upgrade' as the selection of railway level crossings for upgrade may be delayed for various reasons. For each

Behaviour	Risk Modes
Task	T1 – Task mistimed T2 – Task omitted T3 – Task completed inadequately T4 – Inadequate task object T5 – Inappropriate task
Communication	C1 – Information not communicated C2 – Wrong information communicated C3 – Inadequate information communicated C4 – Communication mistimed
Environmental	E1 – Adverse environmental con ditions

FIGURE 2.3 NET-HARMS risk mode taxonomy.

credible risk (i.e. those judged by the analyst to be possible), the analyst should give a description of the form that the risk would take, such as 'site selection process is delayed due to high workload and other external pressures'.

For each credible risk the analyst should also describe the associated consequences. For the risk of 'site selection process is delayed due to high workload and other external pressures' the consequence recorded would be 'site selection is delayed which in turn delays the upgrade program meaning unsafe crossings remain in use'.

After recording the risk mode, risk description and its consequences, analysts should provide a rating of the probability of the risk occurring. An ordinal probability scale of low, medium, or high is used. If the risk has not occurred previously, or the analyst deems it to be unlikely, a low (L) probability is assigned. If the risk has occurred on previous occasions, or the analyst feels that it could occur in certain situations, a medium (M) probability is assigned. Finally, if the risk has occurred on frequent occasions, or the analysts feels like it is likely to occur frequently, a high (H) probability is assigned.

Following the assignment of probability, the analyst assigns a criticality rating for each of the task risks. Again, a scale of low, medium, and high is used to rate risk criticality. Here the analyst should consider how critical the risk is to the overall task/system under analysis. Normally, if the risk would lead to a critical incident (in relation to the task/system in question), it is rated as a highly critical risk.

The final phase of the task risk identification step involves classifying each of the identified task risks into one of the risk categories described in Table 2.1.

TABLE 2.1
Net-HARMS Risk Types

Risk Category	Description
Reputational	Risks that may adversely impact the reputation of stakeholders associated with the task or system under analysis.
Financial	Risks that may adversely impact the financial status of stakeholders associated with the task or system under analysis.
Safety	Risks that may adversely impact the safety of stakeholders associated with the task or system under analysis.
Performance	Risks that may adversely impact the performance the task or system under analysis.
Legal	Risks with legal implications for stakeholders associated with the task or system under analysis.

STEP 5: IDENTIFY EMERGENT RISKS

Once the analyst is satisfied that all task risks have been identified, the task network and risk mode taxonomy are then used in conjunction with the task risks to identify emergent risks. Emergent risks represent new risks that arise as a result of the interaction between the task risks identified during Step 3 and other tasks in the task network (see Figure 2.4 for example). Identifying emergent risks involves identifying which tasks are related via the task network, and then identifying what new risks emerge when task risks from one task influence the performance of related task. For example, if tasks 1 and 2 are related, then analysts should determine what

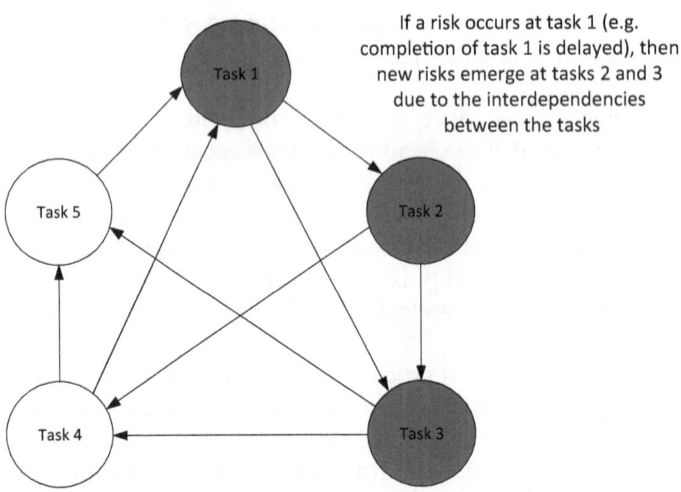

FIGURE 2.4 Emergent risk example.

the impact on task 2 is if each of the task risks for task 1 actually occur. The Net-HARMS taxonomy is used again to support the identification of emergent risks, with the analyst again working through the ten risk modes to consider which are credible should the task in question be influenced by a task risk from a related task. For each credible emergent risk, analysts should again determine and describe the associated consequences. Figure 2.5 provides a visual representation of the identification of emergent risks and how emergent risks relate back to related tasks and task risks.

Next, analysts should provide a rating of the probability of the emergent risk occurring. An ordinal probability scale of low, medium, or high is again used. It is worth noting that the probability of emergent risks is often rated as high as the related task risk is viewed as having occurred, which in turn increases the likelihood that the emergent risk will occur.

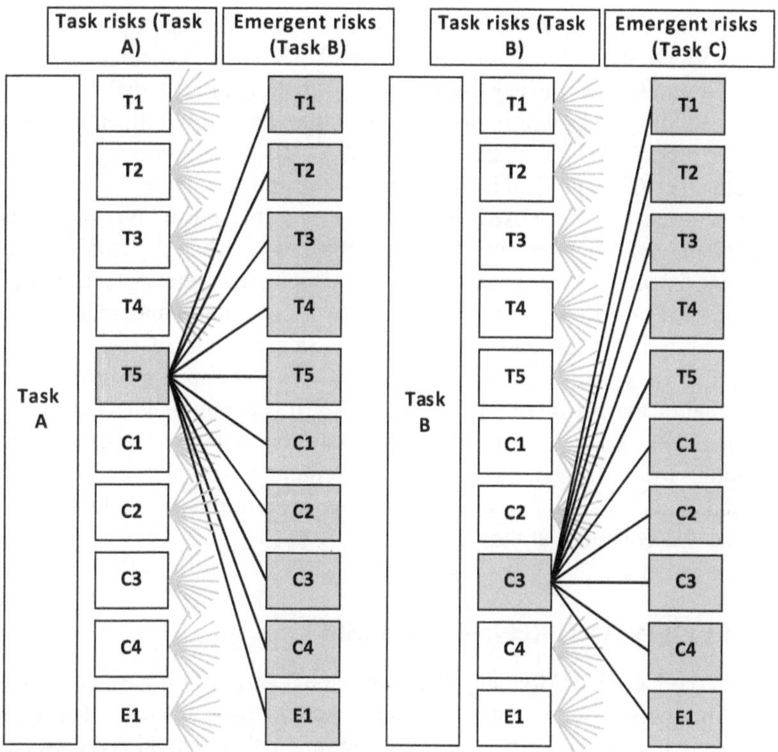

FIGURE 2.5 A generic emergent risk identification example that includes three tasks, Task A, Task B, and Task C. The illustration shows that it is theoretically possible to have 10 emergent risks per task risk, or 100 emergent risks should all 10 task risks be identified for every task. Further, the figure shows second order emergent risks, however Net-HARMS has the capacity to identify third, fourth, and fifth order emergent risks. The trade-off with this level of detail is excessive time performing the analysis.

As with the task risk phase, the analyst also assigns a criticality rating for each of the emergent risks. Again, a scale of low, medium, and high is used to rate risk criticality. Here, the analyst should consider how critical the risk is to the overall task/system under analysis. Normally, if the risk would lead to a critical incident (in relation to the task/system in question), it is rated as a highly critical risk.

The final phase of the emergent risk identification phase involves classifying each of the identified emergent risks into one of the risk categories described in Table 2.1.

STEP 6: IDENTIFY RISK MANAGEMENT STRATEGIES

The final stage in the process is to propose risk management strategies for the identified task and emergent risks. If time constraints are a consideration, then it is useful to focus only on the risks that are rated as high probability and high criticality. Another option to approach the prioritisation of risks is to apply Social Network Analysis (SNA) to the task network (Scott, 1988). SNA is a mathematical and graph theory-based method that reveals important information about social structures and networks, including tasks within task networks that are highly influential and well-connected to other tasks. It is possible to use SNA metrics such as centrality or sociometric status (see Chapter 7) to identify the most prominent nodes within a network (i.e. tasks in the case of Net-HARMS). Thus, it is possible to use SNE to identify key tasks within the task network that should be focussed on when prioritising risks and development risk management strategies. For example, the nodes with a centrality or sociometric status value of the mean for the network plus one standard deviation can be considered to be key tasks within the system (Salmon et al., 2018).

Irrespective of whether the prioritisation of risks is based on probability and criticality, and/or via the use of tools such as SNA, it is recommended that a group of analysts or SMEs be used to identify appropriate risk management strategies. The group should work through the task and emergent risk tables and discuss risk management strategies that can be used to either prevent the risk from occurring or mitigate the consequences of the risk.

STEP 7: REVIEW AND REFINE ANALYSIS

Once the initial Net-HARMS analysis is complete it is useful to have various SMEs review it. The analysis should then be refined based on SME feedback. This typically involves removing identified risks, identifying new risks, or modifying probability and criticality ratings and risk management strategies.

ADVANTAGES

- Net-HARMS has the capacity to go beyond existing sharp-end focussed risk assessment methods to identify risks across overall systems;

- Net-HARMS identifies both task risks and emergent risks. The identification of emergent risks is a considerable advancement over existing risk assessment methods;
- The analysis can be extremely comprehensive, covering risks across the overall system including government and regulatory levels;
- Net-HARMS supports the identification and development of risk management strategies for each of the task and emergent risks identified;
- The Net-HARMS taxonomy prompts analysts for potential task and emergent risks;
- Net-HARMS is based on SHERPA, which, according to the literature has the most supportive reliability and validity evidence (Stanton et al., 2013);
- Owing to the methods structured and systematic approach, Net-HARMS is easy to learn and apply and requires minimal training;
- It is generic, allowing it to be used in any domain;
- Creating the HTA and task network enables analysts to develop an in-depth understanding of the system under analysis; and
- It can be integrated with AcciNet, a systems theory-based incident analysis method (see Chapter 10).

DISADVANTAGES

- Depending on the nature and scope of the analysis, Net-HARMS can be time consuming to apply. The time to complete a Net-HARMS analysis is primarily determined by the number of included tasks, and in particular, the number of relationships between tasks within the task network;
- It is a relatively new method with few published applications;
- The quality of the analysis is highly dependent on the expertise of the analysts involved;
- The analysis incurs high levels of repetition, with risks often being identified multiple times (noting that this is an artefact of the method which was designed to enhance comprehensiveness);
- The requirement to develop the initial HTA and subsequent task network adds significant time to the analysis;
- Due to its infancy, there is little reliability and validity evidence associated with Net-HARMS. The first published study found an acceptable level of validity; however, the task risk phase performed better than the emergent risk phase (Hulme et al. 2021a);
- Some predicted risks and risks management strategies may be unlikely or lack credibility, thus posing a false economy (Stanton et al., 2005); and
- Currently, there is no performance shaping factors taxonomy to support consideration of contextual factors and their impact on task and emergent risks.

RELATED METHODS

Net-HARMS requires an initial HTA of the system under analysis. The HTA is subsequently used to develop a task network for the system under analysis. The Net-HARMS approach and taxonomy was developed based on the SHERPA human error identification method (Embrey, 1986).

APPROXIMATE TRAINING AND APPLICATION TIMES

Net-HARMS is a relatively simple method that requires little training. Indeed, once the procedure is understood by analysts, applying Net-HARMS is relatively straightforward. In a recent reliability and validity study participants were provided with training that lasted around 2 hours (Hulme et al., 2021a); however, a day-long workshop involving multiple opportunities for practice and feedback is recommended. Overall application time depends on the system under analysis and size of the task network; however, Net-HARMS analyses typically require significant time as the analysis is comprehensive and often covers overall systems rather than merely sharp end performance. The emergent risk phase requires the greatest amount of time, and the higher the number of relationships between tasks within the task network, the longer the analysis will take.

RELIABILITY AND VALIDITY

The SHERPA method on which the Net-HARMS risk modes taxonomy is based has a significant body of promising validation evidence associated with it (Stanton and Stevenage, 1998; Stanton and Young, 1999). HTA also has some validation evidence associated with it (Stanton and Young, 1999). Hulme et al. (2021a) recently compared the reliability and validity of Net-HARMS, STPA (see Chapter 4) and EAST-BL (See Chapter 3) when used to identify risks across the railway level crossing design lifecycle. Based on a comparison of the identified risks with a gold standard and an analysis using the Signal Detection Theory (SDT) paradigm (Green and Swets, 1966), Hulme et al. (2021a) found a weak-to-moderate positive correlation coefficient for Net-HARMS, meaning that the method did indeed support relatively novice analysts with the identification of credible system risks, albeit up to a point.

TOOLS NEEDED

Net-HARMS can be conducted using pen and paper; however, analyses are normally undertaken using a Microsoft Excel spreadsheet. The HTA tool can be used for the HTA component and task networks are typically drawn in Microsoft Visio.

CASE STUDY EXAMPLE: RAILWAY LEVEL CROSSINGS

Net-HARMS was used to identify risks arising during the design and operation of railway level crossings in Victoria, Australia. Initially a HTA of the railway level crossing system lifecycle was developed based on various sources of information, including analyses of railway level crossing systems (e.g. Mulvihill et al., 2016; Salmon et al., 2016) and documentation review (e.g. road and rail safety strategy and policy documents, crash investigation reports, relevant academic literature). The resulting HTA described the goals, sub-goals, and operations required when designing, implementing, operating, and removing RLXs in Victoria, Australia. The high-level goals from the HTA are:

1. Select sites for upgrade;
2. Announce upgrade program;
3. Identify user community of RLX;
4. Identify RLX type and infrastructure;
5. Design upgrade;
6. Construction and commissioning of safe and efficient RLX;
7. Manage infrastructure during construction process;
8. Meet legislative requirements around risk management for RLX;
9. Operate railway level crossing;
10. Monitor performance;
11. Check compliance with legislation;
12. Grade separation;
13. Crossing closure;
14. Continuous improvement of standards; and
15. Budget allocated to level crossing program.

The HTA was converted into a task network (Figure 2.6), and a workshop was then held in order to identify the task and emergent risks. This involved four analysts working through the HTA and task networks to identify credible task and emergent risks that might impact the design and operation of railway level crossings.

The task network in Figure 2.6 was created by taking the first sub-goal layer of the HTA of the system and applying the task network relationship rules to each task. This is a relatively high-level task network as it includes broad task descriptions and covers all levels of the system. For example, there are tasks pertaining to budget allocation and tasks around the operation of the RLX system. A particularly interesting feature of task networks is the notion of functional decomposition whereby each task within a network may be divided down further into its own task network. The scope and scale of the analysis was described in step one of the Net-HARMS analysis process above.

For the task risks the analysts used the Net-HARMS taxonomy to identify credible risks that could occur during completion of the railway level crossing design and operation tasks. For each credible task risk the analysts recorded a description of the risk, its consequences, and provided a rating of the probability and criticality of the risk occurring. An extract of the railway level crossing design and operation task risks is presented in Table 2.2.

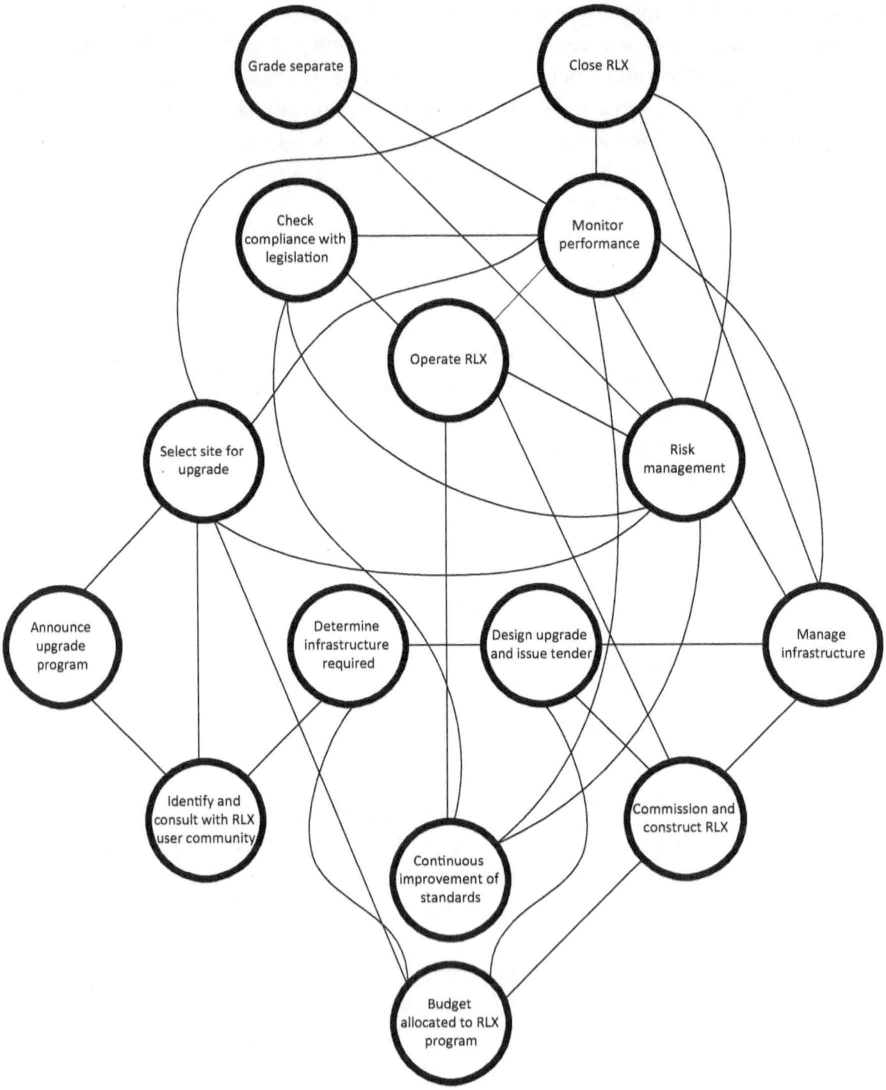

FIGURE 2.6 Task network of railway level crossing system lifecycle.

The analysts then used the Net-HARMS taxonomy and task network to identify credible emergent risks that could occur as a result of the interaction of task risks across the railway level crossing design and operation tasks. This involved examining each set of related tasks in the task network and identifying what the impact of task risks would be on related tasks. For each credible emergent risk the analysts recorded a description of the risk, its consequences, and provided a rating of the probability and criticality of the risk occurring. An extract of the railway level crossing design and operation emergent risks is presented in Table 2.3.

TABLE 2.2
Extract of Railway Level Crossing System Lifecycle Task Risks

Task	Risk Mode	Task Risk Description	Task Risk Consequences
1. Select site for upgrade	T1	Sites selection delayed.	Delay in upgrade of risky RLXs.
	T3	Inappropriate sites are selected for upgrade.	Risky RLXs may not be upgraded/Inappropriate RLXs are upgraded.
	T4	Tools to support site selection are inadequate/inappropriate.	Site selection is misinformed (e.g. ALCAM/Kerang). Inappropriate RLXs are upgraded.
	T5	Site selection process is driven by external pressures.	Site selection is misinformed (e.g. ALCAM/Kerang). Inappropriate RLXs are upgraded.
	C3	Incomplete/inadequate information is communicated regarding site selection.	Upgrade may be inappropriate.
	C4	Site selection communication is delayed.	Delay in upgrade of risky RLXs.
2. Announce upgrade program	T1/C4	Announcement is made too early.	Places time pressure on the site selection process.
	T1/C4	Announcement is delayed.	Delay in upgrade of risky RLXs.
3. Identify and consult with RLX user community	T1	Identify and consult with community too late.	Upgrade and design doesn't take into account end-user need.
	T2	Failure to identify and consult with community.	Upgrade and design don't take into account end-user need.
	T3	Consultation is inadequate (e.g. language and translation of information).	Upgrade and design don't take into account end-user need.
	T4	Consultation methods are inadequate e.g. not suitable for user community.	Upgrade and design don't take into account end-user need.
3. Identify and consult with RLX user community	T5	Consultation methods are inappropriate e.g. sausage sizzle.	Upgrade and design don't take into account all end-user need.
	C1	Information not communicated or gathered.	Upgrade and design don't take into account all end-user need.
	C2	The wrong information is communicated or gathered	End-users needs not understood.
	C3	Consultation methods are inadequate e.g. not suitable for user community.	Upgrade and design don't take into account end-user need.

(*Continued*)

TABLE 2.2 (Continued)

Task	Risk Mode	Task Risk Description	Task Risk Consequences
	C4	Identify and consult with community too late.	Upgrade and design don't take into account end-user need.
4. Determine infrastructure required	T1	Type of upgrade and infrastructure required is determined too late.	Upgrade may not be possible/Delay in upgrade of risky RLXs.
	T3	Determining upgrade and type of infrastructure is done inappropriately.	Wrong upgrade/infrastructure is selected.
	T4	Standards and legislation are inadequate.	Wrong upgrade/infrastructure is selected.
	T5	Inappropriate methods/information are used to determine infrastructure required.	Wrong upgrade/infrastructure is selected. Standards and guidelines are violated
	C2	Wrong information communicated regarding infrastructure and upgrade required.	Design documentation includes wrong information regarding upgrade.
	C3	Information communicated regarding infrastructure and upgrade required is inadequate (e.g. missing detail).	Design documentation includes wrong information regarding upgrade.
	C4	Information communication delayed	Design and tender process is delayed
		Information communicated too early (e.g. before legislation change).	Design decisions do not comply with current standards.
5. Design upgrade and issue tender	T1	Upgrade design delayed.	Delay in upgrade of risky RLXs.
	T1	Tender issue delayed.	Delay in upgrade of risky RLXs.
	T3	Design is inadequate.	Design not fit for purpose.
	T3	Tender is inadequate.	Contractor doesn't understand upgrade requirements.
15. Budget allocated to rail level crossing upgrade program	T1	Budget allocation is delayed.	Delay in upgrade of risky RLXs.
	T2	Budget isn't allocated.	Risky RLXs not upgraded.
	T4	Insufficient budget is allocated	All risky RLXs may not be upgraded.
	C3	Insufficient information communicated to stakeholders regarding upgrade program.	Stakeholders are unclear on upgrade program.
	C4	Budget allocation communication is delayed.	Delay in upgrade of risky RLXs.

TABLE 2.3
Extract of Railway Level Crossing System Lifecycle Emergent Risks

Task	Risk Mode	Task Risk Description	Related Task	Emergent Risk Mode	Emergent Risk Description	Emergent Risk Consequences	P	C
15. Budget allocated to rail level crossing upgrade program	T1	Budget allocation is delayed.	1. Select site for upgrade.	T1	Site selection process is delayed.	Inadequate railway level crossings remain in service for prolonged period.	H	H
	T4	Insufficient budget is allocated.	1. Select site for upgrade.	T3	Site selection is limited to fewer sites.	Inadequate railway level crossings remain in service for prolonged period.	H	H
	T1	Budget allocation is delayed.	6. Commission and construct railway level crossing.	T1	Commissioning and construction of railway level crossing/design upgrade is delayed.	Inadequate railway level crossing remains in service for prolonged period.	H	H
	T2	Budget isn't allocated.	6. Commission and construct railway level crossing.	T2	New railway level crossing/design upgrade is not commissioned.	Inadequate railway level crossing remains in service for prolonged period.	L	H
	T4	Insufficient budget is allocated.	6. Commission and construct railway level crossing.	T3	All required railway level crossings/design upgrades are not implemented.	Inadequate railway level crossings remain in service for prolonged period.	H	H
	C4	Budget allocation communication is delayed.	6. Commission and construct railway level crossing.	T1	Commissioning and construction of railway level crossing/design upgrade is delayed.	Inadequate railway level crossing remains in service for prolonged period.	H	H

RECOMMENDED READING

Dallat, C., Salmon, P. M., & Goode, N. (2018). Identifying risks and emergent risks across sociotechnical systems: The NETworked Hazard Analysis and Risk Management System (NET-HARMS). *Theoretical Issues in Ergonomics Science*, 19:4, 456–482.

Dallat, C., Salmon, P. M., & Goode, N. (2019). Risky systems versus risky people: To what extent do risk assessment methods consider the systems approach to accident causation? A review of the literature. *Safety Science*, 119, 266–279.

Stanton, N. A. (2006). Hierarchical task analysis: Developments, applications, and extensions. *Applied Ergonomics*, 37, 55–79.

REFERENCES

Annett, J., Duncan, K. D., Stammers, R. B., & Gray, M. J. (1971). *Task Analysis: Department of Employment Training Information Paper 6*. London, UK: Her Majesty's Stationary Office (HMSO).

Dallat, C., Salmon, P. M., & Goode, N. (2018). Identifying risks and emergent risks across sociotechnical systems: The NETworked Hazard Analysis and Risk Management System (NET-HARMS). *Theoretical Issues in Ergonomics Science*, 19(4), 456–482.

Dallat, C., Salmon, P. M., Goode, N. (2019). Risky systems versus Risky people: To what extent do risk assessment methods consider the systems approach to accident causation? A review of the literature. *Safety Science*, accepted for publication March 2017.

Embrey, D. E. (1986). SHERPA: A systematic human error reduction and prediction approach. In *Proceedings of the international topical meeting on advances in human factors in nuclear power systems*.

Green, D.M., & Swets, J.A. (1966). *Signal Detection Theory and Psychophysics*. New York: Wiley.

Hulme, A., Stanton, N.A., Walker, G.H., Waterson, P., & Salmon, P.M. (2021a). Testing the reliability and validity of risk assessment methods in Human Factors and Ergonomics. *Ergonomics* in press.

Hulme, A., McLean, S., Dallat, C., Walker, G.H., Waterson, P., Stanton, N.A., & Salmon, P.M. (2021b). Systems thinking-based risk assessment methods applied to sports performance: A comparison of STPA, EAST-BL, and Net-HARMS in the context of elite women's road cycling. *Applied Ergonomics* 91, 103297.

Kirwan, B., Ainsworth, L.K. (1992). *A Guide to Task Analysis*. London, UK.

Mulvihill, C.M., Salmon, P.M., Beanland, V., Lenné, M.G., Read, G.J.M., Walker, G.H., & Stanton, N.A. (2016). Using the decision ladder to understand road user decision making at actively controlled rail level crossings. *Applied Ergonomics* 56, 1–10.

Salmon, P.M., Lenné, M.G., Read, G.J.M., Mulvihill, C.M., Cornelissen, M., Walker, G.H., Young, K.L., Stevens, N., & Stanton, N.A. (2016). More than meets the eye: Using cognitive work analysis to identify design requirements for future rail level crossing systems. *Applied Ergonomics* 53, 312–322.

Salmon, P. M., Read, G. J. M., Walker, G. H., Goode, N., Grant, E., Dallat, C., Carden, T., Naweed, A., & Stanton, N. A. (2018). STAMP goes EAST: Integrating systems ergonomics methods for the analysis of railway level crossing safety management. *Safety Science*, 110:B, 31–46.

Scott, J. (1988). Social network analysis. *Sociology* 22(1), 109–127.

Stanton, N. A. (2006). Hierarchical task analysis: Developments, applications, and extensions. *Applied Ergonomics*, 37, 55–79.

Stanton, N. A. (2014). Representing distributed cognition in complex systems: how a submarine returns to periscope depth. *Ergonomics*, 57(3), 403–418.

Stanton, N. A., Salmon, N. A., & Walker, G. H. (2018). *Systems Thinking in Practice: The Event Analysis of Systemic Teamwork*. Boca Raton, FL: CRC Press.

Stanton, N. A., Salmon, P. M., Baber, C., & Walker, G. (2005). *Human Factors Methods: A Practical Guide for Engineering and Design*. Aldershot, UK: Ashgate.

Stanton, N. A., Salmon, P. M., Rafferty, L., Walker, G. H., Jenkins, D. P., & Baber, C. (2013). *Human Factors Methods: A Practical Guide for Engineering and Design*. 2nd Edition, Aldershot, UK: Ashgate.

Stanton, N.A., & Stevenage, S.V. (1998). Learning to predict human error: Issues of reliability, validity and acceptability. *Ergonomics* 41(11), 1737–1756.

Stanton, N.A., & Young, M.S. (1999). What price ergonomics? *Nature* 399(6733), 197–198.

Svedung, I., & Rasmussen, J. (2002). Graphic representation of accident scenarios: Mapping system structure and the causation of accidents. *Safety Science*, 40(5), 397–417.

3 The Event Analysis of Systemic Teamwork Broken Links (EAST-BL)

BACKGROUND

The Event Analysis of Systemic Teamwork-Broken Links (EAST-BL, Stanton and Harvey, 2017) is an extension to the EAST systems analysis framework that enables it to be used for risk assessment purposes. EAST (Stanton et al., 2013) provides an integrated suite of methods for analysing the performance of complex sociotechnical systems. The framework supports this by providing methods to describe, analyse and integrate three network-based representations of activity: task, social, and information networks.

Task networks are used to provide a summary of the interdependence of activities performed within a system. Social networks are used to describe the human and non-human agents performing tasks within the system and the communications that take place between them during task performance. Information networks describe the information that agents use when performing tasks and how this information is distributed across different tasks and system agents. The EAST 'network of networks' approach is represented in Figure 3.1.

Applying EAST-BL involves breaking the links in EAST task, social, and information networks to identify failures in communication and information transfer and any risks associated with them. According to Stanton and Harvey (2017), broken links represent "failures in communication and information transfer between nodes in the networks and these failures can be used to make predictions about the possible risks within the sociotechnical system" (p. 227).

DOMAIN OF APPLICATION

EAST was originally developed for use in describing and evaluating military and civilian command and control systems (Stanton et al., 2013). It has since been applied in all manner of domains ranging from aviation, road and rail transport, to defence, elite sport, and darknet marketplaces (Stanton et al., 2018).

EAST-BL was originally applied to identify the risk to life associated with a maritime missile attack training system (Stanton and Harvey, 2017); however, the approach is generic and can be applied in any domain. For example, Lane et al. (2019) recently used the EAST-BL approach to identify ways in which to disrupt the trading of illicit identify credentials in darknet marketplaces. Hulme et al. (2021) recently applied EAST-BL to identify risks to performance in elite women's road race cycling.

DOI: 10.1201/9780429281624-3

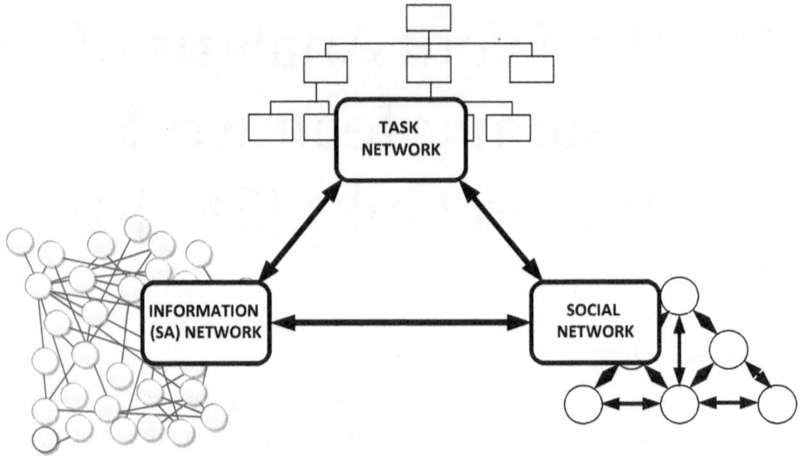

FIGURE 3.1 EAST network of network approach for systems analysis.

PROCEDURE AND ADVICE

A flowchart depicting the EAST-BL procedure is presented in Figure 3.2. Step-by-step guidance for the EAST-BL process is presented below. For a more detailed description of network construction the reader is referred to Chapter 7.

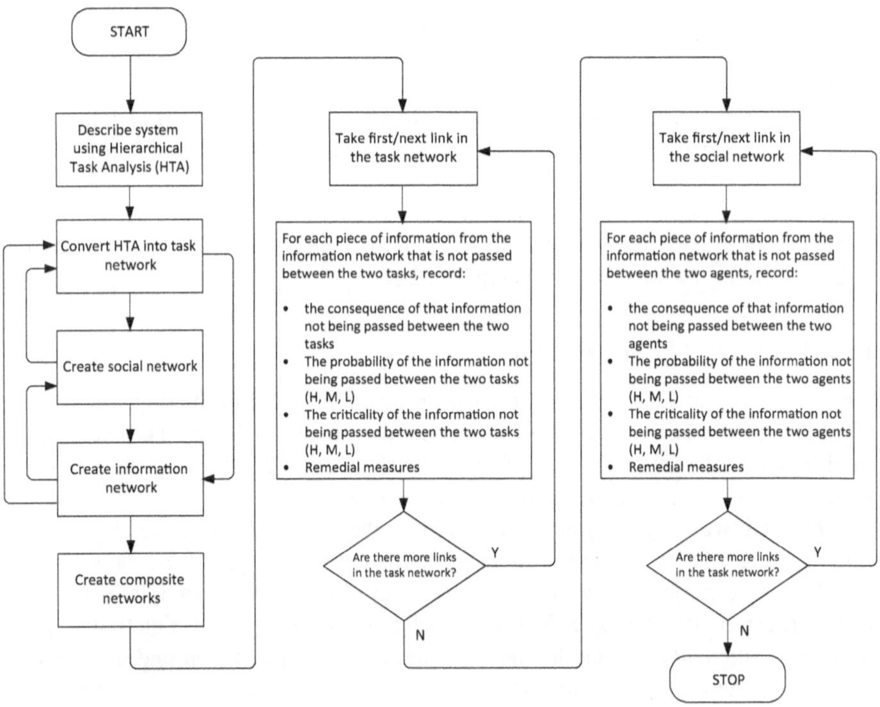

FIGURE 3.2 EAST-BL flowchart.

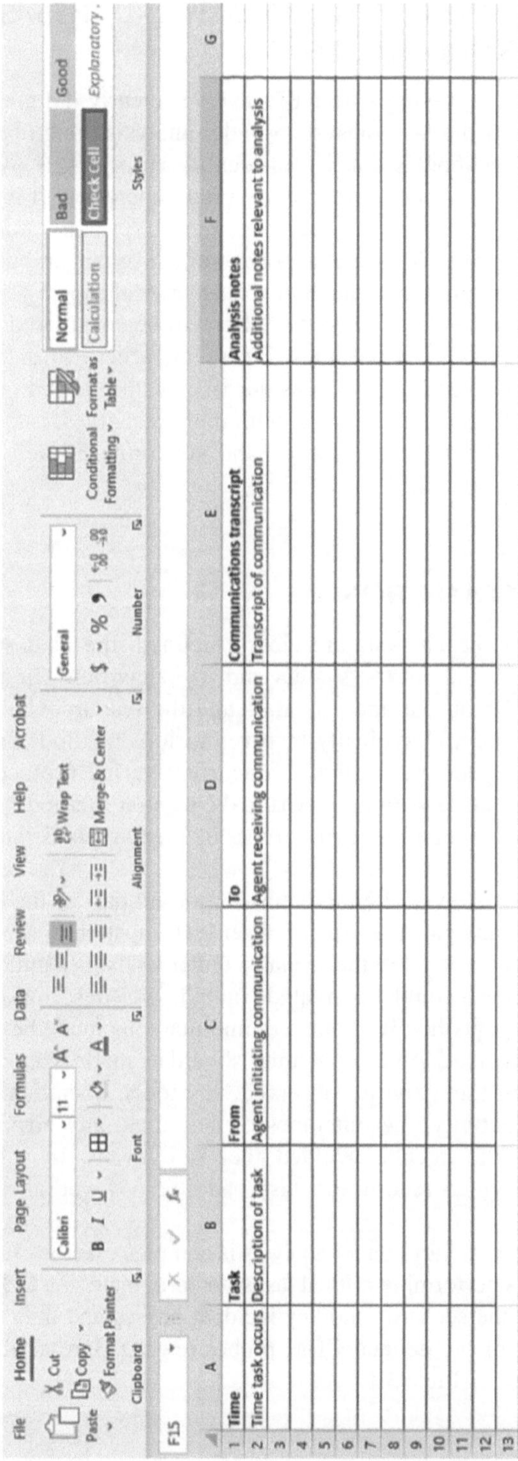

FIGURE 3.3 Generic event transcript.

STEP 1: DEFINE AIMS OF THE ANALYSIS AND THE TASK OR SYSTEM UNDER ANALYSIS

The first step in an applying EAST-BL involves clearly defining the aims of the analysis along with the task or system under analysis. One of the key strengths of the EAST-BL method is that it can identify risks across overall systems, as opposed to risks only at the sharp end of system operation. It is therefore recommended that EAST-BL analyses focus on overall systems rather than front-line tasks or operations only. This includes beyond the organisational level up to and including regulatory and government levels. A critical aspect of this step involves clearly defining the levels of the system that will be considered during the analysis. For example, based on the aims, resources, access to data and time available it may be only necessary to consider risks within a particular organisation. Alternatively, on larger scale projects with more resources, it may be appropriate to consider risks also at the regulatory and government levels of the system in question.

STEP 2: DATA COLLECTION

Once the aims of the analysis are clearly defined, the next step involves collecting targeted data about the system and its behaviour. The specific data collected is dependent on the analysis aims and the resources available; however, data collection for EAST typically involves various methods including observations, recording of communications, concurrent verbal protocols, structured or semi-structured interviews (e.g. the Critical Decision Method), task walkthrough analyses and documentation review (e.g. incident reports, standard operating procedures).

The observation step is often the most important part of the EAST procedure. Typically, several analysts are used to observe the system or scenario under analysis. All activities involved in the scenario under analysis should be recorded and an activity timeline should be created, including a description of the activity undertaken, the agents involved, any communications made between agents, and the technology involved. Additional notes should be made where required, including the purpose of the activities observed, any tools, documents, or instructions used to support activity, the outcomes of activities, any errors made, and any information that the agents involved feel is relevant. In addition, it is normal practice to video-record the task, and record verbal transcripts of all communications.

Once the task or scenario under analysis is complete, it is useful for 'key' human agents (e.g. agents performing critical tasks) to participate in a CDM interview. This involves dividing the scenario into key incident phases and then interviewing each agent using a set of pre-defined CDM probes (e.g. O'Hare et al., 2000). A set of probes is presented in Table 3.1.

TABLE 3.1
CDM Probes

Goal Specification	What Were You Aiming to Accomplish Through This Activity?
Assessment	Suppose you were to describe the situation at this point to someone else. How would you summarise the situation?
Cue Identification	What features were you looking for when you formulated your decision?
	How did you know that you needed to make the decision? How did you know when to make the decision?
Expectancy	Were you expecting to make this sort of decision during the event?
	Describe how this affected your decision-making process.
Options	What courses of action were available to you? Were there any other alternatives available to you other than the decision you made?
	How/why was the chosen option selected? Why were the other options rejected?
	Was there a rule that you were following at this point?
Influencing factors	What factors influenced your decision-making at this point?
	What was the most influential factor/piece of information that influenced your decision-making at this point?
Situation Awareness	What information did you have available to you at the time of the decision?
Situation Assessment	Did you use all the information available to you when formulating the decision?
	Was there any additional information that you might have used to assist in the formulation of the decision?
Experience	What specific training or experience was necessary or helpful in making this decision?
	Do you think further training is required to support decision-making for this task?
Mental models	Did you imagine the possible consequences of this action?
	Did you create some sort of picture in your head? Did you imagine the events and how they would unfold?
Decision making	How much time pressure was involved in making the decision? How long did it actually take to make this decision?
Conceptual	Are there any situations in which your decision would have turned out differently?
Guidance	Did you seek any guidance at this point in the task/incident? Was guidance available?
Basis of choice	Do you think that you could develop a rule, based on your experience, which could assist another person to make the same decision successfully?

(Adapted from O'Hare et al, 2000 and Crandall et al., 2006.)

STEP 3: TRANSCRIBE DATA

Once all data is collected, it should be transcribed in order to make it usable during the EAST analysis phase. An event transcript should then be constructed (see Figure 3.3). This should describe the scenario over a timeline, including descriptions of activity, the agents involved, any communications made, and the technology used. In order to ensure the validity of the data, the event transcript should be reviewed by as many subject matter experts as possible.

STEP 4: CONSTRUCT A HIERARCHICAL TASK ANALYSIS FOR THE TASK/SCENARIO/SYSTEM UNDER ANALYSIS

Once the data is collected, transcribed, and verified, the next step involves creating a Hierarchical Task Analysis (HTA; Stanton, 2006) for the task/scenario/system under analysis. HTA works by decomposing systems and behaviour into a hierarchy of goals, subordinate goals, operations, and plans. In doing so, it focuses on "what an operator is required to do, in terms of actions and/or cognitive processes to achieve a system goal" (Kirwan and Ainsworth 1992, p. 1). HTA outputs specify the overall goal of a system, the sub-goals to be undertaken to achieve this goal, the operations required to achieve each of the sub-goals specified and the plans that trigger engagement and achievement of the different goals and operations. The plans component specifies the sequence, and under what conditions different sub-goals must be achieved in order to satisfy the requirements of a super-ordinate goal. For the full HTA procedure see Chapter 5. Once the HTA is complete it is important to have it reviewed by appropriate subject matter experts. Following this, the HTA is often used to identify task or scenario phases for which task, social, and information networks will be constructed.

STEP 5: CREATE TASK NETWORK

The first analysis step involves constructing a task network. Task networks (see Figure 3.4) are used to represent HTA outputs in the form of a network which shows key tasks and the relationships between them (Stanton et al., 2013). Constructing the task network involves identifying high level tasks, typically from the first layer of

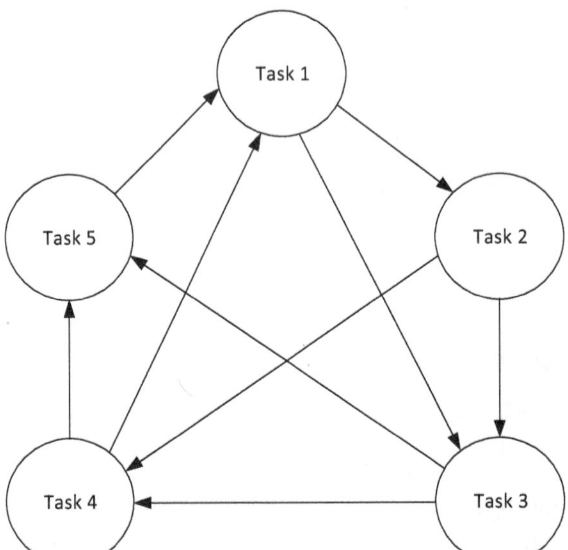

FIGURE 3.4 Task network example.

sub-goals in the HTA, and the relationships between them and creating a network to represent this. Relationships between tasks are included in the task network when:

1. Tasks are undertaken sequentially e.g. task 3 is undertaken after completion of task 2;
2. Task are undertaken together e.g. tasks 1 and 2 are undertaken together;
3. The outcomes of one task influence the conduct of another e.g. the outcomes of task 4 influence how task 5 is undertaken; or
4. The conduct of one task is dependent on completion of the other e.g. task 5 cannot be undertaken until task 4 is completed.

STEP 6: CREATE SOCIAL NETWORK

Social Network Analysis (SNA; Driskell and Mullen, 2004) is used to describe and analyse the relationships (e.g. communications) between the agents involved in the task/scenario/system under analysis. This involves first creating a social network matrix showing the relationships between agents in terms of interactions (e.g. communications) and the frequency of interactions (e.g. number of communications between agents). Following this a social network diagram is constructed which provides a visual representation of the social network (see Figure 3.5). Typically, the direction (i.e. from actor A to actor B) and frequency of communications is depicted; however, the type and content of associations should also be recorded. It is important to ensure that SNAs are undertaken for the same task or scenario phases that the task and information networks are used to describe.

STEP 7: CREATE INFORMATION NETWORKS

The next step of the initial EAST analysis involves constructing information networks for the task/scenario/system under analysis. Information networks can be constructed either manually or via the Leximancer software tool using data derived from communications transcripts or concurrent verbal protocols provided by participants as they perform the task or scenario under analysis. If this data is not available, however, they can also be developed based on cognitive task analysis interview data, procedures, or accident investigation reports.

Information networks depict the information or concepts underlying situation awareness and the relationships between them. For example, Figure 3.6 shows the information network resulting from a driver's verbal protocol transcript extract "the red car behind is slowing down". As shown in Figure 3.6, this produces a network comprising the linked concepts 'Car', 'Red', 'Behind' and 'Slowing' as in the car 'is' red, it is located 'behind' the participant's vehicle and it 'is' slowing down.

Manual construction of the networks thus involves performing a content analysis on verbal transcript or communications log data to identify concepts and the relationships between them. It is often useful to have a second analyst also construct an information network (or at least a portion of it) from the same data for the purposes of reliability testing.

The Leximancer thematic analysis software can be used to auto-construct information networks. Leximancer uses text representations of natural language to

From \ To	Agent A	Agent B	Agent C	Agent D
Agent A	-	20	2	10
Agent B	17	-	4	8
Agent C	1	5	-	8
Agent D	10	10	1	-

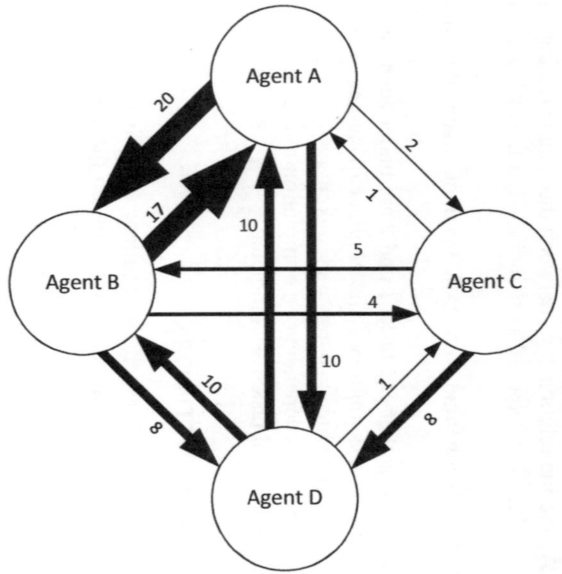

FIGURE 3.5 Simple social network matric and associated social network diagram.

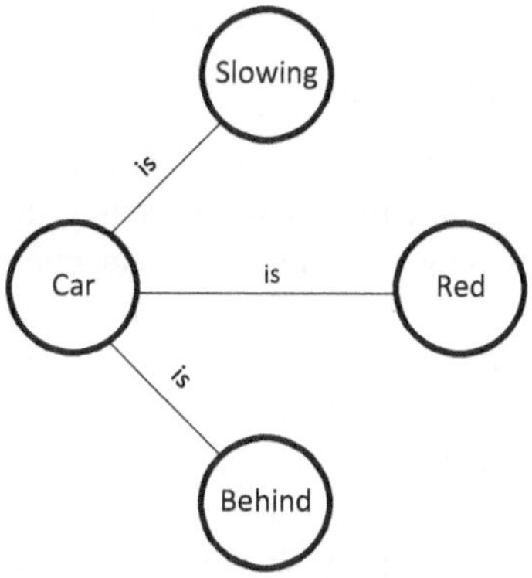

FIGURE 3.6 Simple example information network for concurrent verbal protocol extract "the red car behind is slowing down".

interrogate verbal transcripts and identify themes, concepts, and the relationships between them. The software does this by using algorithms linked to an in-built thesaurus and by focusing on features within the verbal transcripts such as word proximity, quantity, and salience. Leximancer thus automates the content analysis procedure by processing verbal transcript data through five stages: 1) conversion of raw text data, 2) concept identification, 3) thesaurus learning, 4) concept location, and 5) mapping (i.e. creation of network). The output is a concept map, similar to an information network, which shows concepts derived from the verbal transcript and the relationships between them reflected within the verbalisations.

STEP 8: CREATE COMPOSITE NETWORKS

Composite networks are used to explore the relationships between tasks, agents, and information and are constructed by combining the different networks. For example, a task by agent network can be constructed by combining the task and social network to show which tasks are undertaken by which agents. This involves assigning a colour to the different agents within the social network and shading each node within the task network to show which agent performs that task. Useful composite networks to construct include:

- Task by agent network (task-social network showing which tasks are performed by different agents);
- Information by agent network (social-information network showing which agents use what information);
- Task and associated information network (task-information network showing what information is used when performing each task and
- Fully composite network showing information by agents and related to tasks (combined task, social, and information network).

For EAST-BL analyses it is not necessary to produce composite networks; however, it is useful to construct task-information and social-information networks as these show what information is passed between tasks and between agents.

STEP 9: BREAK LINKS IN TASK NETWORK AND IDENTIFY ASSOCIATED RISKS

Once the networks are finalised, the task network is first subjected to the EAST-BL process. This involves systematically breaking each of the relationships between the tasks in the task network and identifying what risks emerge when the relevant information from the information network is not transferred between tasks. For example, in a driving task network, the two tasks 'assess current speed' and 'drive within speed limit' would be related. Applying EAST-BL, the analyst would break the link, and then determine what the outcome would be if relevant information was not transferred between the two tasks. In this case, if the link between the two tasks was broken, the information 'current speed' would not be available when performing the task 'drive within speed limit'. A consequence of this may be that the driver exceeds the speed limit as they are not aware of their current speed. It is important to note that when applying EAST-BL, analysts do not need to justify breaking any of the relationships, nor give reasons why the relationship is broken. Rather, the process involves systematically breaking every relationship in the task and social network, and identifying the associated risks. A task network EAST-BL proforma with example is presented in Table 3.2 As a first task during step 9 it is useful to construct the task network EAST-BL proforma by identifying all related tasks in the task network and then information from the information network that is passed between the related tasks. This provides a table of related tasks and associated information that the analyst can systematically work through to identify risks.

STEP 10: BREAK LINKS IN SOCIAL NETWORK AND IDENTIFY ASSOCIATED RISKS

Once the task network breaks have been assessed, the social network is subjected to the EAST-BL process. This involves systematically breaking each of the relationships between the agents in the social network and identifying what risks emerge when the relevant information from the information network is not transferred between agents. It is important to note here that agents can be both human and non-human in nature. For example, in a driving social network, the two agents' 'driver' and 'traffic light' would be related, as they interact when drivers negotiate signalised intersections. Applying EAST-BL, the analyst would break the relationship between both agents

TABLE 3.2
Task Network EAST-BL Template

From (Task)	To (Task)	Information Not Transferred	Outcome	Remedial Measure
Assess current speed	Drive within speed limit	Current speed	Driver could exceed the post speed limit	Intelligent speed adaptation system Speed limit broken alert

and then determine what the outcome would be if relevant information was not transferred between them. In this case, if the relationship between the two agents was broken, the information 'traffic light status' would not be transferred to the driver when negotiating the intersection. A consequence of this may be that the driver proceeds through a red signal. Again, it worth noting that, when applying EAST-BL, analysts do not need to justify breaking any of the relationships or give reasons why the relationship is broken. Rather, the process involves systematically breaking every relationship in the task and social network and identifying the associated risks. A social network EAST-BL proforma with example is presented in Table 3.3.

STEP 11: IDENTIFY REMEDIAL MEASURES

The final stage in the process is to propose risk management strategies for the identified risks. It is recommended that a group of analysts or SMEs be used to identify appropriate risk management strategies. The group should work through the task and social network risks and discuss risk management strategies that can be used to either prevent the risk from occurring or mitigate the consequences of the risk. Normally, risk management strategies comprise suggested changes to the design of the process or system. According to Stanton (2005), remedial measures proposed following risk assessment applications are normally proposed under the following four categories:

1. Equipment (e.g. redesign or modification of existing equipment);
2. Training (e.g. changes in training provided);
3. Procedures (e.g. provision of new, or redesign of old, procedures); and
4. Organizational (e.g. changes in organisational policy or culture).

STEP 12: REVIEW AND REFINE ANALYSIS

Once the initial EAST-BL analysis is complete it is useful to have various SMEs review it. The analysis should then be refined based on SME feedback. This typically involves removing identified risks, identifying new risks, or adding or modifying risk management strategies.

TABLE 3.3
Social Network Broken Links Template

From (Agent)	To (Agent)	Information Not Transferred	Outcome	Remedial Measure
Driver	Traffic lights	Traffic light status.	Driver is not aware that the traffic lights are on red and may proceed through a red light	Vehicle to Infrastructure communications (i.e. traffic lights communicate with vehicle and vehicle alerts the driver or stops automatically)

ADVANTAGES

- EAST-BL can be used to identify risks across overall work systems;
- As every link in the networks is systematically broken, the process requires less subjective judgement than other risk assessment methods;
- EAST-BL can be used to examine both task to task risks and agent to agent risks;
- The analysis is extremely comprehensive, covering risks across all interactions with task and social networks;
- EAST-BL analyses provide risk management strategies for each of the risks identified;
- The method is based on EAST which is a highly popular systems analysis framework that has been applied in many domains (Stanton et al., 2018);
- Whilst EAST can be difficult to learn and apply, the EAST-BL process is easy to learn and apply, and requires minimal training;
- The EAST-BL method is generic, allowing it to be used in any domain; and
- Conducting the HTA and task, social, and information networks enable analysts to develop an in-depth understanding of the system under analysis.

DISADVANTAGES

- As it requires an initial EAST analysis, EAST-BL can be time consuming to apply;
- The systematic breaking of all links within the task and social network can become laborious for large analyses;
- EAST analyses can become large, complex, and unwieldy, especially when applied to overall work systems;
- There is little reliability and validity evidence associated with EAST or EAST-BL.
- Some predicted risks and risks management strategies may be unlikely or lack credibility, thus posing a false economy (Stanton, 2005); and
- There is currently no performance shaping factors taxonomy to support consideration of contextual factors and their impact on risks.

RELATED METHODS

EAST-BL requires an initial EAST analysis comprising task, social, and information networks. EAST typically requires an initial HTA of the system under analysis. This is then used to develop the EAST task network.

APPROXIMATE TRAINING AND APPLICATION TIMES

EAST-BL is a relatively simple method that requires little training. In a recent reliability and validity study, participants were provided with training that lasted around two hours and were able to achieve acceptable levels of validity (Hulme et al., 2021). The training time for EAST, however, is longer. Typically, EAST is trained via full day workshops lasting up to 7 hours in duration. Application time depends on the system under analysis; however, EAST-BL analyses typically require significant time as the analysis covers overall systems rather than merely sharp-end performance.

RELIABILITY AND VALIDITY

Hulme et al. (2021) recently compared the reliability and validity of Net-HARMS (see Chapter 2), STPA (see Chapter 4) and EAST-BL when used to identify risks across a railway level crossing design lifecycle. Based on a comparison of the identified risks with a gold standard and an analysis using the Signal Detection Theory (SDT) paradigm (Green and Swets 1966), Hulme et al. (2021) found a weak to moderate positive correlation coefficient for EAST-BL, meaning that the method did indeed support relatively novice analysts with the identification of credible system risks, albeit up to a point.

TOOLS NEEDED

EAST and EAST-BL can be conducted using pen and paper; however, EAST-BL analyses are normally undertaken using a Microsoft Excel spreadsheet. The HTA tool can be used for the HTA component and task, social, and information networks are typically drawn in Microsoft Visio. SNA tools such as SOCNET Viewer or Agna can be used to construct and analyse the task, social, and information networks.

CASE STUDY EXAMPLE: DARKNET MARKETPLACES

The darknet is a layer of the Internet that cannot be indexed and searched through popular search engines such as Google, and cannot be accessed via conventional browsers such as Google Chrome and Mozilla Firefox. Within the darknet various marketplaces exist to facilitate the trade of both legitimate and illicit products and services. A key feature of the darknet is that applications enable Internet traffic to route through random pathways in order to disguise their point of origin (Bradbury, 2014). This allows users to communicate and trade without their real identity becoming known to authorities, reducing the likelihood of detection and prosecution (Martin, 2014). Consequently, darknet marketplaces are used as a means to trade illicit goods including drugs, firearms, child pornography, and stolen identity credentials (Bhaskar et al., 2019; Chen, 2012; Haasio et al., 2020; Europol, 2017a; Lane et al., 2019; Rhumorbarbe et al., 2016). Transactions span continents (Broséus et al. 2017) and market administrators have been known to earn millions of dollars in illicit profits (US Department of Justice, 2017).

Darknet markets represent a significant challenge for law enforcement agencies. Many marketplaces of different sizes exist, owners, vendors and sellers remain anonymous, and marketplaces are agile and constantly evolving. These marketplaces provide users with access to a global network of suppliers selling illicit goods, presenting novel challenges compared with traditional offline illicit trading. Of the known successful interventions to date, most have involved the closure of marketplaces or the arrest of key marketplace stakeholders. For example, Van Buskirk et al. (2014) describe three significant disruptions: the FBI's seizure of Silk Road in 2013, the seizure of multiple marketplaces via Operation Onymous in 2014, and the closure of the Evolution marketplace in 2015. Since these occurrences, Operation Bayonet saw the closure of AlphaBay, the largest marketplace to have emerged, with users drawn to a second marketplace, Hansa Market, under the control of law enforcement (Europol, 2017b). Despite successful interventions such as these, darknet marketplaces have proven to be resilient through avenues such as the rapid creation of new marketplaces or the return of vendors (Van Buskirk et al., 2014; Europol, 2017a). Novel forms of intervention are required along with research exploring different mechanisms for disrupting marketplace trading (Europol, 2017a; Lane et al., 2019).

Recent research has explored the extent to which HFE methods in particular systems HFE methods, can be used to inform the development of strategies to diminish the functioning of systems that have been created to achieve illicit ends (Lane et al., 2019; Salmon et al., 2018). Lane et al. (2019) used the Event Analysis of Systemic Teamwork (EAST; Stanton et al., 2013) to analyse the processes involved when

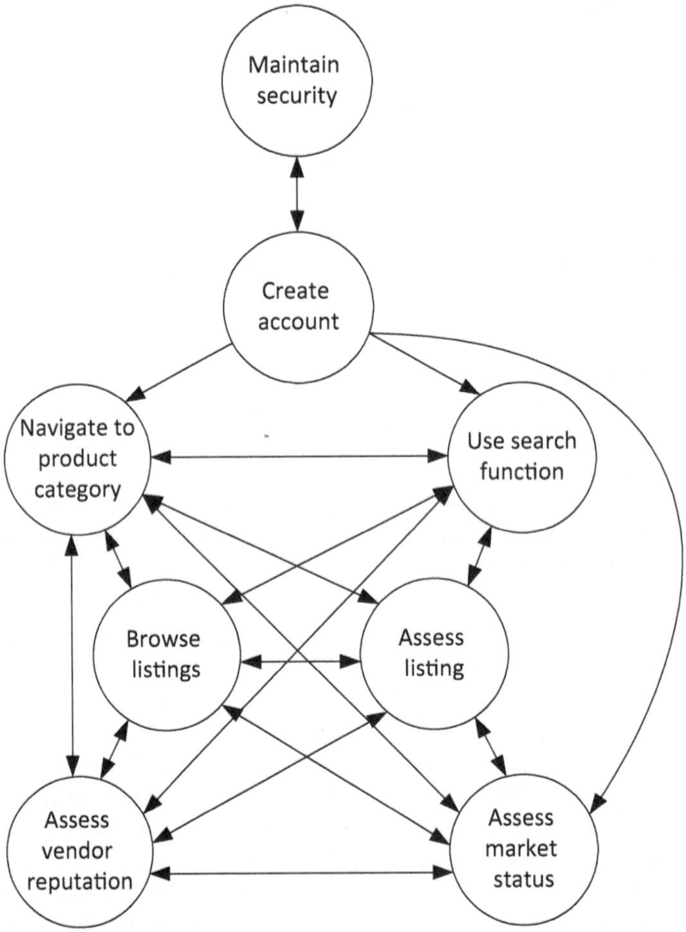

FIGURE 3.7 Task network for DreamMarket trading task 'Create account and search listings'. (Adapted from Lane et al., 2019).

buying and selling identity credentials on DreamMarket (a darknet marketplace that has since been closed). Lane et al. (2019) subsequently applied the EAST-Broken-Links approach (EAST-BL; Stanton and Harvey, 2017) to identify strategies for disrupting trading. A task network representing the key phase of creating an account and searching listings is presented in Figure 3.7. A social network for the same tasks is presented in Figure 3.8. An extract of the task-to-task EAST-BL analysis is presented in Table 3.4. An extract of the agent-to-agent EAST-BL analysis is presented in Table 3.5.

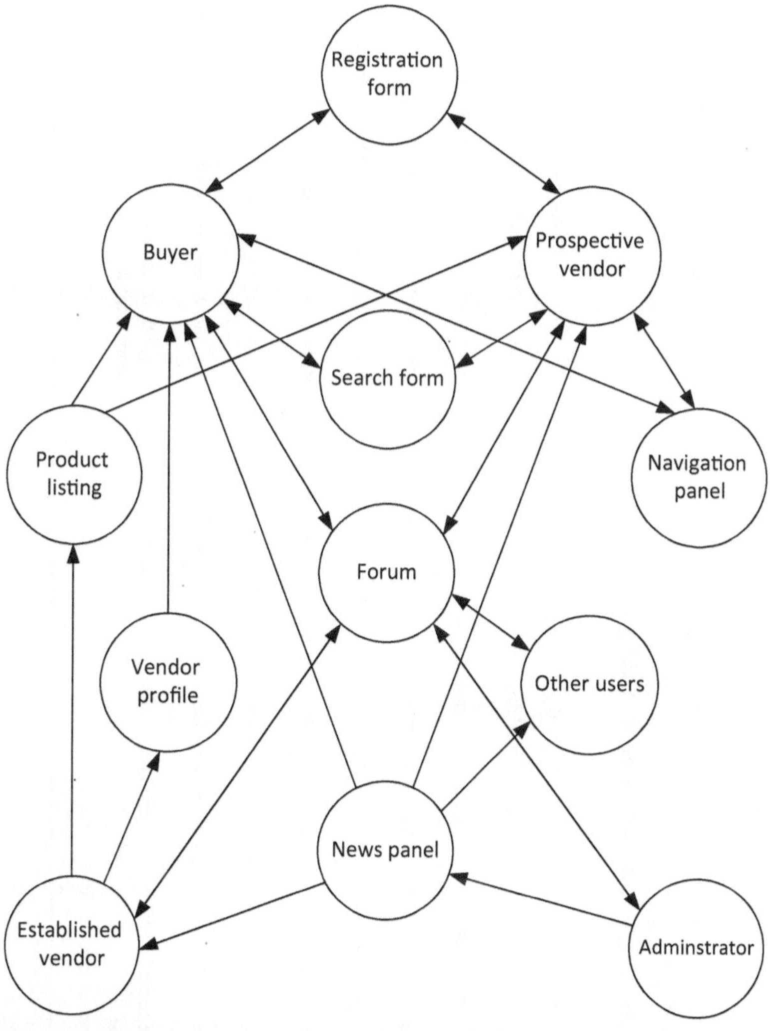

FIGURE 3.8 Social network for DreamMarket trading task 'Create account and search listings'. (Adapted from Lane et al., 2019).

TABLE 3.4
Extract of Task Network (Task to Task) EAST-BL Analysis

From (Task)	To (Task)	Information Not Communicated	Outcome	Intervention Strategy
Negotiate solution	Create formal dispute	*Order status*	Order status does not inform whether a formal dispute is necessary	Prevent order status from being communicated
Negotiate solution	Create formal dispute	*Order details*	Order details are not communicated	Prevent order details from being communicated
Negotiate solution	Create formal dispute	*Username*	Unclear which users are involved	Prevent username from displaying in messaging service
Negotiate solution	Create formal dispute	*Solution*	Solution unknown; unclear how to proceed	Prevent messages from displaying in messaging service
Negotiate solution	Create formal dispute	*Dispute message*	Message is not communicated	Prevent messages from displaying in messaging service
Create formal dispute	Negotiate solution	*Tracking number*	No verification that order was shipped	Prevent users from accessing tracking information
Create formal dispute	Negotiate solution	*Solution*	No solution communicated	Prevent messages from displaying in messaging service
Create formal dispute	Negotiate solution	*Dispute message*	Previous messages about dispute not communicated	Prevent messages from displaying in messaging service

Source: Lane et al., 2019

TABLE 3.5
Extract of Social Network (Agent to Agent) EAST-BL Analysis

From (Agent)	To (Agent)	Information Not Communicated	Outcome	Intervention Strategy
Vendor	Product listing	*Packaging*	Product listing does not describe packaging	Prevent vendor from describing packaging on listing
Vendor	Product listing	*Product qualities*	Product listing does not accurately describe product qualities	Prevent vendor from describing product qualities on listing
Vendor	Product listing	*Price*	Product listing does not display price	Prevent vendor from entering price on product listing
Vendor	Product listing	*Escrow option*	Product listing does not display escrow option	Prevent vendor from selecting whether escrow is available
Vendor	Product listing	*Product image*	Product listing does not display image	Prevent vendor from uploading product image
Vendor	Product listing	*Product description*	Product listing does not have description	Prevent vendor from typing product description
Vendor	Product listing	*Title*	Product listing does not have title	Prevent vendor from entering title for listing
Vendor	Product listing	*Shipping details*	Product listing does not have shipping details	Prevent vendor from selecting shipping details for listing
Forum	Buyer	*Username*	User account associated with forums posts not communicated	Prevent username from appearing with user forum posts
Forum	Buyer	*Product reviews*	Product reviews not communicated	Prevent product reviews from displaying on forum
Forum	Buyer	*Product description*	Product description not communicated	Prevent product descriptions from displaying on forum
Forum	Buyer	*Title*	Title not communicated	Prevent product titles from displaying on forum
Forum	Buyer	*Vendor reviews*	Vendor reviews not communicated	Prevent vendor reviews from displaying on forum
Forum	Buyer	*Product adverts*	Product adverts not communicated	Prevent product adverts from displaying on forum

Source: Lane et al., 2019

RECOMMENDED READING

Lane, B. R., Salmon, P. M., Cherney, A., Lacey, D., & Stanton, N. A. (2019). Using the Event Analysis of Systemic Teamwork (EAST) broken-links approach to understand vulnerabilities to disruption in a darknet market. *Ergonomics.*

Stanton, N. A., & Harvey, C. (2017). Beyond human error taxonomies in assessment of risk in sociotechnical systems: A new paradigm with the EAST 'Broken-Links' approach. *Ergonomics*, 60, 2.

Stanton, N. A., Salmon, N. A., & Walker, G. H. (2018). *Systems Thinking in Practice: The Event Analysis of Systemic Teamwork.*Boca Raton, FL: CRC Press.

Stanton, N. A., Salmon, P. M., Rafferty, L., Walker, G. H., Jenkins, D. P., & Baber, C. (2013). *Human Factors Methods: A Practical Guide for Engineering and Design.* 2nd Edition, Aldershot, UK: Ashgate.

REFERENCES

Bhaskar, V., Linacre, R., & Machin, S. (2019). The economic functioning of online drug markets. *Journal of Economic Behavior & Organization*, 159, 426–441.

Bradbury, D. (2014). Unveiling the dark web. *Network Security*, 4, 14–17.

Broséus, J., Rhumorbarbe, D., Morelato, M., Staehli, L., & Rossy, Q. (2017). A geographical analysis of trafficking on a popular darknet market. *Forensic Science International*, 277, 88–102. doi:10.1016/j.forsciint.2017.05.021

Chen, H. (2012). Dark Web research overview. In *Dark Web*, New York: Springer.

Crandall, B., Klein, G. A., & Hoffman, R. R. (2006). *Working Minds: A Practitioner's Guide to Cognitive Task Analysis.* Cambridge: Mit Press.

Driskell, J. E., & Mullen, B. (2004). Social network analysis. In *Handbook of Human Factors and Ergonomics Methods* (pp. 565–571). CRC Press.

Europol. (2017a). Drugs and the darknet: perspectives for enforcement, research and policy. https://www.europol.europa.eu/publications-documents/drugs-and-darknet-perspectives-for-enforcement-research-and-policy, accessed 27 August 2019.

Europol. (2017b). Massive blow to criminal dark web activities after globally coordinated operation. https://www.europol.europa.eu/newsroom/news/massive-blow-to-criminal-dark-web-activities-after-globally-coordinated-operation, accessed 27 August 2019.

Haasio, A., Harviainen, J. T., & Savolainen, R. (2020). Information needs of drug users on a local dark Web marketplace. *Information Processing & Management*, 57, 2, 1–13.

Hulme, A., McLean, S., Dallat, C., Walker, G. H., Waterson, P., Stanton, N. A., & Salmon, P. M. (2021). Systems thinking-based risk assessment methods applied to sports performance: A comparison of STPA, EAST-BL, and Net-HARMS in the context of elite women's road cycling. *Applied Ergonomics*, 91, 103297.

Kirwan, B., & Ainsworth, L. K. (Eds.). (1992). *A Guide to Task Analysis: The Task Analysis Working Group.* CRC Press.

Lane, B. R., Salmon, P. M., Cherney, A., Lacey, D., & Stanton, N. A. (2019). Using the Event Analysis of Systemic Teamwork (EAST) broken-links approach to understand vulnerabilities to disruption in a darknet market. *Ergonomics*, 62(9), 1134–1149.

Martin, J. (2014). *Drugs on the Darknet: How Cryptomarkets are Transforming the Global Trade in Illicit Drugs.* Palgrave Macmillan, UK: Palgrave Pivot.

Rhumorbarbe, D., Staehli, L., Broséus, J., Rossy, Q., & Esseiva, P. (2016). Buying drugs on a darknet market: A better deal? Studying the online illicit drug market through the analysis of digital, physical and chemical data. *Forensic Science International* 267, 173–182.

Salmon, P. M., Carden, T., & Stevens, N. J. (2018). Breaking Bad Systems: Using Work Domain Analysis to Identify Strategies for Disrupting Terrorist Cells. *Proceedings of the Chartered Institute of Ergonomics and Human Factors Annual Meeting 2018*, Daventry, UK.

Stanton, N. A. (2006). Hierarchical task analysis: Developments, applications, and extensions. *Applied Ergonomics*, 37, 55–79.

Stanton, N. A., & Harvey, C. (2017). Beyond human error taxonomies in assessment of risk in sociotechnical systems: A new paradigm with the EAST 'Broken-Links' approach. *Ergonomics*, 60, 2.

Stanton, N. A., Salmon, N. A., & Walker, G. H. (2018). *Systems Thinking in Practice: The Event Analysis of Systemic Teamwork.* Boca Raton, FL: CRC Press.

Stanton, N. A., Salmon, P. M., Baber, C., & Walker, G. (2005). *Human Factors Methods: A Practical Guide for Engineering and Design*. Ashgate, Aldershot, UK.

Stanton, N. A., Salmon, P. M., Rafferty, L., Walker, G. H., Jenkins, D. P., & Baber, C. (2013). *Human Factors Methods: A Practical Guide for Engineering and Design.* 2nd Edition, Aldershot, UK: Ashgate.

Stanton, N. A., & Young, M. S. (1999). *Guide to Methodology in Ergonomics: Designing for Human Use*. CRC Press.

US Department of Justice. (2017). Acting Manhattan U.S. Attorney Announces Forfeiture of $48 Million from Sale of Silk Road Bitcoins. https://www.justice.gov/usao-sdny/pr/acting-manhattan-us-attorney-announces-forfeiture-48-million-sale-silk-road-bitcoins

Van Buskirk, J., A. Roxburgh, M. Farrell, and L. Burns. 2014. The closure of the silk road: what has this meant for online drug trading. *Addiction* 109(4), 517–518. doi:10.1111/add.12422

4 The Systems Theoretic Process Analysis (STPA) Method

BACKGROUND

The Systems Theoretic Accident Model and Processes (STAMP) (Leveson, 2004) is a model of accident causation which has an associated proactive risk assessment method, the Systems Theoretic Process Analysis (STPA) (Leveson, 2011; Leveson and Thomas, 2018), and a retrospective incident analysis method, the Causal Analysis based on Systems Theory (CAST, See Chapter 11) (Leveson, 2004; Leveson et al., 2009). STAMP is based on systems and control theory and suggests that adverse events occur when interactions between system components are not controlled through managerial, organisational, physical, operational, and manufacturing-based controls. According to Leveson (2004), safety risks are managed through a hierarchy of controls and feedback mechanisms, and adverse events emerge when behaviour and emergent properties are not adequately controlled. STAMP therefore views safety as an issue of control and one that should be managed through a control structure that has the goal of enforcing constraints on the behaviour of key agents and factors from across the system.

STPA (Leveson, 2011; Leveson and Thomas, 2018) is a risk assessment method that is used to forecast instances where the control and feedback mechanisms in the control structure model could fail, with the outputs supporting the proactive development of risk controls. STPA uses an unsafe control actions taxonomy that is applied to a control structure model of the system in question to identify credible risks. The control structure model is used to describe the control and feedback mechanisms that exist within the system during both system development and operation. A generic control structure is presented in Figure 4.1. The control structure takes inspiration from Rasmussen's (1997) risk management framework and incorporates a series of hierarchical system levels to represent the structure of the overall work system. Levels of the work system include congress and legislatures at the top (i.e. government and policy) all the way down to the operations management and operating processes that occur at the bottom. Accordingly, systems are viewed as comprising interrelated components that maintain a state of dynamic equilibrium through feedback loops of control and information across all levels (Leveson 2004). Each level therefore includes a description of the relevant agents and organisations that play a role in system design or operation. Control and feedback mechanisms are included to show what controls are enacted down the

DOI: 10.1201/9780429281624-4

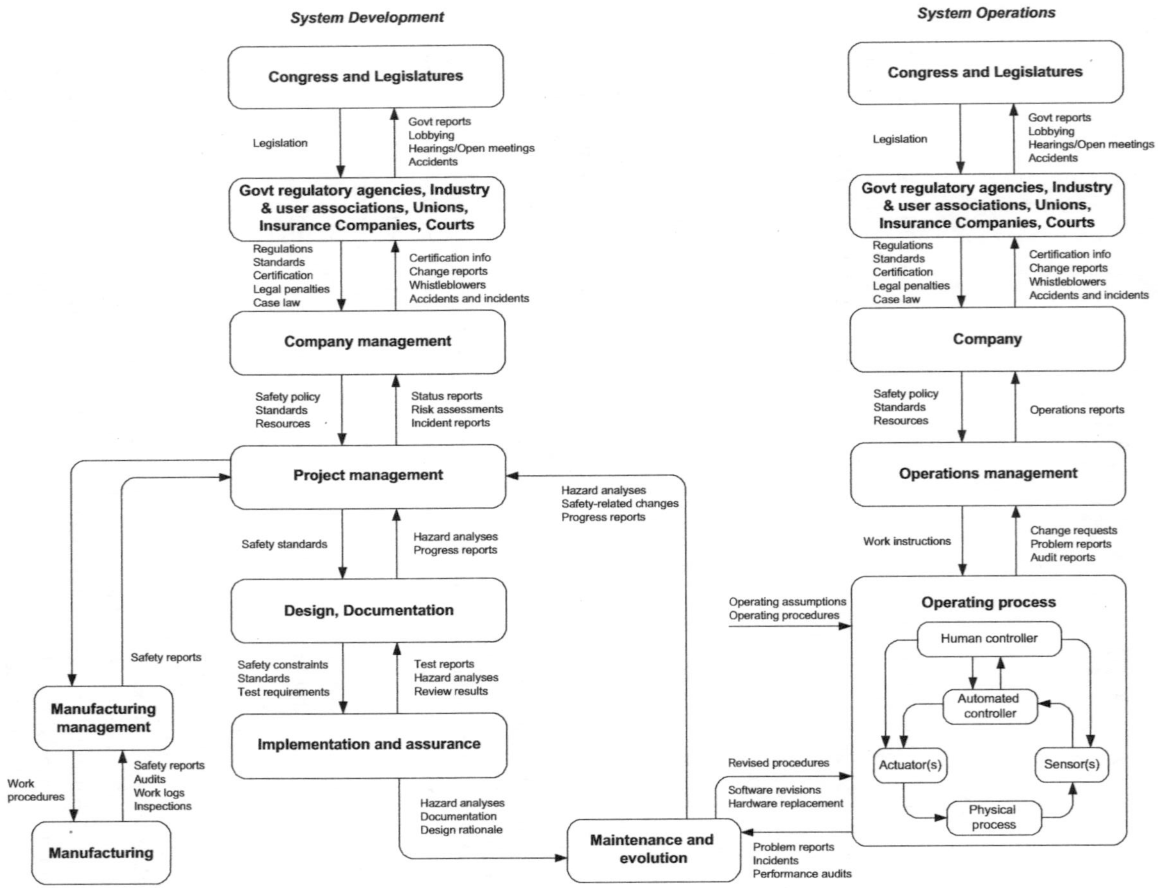

FIGURE 4.1 Generic control structure. (Adapted from Leveson, 2004). On the left of the control structure hierarchy is system development, and on the right is system operation.

hierarchy and what information about the status of the system is sent back up the hierarchy. Within Figure 4.1, the arrows flowing down the hierarchy represent controls (or reference channels) and the arrows flowing up the hierarchy represent feedback loops (or measuring channels). In other words, each entity has control and authority over the entities immediately below it, and each entity is likewise subject to control and authority from the entities immediately above.

As a point of reference, there are other popular hazard identification and risk assessment methods used in research and practice, including Failure Modes and Effect Analysis (FMEA) (Rausand and Hoyland, 2003), the Bowtie method, Hazard and Operability study (HAZOP) (Adhitya et al., 2009) and many other approaches predicated on Fault Tree Analysis (FTA), Event Tree Analysis (ETA) (de Ruijter and Guldenmund, 2016) and Cause-Consequence diagrams (Nielsen, 1971). A study by Sulaman et al. (2019) compared FMEA, one of the most widely applied risk assessment methods, to STPA. The main differences between the two methods were as follows: FMEA is based on reliability theory whereas STPA is based on systems and control theory; FMEA typically uses a physical component diagram whereas STPA uses a model of the entire system (or sub-system of interest) that consists of a functional control diagram (i.e. Figure 4.1); STPA also considers safety as a theoretic systems control problem rather than, exclusively, a component failure problem; and, unlike FMEA, STPA also considers the cyclical interactions between various system components across multiple system levels and was designed with non-linear rather than linear interactions in mind (Sulaman et al., 2019). Whilst the purpose of this chapter is not to delve deeply into a methods comparison, it can nevertheless be useful to understand such differences and why they point to STPA as a contemporary systems-based risk assessment approach.

STPA works by considering each of the control and feedback mechanisms described in the control structure along with an unsafe control actions taxonomy comprising the following four failure modes (Leveson, 2011; Leveson and Thomas, 2018):

1. Control or feedback action is not provided or followed;
2. An unsafe (incorrect) control or feedback action is provided;
3. Control or feedback action is provided too early or too late (wrong time or sequence); and
4. Control or feedback action is stopped too soon or applied too long (for continuous control actions, not discrete ones).

The output therefore includes a description of potential control and feedback failures and their consequences which is used to support the identification, development, and implementation of appropriate risk controls. Given that an introduction to STAMP and STPA has now been provided, the remainder of this chapter will integrate the two concepts and use the terminology 'STAMP-STPA' when referring to the proactive risk assessment method.

DOMAIN OF APPLICATION

STAMP-STPA is a generic approach and can be applied in any domain for risk assessment purposes. Recent applications of STAMP-STPA have taken place in areas such as sport (Hulme et al., 2021a), cybersecurity (Lane et al., 2020), maritime (Chaal et al., 2020), healthcare (Bas 2020), rail (Yang et al., 2019), aviation (Castilho et al., 2018), road transport (Mahajan et al., 2017), and space vehicle launch missions (Rising and Leveson 2018). STAMP-STPA is therefore a versatile approach that can be applied in various domains and equally to different problem types, whether it be traditional safety, security, privacy, performance, and other system specific properties. Given its connection to STAMP, STPA can account for design errors, software flaws, component interactions, cognitively complex human decision-making errors, and social, organisational and management factors that could conceivably lead to a loss event in nearly any domain (Leveson 2011). The case study example shown later in this chapter demonstrates the application of STAMP-STPA in the elite sports context to identify potential risks that could conceivably threaten cycling team performance (Hulme et al., 2021a).

PROCEDURE AND ADVICE

A flowchart depicting the STAMP-STPA procedure is presented in Figure 4.2. Step-by-step guidance is presented below.

STEP 1: DEFINE AIMS OF THE ANALYSIS AND THE TASK OR SYSTEM UNDER ANALYSIS

The first step in an applying STAMP-STPA involves clearly defining the aims of the analysis along with the system, task, or scenario under analysis and any analysis boundaries. Defining the boundaries of a given analysis is important as project constraints will dictate how deep the analysis can go in terms of the parts of the system considered. It may be that an analysis is limited to the organisational level or a particular work process only. Further, it is worth asking what kinds of losses will the analysis aim to prevent, and whether the focus is on traditional safety risks or performance risks (Leveson & Thomas 2018). Potential adverse incidents and losses can mean different things to different stakeholders within the system. It is during step 1 that analysts should be clear about what the STAMP-STPA analysis is intended to achieve, and to whom the outputs are applicable. The case study example presented later in this chapter clearly defines the STAMP-STPA performance endpoint which assists with the identification of potential hazards and risks. Examples from the literature are provided below to illustrate the idea of defining system boundaries when developing control structure models of sociotechnical systems – whether this be to support a STAMP-CAST (reactive) or STAMP-STPA (proactive) analysis.

In a healthcare study applying STAMP to model the medication administration system, Canham et al. (2018) included the operating process (i.e. the patient monitoring and observation cycle), hospital ward, and hospital care sub-systems *within* the broader healthcare system as their unit(s) of analysis. Whilst the authors did not go

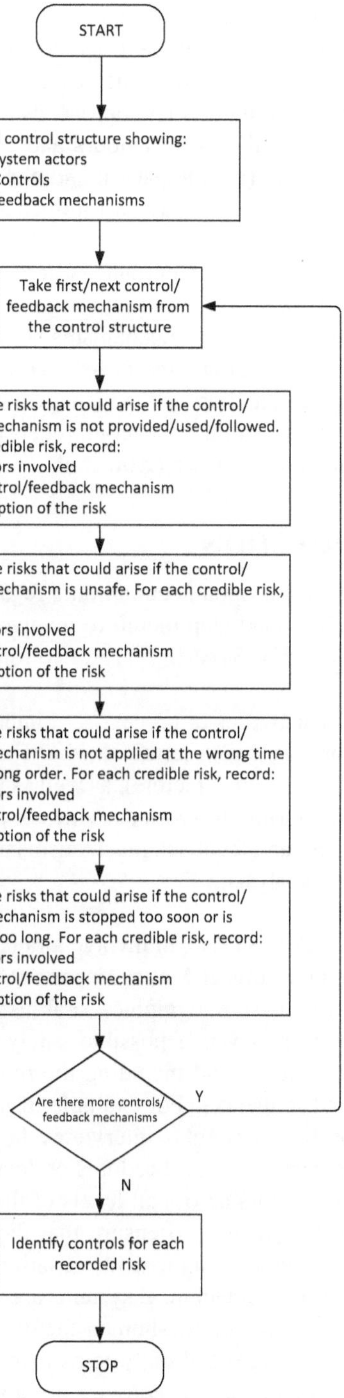

FIGURE 4.2 STAMP-STPA analysis process.

beyond the boundary of the hospital management level, their study was nevertheless useful as the control structure model identified how the flow of information between various stakeholders and workers within the hospital system influenced patient safety (Canham et al., 2018). Most modelling approaches up to that point had been reductive and linear, focussing only on behaviours and cause-effect relationships at the 'sharp end' of patient safety (i.e. the patient and nurse). A second example is found in work conducted by Kim et al. (2016) who applied STAMP to better understand the cause of the Korean Sewol ferry incident (ship sinking). The control structure model included both the system development and system operation hierarchies, and also incorporated factors at the government and regulatory levels. Particular emphasis was placed on control and feedback mechanisms between the ship-owning company, vessel design, shipyard operations, the classification society, the Korean shipping association and coastguard, and operation management levels of the system. For this reason, the analysis was systemic and comprehensive as all the relevant factors, agents and organisations were shown (Kim et al., 2016).

STEP 2: DATA COLLECTION

STAMP-STPA is dependent upon accurate data regarding the system, task, or scenario under analysis. The next step therefore involves collecting data regarding the system, task, or scenario. For STAMP-STPA, two forms of data are required:

1. Data on the work activities or processes of interest.
2. Data on the work system in terms of who/what resides within the system (i.e. human and non-human factors), as well as who/what shares the responsibility for safety during the work activities or processes.
3. Data on the control and feedback mechanisms used to manage safety within the system under analysis.

Data collection for STAMP-STPA can involve a range of field-based data activities, including interviews with Subject Matter Experts (SMEs) for the domain, surveys, questionnaires, observations, ethnographies, oral histories, workplace walkthroughs and logs, and document reviews. If possible, analysts are encouraged to consider all levels of the system, up to and including the regulatory, government, and even international levels and influences. For example, if using interviews as the primary data collection approach, it is useful to interview relevant actors from all levels of the system. Likewise, when reviewing procedures or documents, this should include the procedures for different actors across all levels of the system. The reason for being so comprehensive at this stage is to ensure that the control structure is complete and exhaustive. As is the case with incident causation and analysis, it is now widely accepted that risks within sociotechnical systems are also created due to the complex interactions between human and non-human factors across all levels (Dallat et al., 2019). Often, risks at the sharp end of systems that manifest in the behaviours and actions of frontline workers and technologies are a symptom of a deeper, underlying systemic cause or set of causes (Dallat et al., 2019). Consequently, if the data collection activities used to develop the initial control structure model consider all system levels, then the analyst can have confidence that their STAMP-STPA analysis will identify many critical risks, and ultimately, effective risk controls.

STEP 3: CONSTRUCT AND VERIFY THE CONTROL STRUCTURE

Once data collection is complete, the analyst should first develop a control structure showing the hierarchical system levels to be considered, relevant actors and organisations involved in the work system, and the controls and feedback mechanisms used to manage safety. As shown in Figure 4.1, the generic control structure specifies a set of hierarchical levels for both system design and operation; however, these are flexible and can be modified depending on the system, aims of the analysis, and the resources available.

Building the control structure first involves specifying the levels to be used (step 1), identifying relevant actors and organisations at each level, and then specifying the controls and feedback mechanisms that are currently in place to manage safety. The data collected during step 2 should be used for this purpose; however, the process is iterative and often involves the identification of new data requirements. Development of the control structure usually begins at a very abstract level and is gradually refined to capture more detail about the system (Leveson & Thomas 2018). For example, in cases where more than one agent or organisation is responsible for safety and performance, it can be useful to collapse these into a single entity where necessary. Likewise, instead of listing each and every control mechanism, it can be useful to start at a more abstract level and refine from there. This may mean breaking down an initial control or feedback description into multiple further controls or feedback mechanisms. There is also the matter of determining whether the control structure in question requires a strict linear hierarchy, or in cases where control structures and sub-systems are detailed, a hierarchy that combines both linear and horizontal control and feedback mechanisms in more complex configurations (Figure 4.3).

Once developed, it is useful to subject the control structure to a validation exercise whereby it is reviewed and refined by appropriate SMEs. This can take on various forms, including an SME workshop or Delphi study (e.g. Salmon et al., 2016; Hulme et al., 2017). It can be useful to supplement the control structure model with further information about the included controllers such as their description, purpose, responsibilities, actions, and process models.

STEP 4: IDENTIFY POTENTIAL CONTROL FAILURES

STPA works by considering each of the control and feedback mechanisms described in the control structure along with a control/feedback failure taxonomy comprising the following four failure modes (Leveson, 2011):

1. Control action/Feedback mechanism is not provided or followed.
2. An unsafe control action/Feedback mechanism is provided.
3. Control action/Feedback mechanism is provided too early or too late.
4. Control action/Feedback mechanism is stopped too soon or applied too long.

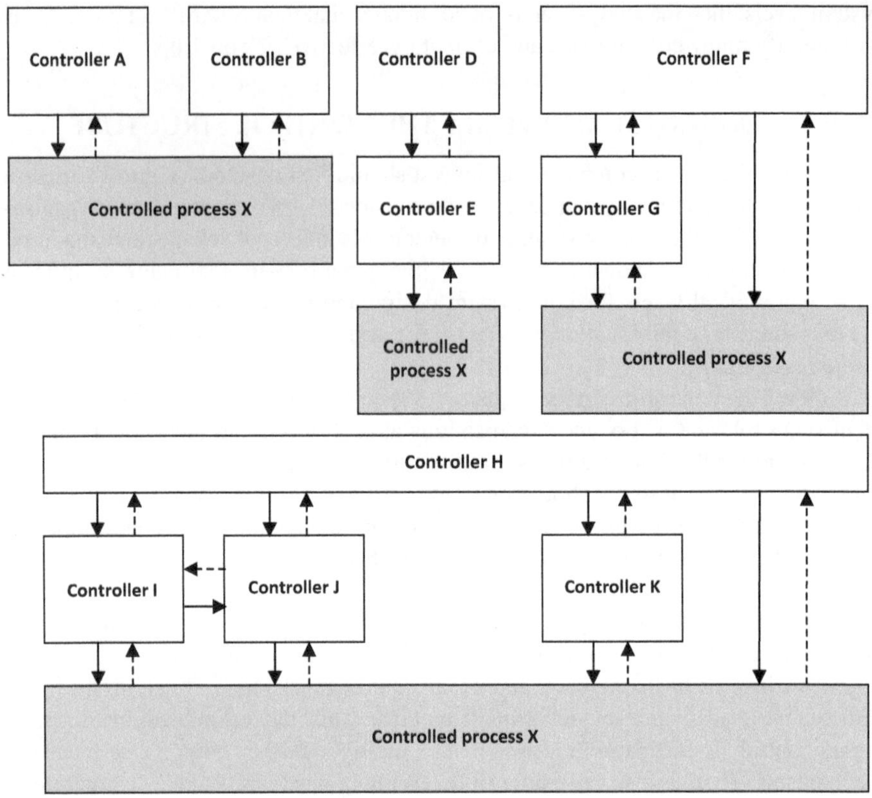

FIGURE 4.3 A comparison of different control structure types. (Adapted from Leveson and Thomas (2018)). Controller A and B both control and monitor shared control process X and do not interact with one another. Controller D and E are integrated linearly, the effects of controller D of which are exerted through controller E on process X. Control and feedback mechanisms may also bypass levels as shown in the control structure including controller F, controller G, and process X. In this case, emergency situations or unusual circumstances may mean that controller F can directly control and demonstrate authority over process X, thus circumventing controller G. Finally, more complex control structures as is the case between controller H, controller I, controller J, and controller K are possible. Controller K may operate independent of controller I and J, which, both exert their influence over process X whilst being controlled by controller H.

During this step, the analyst takes each control and considers whether there are credible risks associated with each of the four failure modes. To illustrate this step of the analysis with an example, a simplified control structure model of a rail maintenance and repair system is shown in Figure 4.4. In this system a rail track maintenance engineer is tasked with repairing a damaged section of track and reports to the workplace health and safety (WHS) manager above them. A control mechanism within this system is 'standard operating procedures (SOPs)' linking the maintenance subdivision WHS manager to the trackside engineer. These SOPs mandate the safe operation of equipment and technologies, in this case a hydraulic weld trimmer for

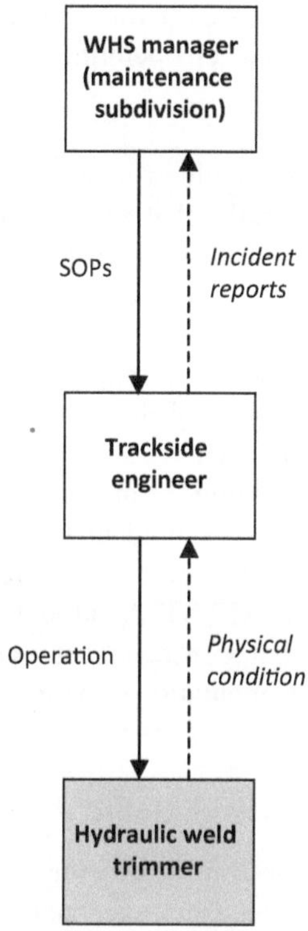

FIGURE 4.4 Example rail maintenance and repair system showing linear hierarchy of control and feedback.

removing excess weld material from hot-rolled steel on the track line. It is assumed that the engineer has the necessary safety certification and can show evidence of understanding the purpose and utility of machine operating SOPs. The trackside engineer has control over the weld trimmer, which in turn provides feedback to the engineer, mostly around its physical condition (e.g. need for new shear blades). The engineer communicates critical information about the status of trackside operations back to the WHS manager, including the reporting of any incidents or issues with metalworking guillotines that may create risks if in poor working order.

Credible risks for the SOP control mechanism in Figure 4.4 could include:

1. The SOP is not followed as required leading to unsafe behaviours when operating the weld trimmer during trackside repair activities;
2. The SOP requires updating in line with company policy leading to unsafe behaviours when operating the weld trimmer during trackside repair activities;
3. The SOP is followed too late (i.e. part way through the repair process) leading to unsafe behaviours when operating the weld trimmer during trackside repair activities; and

4. Compliance with the SOP ceases before the full repair task is completed leading to unsafe behaviours when operating the weld trimmer during track-side repair activities.

For each credible risk, the analyst should provide a description in the STAMP-STPA output table (Table 4.1). It is essential that each potential risk description contains the context in which the unsafe control action occurred (Leveson and Thomas 2018). The context in the above example is trackside repair when operating a hydraulic weld trimmer. The *source* is the WHS manager, the *type* of unsafe control action is whether or not the SOPs are followed/safe/applied/stopped, the *control action* is the SOPs for machine operation, and the *context* is trackside repair with a metalworking guillotine. A relatively high-level context is used for illustrative purposes here, however the context can be highly specific, outlining the precise actions when applying a trimmer to the track line for example. Analysts should aim to be as comprehensive as possible when identifying potential hazards, however if a failure mode is deemed not be applicable, 'N/A' can be used. The STAMP-STPA output therefore includes a description of potential control failures and their consequences which is used to support the identification, development, and implementation of appropriate risk controls in Step 6.

STEP 5: IDENTIFY POTENTIAL FEEDBACK FAILURES

Once all of the controls within the control structure have been assessed, the process should be repeated for all of the feedback mechanisms included in the model. The analyst examines each feedback mechanism and considers whether there are credible risks associated with each of the four STPA failure modes. For example, for the feedback mechanism 'report safety incident' in the rail maintenance and repair system (Figure 4.4), the analyst would consider whether any of the four failure modes are credible and would create a safety risk. Credible risks for the report safety incident feedback mechanism could include:

1. Incidents during trackside repair activities are not reported and so the organisation does not identify a safety issue and implement appropriate safety controls;
2. The incident reporting form around trackside repair activities is poorly designed leading to incorrect or misleading information being included in reports;
3. There is a delay in the reporting of incidents during trackside repair activities and further similar incidents occur before appropriate risk controls are implemented; and
4. The reporting of incidents during trackside repair activities occurs initially but then ceases. As a result, the organisation does not identify safety issues and does not implement appropriate safety controls.

As with Step 4, for each credible risk, the analyst should provide a description in the STAMP-STPA output table (Table 4.1). Up to this point in the process, we have

TABLE 4.1
Extract of STPA Cycling Risk Assessment Analysis

From (Agent/ Factor)	To (Agent/ Factor)	Control Action/ Feedback (from Control Structure Model)	Action or Feedback Required But Not Provided	Unsafe Action or Feedback Provided	Incorrect Timing or Order	Stopped too Soon/ Applied Too Long
Rider	Bicycle	Positional control	Rider does not maintain appropriate position in the peloton	Rider collides with other riders in the peloton	Rider inputs controls/ moves too early; collides with other riders in peloton	Rider performs excessive/ delayed control inputs; collides with other riders in peloton
Rider	Bicycle	Mechanical operation	Rider fails to perform required inputs (e.g. gear change); performs sub-optimally	Rider selects/uses the wrong gear; overexerts leading to fatigue/injury	Rider makes required inputs too early/ too late; performs sub-optimally	Rider inputs are inadequate (i.e. stopped too soon) and performance is sub-optimal
Bicycle	Rider	Physical/haptic feedback	Absence of mechanical feedback to the rider; bicycle control compromised; rider crashes/ falls/collides with other riders	N/A	N/A	N/A
Rider	Teammates	Behaviour of other teams	Rider fails to communicate key information regarding opponent teams' performance	Communication occurs at unsafe time leading to distraction/ overload	Rider informs team mates too late to respond/capitalise	Communication about other teams' behaviours is not fully delivered
Director Sportif (DS)	Rider	Tactics	Riders are unaware of race plan/tactics; performs sub-optimally	Tactics are inappropriate leading to sub-optimal performance	Tactics are provided and enacted too late leading to sub-optimal performance	Multiple (possibly conflicting) tactics are provided leading to confusion and a failure to enact the race plan

(Continued)

TABLE 4.1 (Continued)

From (Agent/ Factor)	To (Agent/ Factor)	Control Action/ Feedback (from Control Structure Model)	Action or Feedback Required But Not Provided	Unsafe Action or Feedback Provided	Incorrect Timing or Order	Stopped too Soon/ Applied Too Long
Director Sportif (DS)	Mechanic	Instructions	Mechanic of unaware of requirements and team needs; does not fulfil role leading to sub-optimal performance	Mechanic is instructed to break the rules when setting up/repairing bikes leading to disqualification	Mechanic instructed to repair/service at inopportune time	Mechanic is given multiple instructions leading to overload/confusion; inappropriate response
Commissaire	Director Sportif (DS)	Race rules and regulations	DS does not fully understand race rules and regulations leading to illegal behaviours	Commissaire distracts the DS from driving task leading to crash with rider/convoy vehicle	Rules are enforced too late; they are violated/ the team penalised	Commissaire is too strict in rule interpretation creating/biased/unfair penalties

Source: Adapted from Hulme et al., 2021a

defined the purposes of the analysis, collected data, constructed a control structure model of the system, and have identified unsafe control and feedback actions.

STEP 6: REVIEW AND REFINE ANALYSIS

Once the draft STAMP-STPA analysis is complete, it is useful to have various SMEs review it. This can be achieved via survey and/or a working group. It is best practice to use SMEs who have extensive knowledge of the system and the work in question. The STAMP-STPA analysis should be updated and finalised based on the feedback provided by the SMEs. Specific revisions at this stage can target the risk descriptions and wording to enhance their meaning and relevance.

STEP 7: IDENTIFY RISK CONTROLS

Once the analyst is satisfied that all credible risks associated with control and feedback failures have been identified, they should be used to support the development of appropriate risk controls. This typically involves a workshop or group session where each risk is considered, and potential controls are generated. It is also useful at this stage to develop an action plan which includes a description of each control, the criteria for evaluating the success of each control, responsibilities for implementing and monitoring each control and a review timeline.

ADVANTAGES

- The control structure model provides a useful description of the system, including relevant actors and organisations and the control and feedback mechanisms used to maintain system safety;
- STAMP-STPA is based upon a sound theoretical model and is aligned with state-of-the-art accident causation theory;
- STAMP-STPA enables the proactive identification of control and feedback failures across the overall system of work, up to and including government and even international levels;
- The consideration of feedback failures and their associated risks is novel and is a key strength over and above other risk assessment methods;
- The use of the unsafe control actions taxonomy potentially enhances reliability and validity;
- The output offers an exhaustive analysis of potential risks within the work system (or sub-system) under analysis;
- STAMP-STPA is a generic approach which has been applied across many domains;
- STAMP-STPA focuses on systematic improvements rather than on blaming individuals; and
- The control structure is a useful approach in its own right, and provides an in-depth overview of the stakeholders who shared the responsibility for safety and accident causation in a particular system.

DISADVANTAGES

- STAMP-STPA can be time consuming to apply, involving the development of a system control structure followed by the identification of control and feedback failures;
- The analysis does not explicitly consider emergent risks that arise when control failures interact with one another;
- The quality of the outputs is highly dependent on the expertise of the analysts;
- There is little reliability and validity evidence associated with STAMP-STPA, albeit one study has recently examined this (see reliability and validity section below);
- STAMP-STPA does not provide a method to support identification and development of corrective measures; these are based on the judgement of the analyst; and
- The outputs can be complex, large, and unwieldy.

RELATED METHODS

STAMP-STPA is part of an integrated safety management approach, the other component of which includes CAST, a retrospective incident analysis method (see Chapter 11). Data collection for STPA can include a range of methods such as structured or semi-structured interviews (e.g. the Critical Decision Method), direct observation, walkthrough analysis and documentation review (e.g. incident reports, standard operating procedures).

APPROXIMATE TRAINING AND APPLICATION TIMES

STAMP-STPA is relatively simple to learn and apply but can become time consuming when applied to complex scenarios or full work systems. Development of the control structure in particular can be time consuming, with timescales of around one to two weeks for data collection and construction of the control structure model, and further time required for model validation. Identification control and feedback failures can also be time consuming and is dependent on the number of control and feedback mechanisms identified as well as availability of data and information sources.

RELIABILITY AND VALIDITY

Hulme et al. (2021b) tested the reliability and validity of STAMP-STPA, EAST-BL (Chapter 3), and Net-HARMS (Chapter 2) when used by analysts to identify the risks across a railway level crossing design lifecycle. Hulme et al. (2021b) report a weak to moderate positive correlation coefficient for STAMP-STPA, meaning that the method did indeed support relatively novice analysts with the identification of credible system risks, albeit up to a point. Many reasons were given explaining the results, covering methodological and practical considerations. For example, one of the key recommendations following the study was a need for further training in the use of the STAMP-STPA method. The development of the control structure model and the application of the unsafe control actions taxonomy takes time to learn and apply. Finally, there is always the consideration of how much domain expertise is required when applying safety methods, including STAMP-STPA. In the reliability and validity study (Hulme et al., 2021b), analysts were expected to identify risks across a railway level crossing system lifecycle, but all did not have experience of working in rail safety, hence the stability and accuracy of the methods may have been affected as a result. Accordingly, there is a need for further reliability and validity studies of STAMP-STPA in both a diverse range of analyst samples and sociotechnical system domains.

TOOLS NEEDED

STAMP-STPA can be conducted using pen and paper; however, the control structure is normally created in a software drawing package such as Microsoft Visio and the output table is typically created in Microsoft Word or Excel.

CASE STUDY EXAMPLE: ELITE WOMEN'S ROAD CYCLING

Hulme et al. (2021a) applied STAMP-STPA to identify safety and performance risks for an elite women's cycling road race team who were competing in the Australian Subaru National Road Series (NRS). Before describing the results of the study, and for the reader unfamiliar with the context forthcoming, a useful starting point is to provide an overview of elite road bike racing.

Australian NRS road races typically range in length from 50–130km, and involve teams of four to five riders. The riders from multiple teams form a 'peloton' of cyclists that negotiate the race route together. Each team has a 'protected rider' or classification rider with the remaining riders acting in a 'domestique' or support capacity. The cycling team's goal is to ensure that their protected rider finishes each stage of the race in the best position and time possible. Domestiques employ various tactics to assist their protected rider, such as allowing the protected rider to ride behind them ('on their wheel') to reduce their physical workload or 'leading them out' to initiate a sprint finish. Throughout the race, riders may attempt to 'attack' and break away from the peloton and other teams' riders may attempt to chase them down and bring them back to the peloton. Alternatively, the peloton may stay together, resulting in a 'bunch sprint' finish. A convoy of vehicles comprising the Commissaries, race officials, medical support, neutral spares, the media, and team support vehicles follow behind the peloton. The Commissaries control the race and following convoy, communicating with the cyclists via loudspeaker and other vehicles in the convoy via UHF radio. Medical support vehicles follow behind the Commissaries and aid in the event of a crash or injury. Each cycling team has a support vehicle which typically includes the Director Sportif (DS), a mechanic, spare parts (e.g. wheels), bicycle tools, and food and drink for the riders. Each support vehicle is permitted to provide mechanical or nutritional support to team riders and can offer tactical guidance and encouragement through verbal exchanges. Whilst there are many more aspects to elite bike racing that could be discussed, the above description contains the essential agents and activities for making sense of the upcoming STAMP-STPA analysis.

As part of a study comparing three risk assessment methods, STAMP-STPA was used to identify the potential control and feedback failures that could adversely impact the racing team's performance and safety during NRS races (Hulme et al., 2021a). A control structure model of the cycling race system was developed by three analysts (Figure 4.5). One analyst then applied the unsafe control actions taxonomy to identify credible risks to performance and safety.

Before applying STAMP-STPA, it was necessary to outline the endpoint of the study and how to explicitly define 'performance' in the elite cycling context. As such, cycling team performance was defined as the ability of cycling team members to

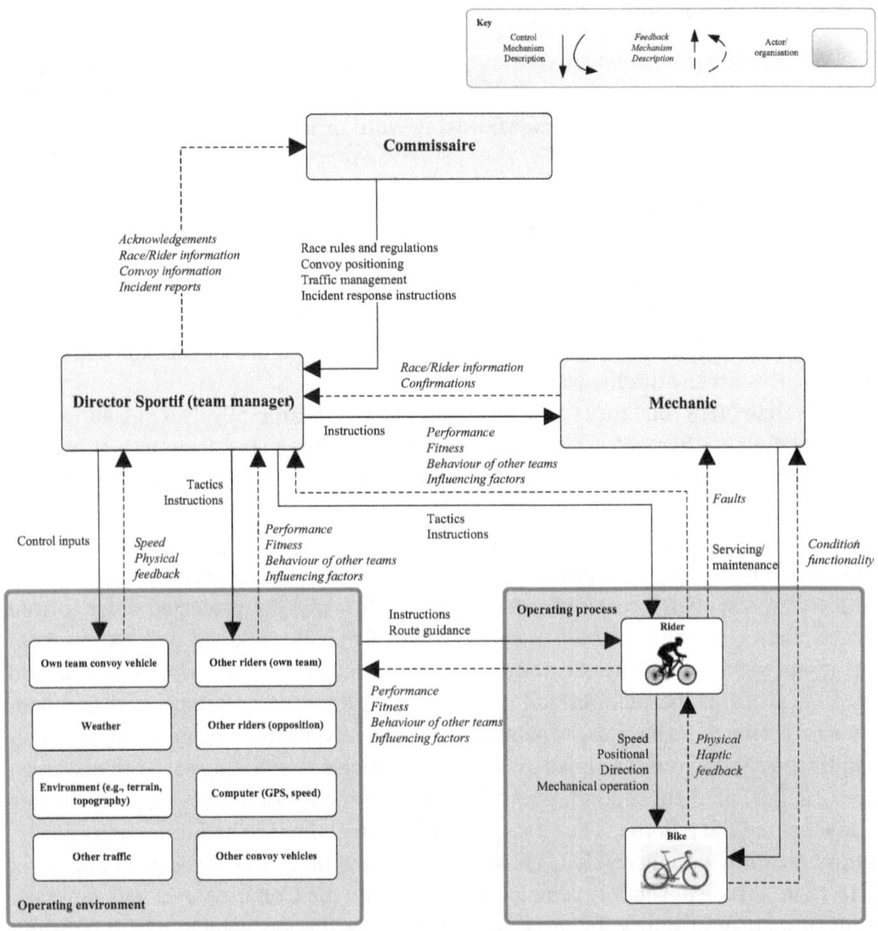

FIGURE 4.5 Control structure model of the cycling race system. (Reprinted with permission from Hulme et al., 2021a). Control mechanisms are indicated by the downward flowing solid relationships between agents/system levels, whereas the feedback mechanisms are indicated by the upward flowing broken relationships and italicised descriptions.

work cooperatively to achieve the shared goal of protecting the classification rider and winning the race. Risk credibility was determined based on the potential of a given risk to detrimentally affect the ability of the cycling team to achieve their shared goal of winning the race.

RESULTS

A total of 211 risks were identified following the application of STAMP-STPA to the control and feedback mechanisms presented in Figure 4.5. Broken down by the four failure modes: 48 risks were identified for 'action or feedback required but not provided'; 47 risks were identified for 'unsafe action or feedback provided'; 59 risks were identified for 'incorrect timing or order'; and 57 risks were identified for 'stopped too soon/applied too long' (Hulme et al., 2021a). Table 4.1 presents an excerpt of the STAMP-STPA analysis showing a selection of risks identified across the four categories. A description of the control and feedback actions between agents from the control structure model, and the risks that were deemed to be credible are presented.

Regarding the types of the risks identified, STAMP-STPA generated useful insights about how performance could be degraded in the elite cycling race system. This suggests that traditional risk assessment methods developed specifically to identify risks in safety-critical domains and work contexts, such as STAMP-STPA, can indeed produce relevant outputs when used to identify risks to athlete, team, and system performance. For example, Table 4.1 includes a range of different conditions and scenarios that could hinder and impair optimal cycling team performance. The STPA analysis contains risk descriptions that are relevant to, and can be used by, all stakeholders within the cycling race system, including the riders, the DS, and support crew. Importantly, the findings also provide further evidence that contemporary safety and accident causation theory apply equally well to sports performance. In the complete study by Hulme et al. (2021a), risks were identified across the sports system relating to equipment and technology (e.g. bikes, power meters, GPS devices), nutrition and ergogenic aids, race planning, tactics and strategy, specific features of the terrain and topography, rider behaviours (e.g. attacking, surging), opposing team behaviours, and race management and governance.

Of further consideration is the need for, and use of, the control structure modelling technique as a basis to conduct a STAMP-STPA analysis. The control structure in Figure 4.5 described a range of different control and feedback actions among key agents within the cycling race system. As a result, the corresponding risk descriptions across the four failure modes in the unsafe control actions taxonomy were also specific to the control and feedback action(s) being examined. The value and utility of a given STAMP-STPA output is therefore largely dependent on the quality and face validity of the control structure model, including the fidelity of the control and feedback actions described. Given the relatively high number of risks described, the next step of the process would involve collaborating with the cycling team to develop appropriate remedial measures and intervention strategies to address potential risks and optimise subsequent race performances. Alternatively, and depending on the expectations and goals of the project and overall clarity of the risk descriptions, researchers and practitioners may have the option of handing over their STAMP-STPA analyses to decision-makers and key stakeholders who can prioritise and/or incorporate findings into future planning activities. In the case of the cycling risk analysis, key lessons and insights could be extracted by the riders and the DS and applied to the next race in the NRS series to enhance team performance moving forwards.

REFERENCES

Adhitya, A., Srinivasan, R., Karimi, I.A. (2009). Supply chain risk identification using a HAZOP-based approach. *AIChE Journal: Process Systems Engineering* 55(6), 1447–1463.

Bas, E. (2020). STPA methodology in a socio-technical system of monitoring and tracking diabetes mellitus. *Applied Ergonomics* 89, 103190.

Canham, A., Jun, G.T., Waterson, P., & Khalid, S. (2018). Integrating systemic accident analysis into patient safety incident investigation practices. *Applied Ergonomics* 72, 1–9.

Castilho, D.S., Urbina, L.M., & de Andrade, D. (2018). STPA for continuous controls: A flight testing study of aircraft crosswind takeoffs. *Safety Science* 108, 129–139.

Chaal, M., Banda, O.A.V., Glomsrud, J.A., Basnet, S., Hirdaris, S., & Kujala, P. (2020). A framework to model the STPA hierarchical control structure of an autonomous ship. *Safety Science* 132, 104939.

Dallat, C., Salmon, P.M., & Goode, N. (2019). Risky systems versus risky people: To what extent do risk assessment methods consider the systems approach to accident causation? A review of the literature. *Safety Science* 119, 266–279.

de Ruijter, A. & Guldenmund, F. (2016). The bowtie method: A review. *Safety Science* 88, 211–218.

Hulme, A., McLean, S., Dallat, C., Walker, G.H., Waterson, P., Stanton, N.A., & Salmon, P.M. (2021b). Systems thinking-based risk assessment methods applied to sports performance: A comparison of STPA, EAST-BL, and Net-HARMS in the context of elite women's road cycling. *Applied Ergonomics* 91, 103297.

Hulme, A., Salmon, P.M., Nielsen, R.O., Read, G.J.M., & Finch, C.F. (2017). From control to causation: Validating a 'complex systems model' of running-related injury development and prevention. *Applied Ergonomics* 65, 345–354.

Hulme, A., Stanton, N.A., Walker, G.H., Waterson, P., & Salmon, P.M. (2021a). Testing the reliability and validity of risk assessment methods in Human Factors and Ergonomics. *Ergonomics* in press.

Kim, T., Nazir, S., & Øvergård, K.I. (2016). A STAMP-based causal analysis of the Korean Sewol ferry accident. *Safety Science* 83, 93–101.

Lane, B.R., Salmon, P.M., Desmond, D., Cherney, A., Carley, A., Hulme, A., & Stanton, N.A. (2020). Out of control? Using STAMP to model the control and feedback mechanisms surrounding identity crime in darknet marketplaces. *Applied Ergonomics* 89, 103223.

Leveson, N.G. (2011). *Engineering a Safer World: Systems Thinking Applied to Safety*, Cambridge, MA: The MIT Press.

Leveson, N.G. & Thomas, J.P. (2018). *STPA Handbook*. Cambridge, MA: The MIT Press.

Leveson, N.J. (2004). A new accident model for engineering safer systems. *Safety Science* 42(4), 237–270.

Leveson, N.J., Dulac, N., Marais, K., & Carroll, J. (2009). Moving beyond normal accidents and high reliability organizations: A systems approach to safety in complex systems. *Organization Studies* 30(2–3), 227–249.

Mahajan, H.S., Bradley, T., & Pasricha, S. (2017). Application of systems theoretic process analysis to a lane keeping assist system. *Reliability Engineering & System Safety* 167, 177–183.

Nielsen, D.S. (1971). *The Cause/Consequence Diagram Method as a Basis for Quantitative Accident Analysis*, Risø National Laboratory: Roskile, Denmark.

Rasmussen, J. (1997). Risk management in a dynamic society: A modelling problem. *Safety Science* 27(2–3), 183–213.

Rausand, M. & Hoyland, A. (2003). *System Reliability Theory: Models, Statistical Methods, and Applications*, Hoboken, NJ: John Wiley & Sons.

Rising, J.M., Leveson, N.G. (2018). Systems-theoretic process analysis of space launch vehicles. *Journal of Space Safety Engineering* 5(3–4), 153–183.

Salmon, P.M., Read, G.J.M., & Stevens, N.J. (2016). Who is in control of road safety? A STAMP control structure analysis of the road transport system in Queensland, Australia. *Accident Analysis and Prevention* 96, 140–151.

Sulaman, S.M., Beer, A., Felderer, M., & Höst, M. (2019). Comparison of the FMEA and STPA safety analysis methods–a case study. *Software Quality Journal* 27(1), 349–387.

Yang, P., Karashima, R., Okano, K., & Ogata, S. (2019). Automated inspection method for an STAMP/STPA-fallen barrier trap at railroad crossing. Procedia Computer Science 159, 1165–1174.

Section III

Systems Analysis and Design Methods

5 Hierarchical Task Analysis (HTA)

BACKGROUND AND APPLICATIONS

Though various task analysis methods exist, Hierarchical Task Analysis (HTA; Annett et al., 1971) is arguably the commonly applied. The name is slightly misleading, however, as HTA does not focus exclusively on tasks. Rather it focuses on goals (an objective or end state) and decomposes them hierarchically to identify sub-goals and requisite operations. The task focussed name is also misleading in terms of what can be analysed with HTA. HTA can be applied at scale wherever the system boundary is drawn.

HTA has its roots in the scientific management movement of the early 20th century (Stanton, 2006). Scientific management methods were used to describe and analyse tasks in a way that supported the development of more efficient work processes. The focus was on how the work was performed physically, what was required to do the work, and how the work could be enhanced (Stanton, 2006). Whilst this approach endured until the mid-20th century, HTA was developed in the 1960s in response to the changing nature of work. Work tasks were becoming more cognitive in nature which in turn created the need for methods which could describe both the physical and cognitive aspects of work (Annett, 2004). At the time HTA provided a significant advancement over scientific management methods which largely focussed on the physical and observable aspects of behaviour.

HTA works by decomposing systems and behaviour into a hierarchy of goals, subordinate goals, operations, and plans. This sounds anathema to a systems approach but there is an important subtlety. HTA focuses on "what an operator [...] is required to do, in terms of actions and/or cognitive processes to achieve a system goal" (Kirwan and Ainsworth 1992, p. 1). It can also be used to focus on what a system is required to do in terms of the operations undertaken by multiple stakeholders to achieve systems goals. It is important to note here that an 'operator' may be human or non-human (e.g. system artefacts such as equipment, devices, documentation, and interfaces). It is also important to note the focus on outcomes.

HTA outputs specify the overall goal of a particular task/scenario/system (system in the present context), the sub-goals to be undertaken to achieve this goal, the operations required to achieve each of the sub-goals specified and the plans that trigger engagement and achievement of the different goals and operations. The plans

DOI: 10.1201/9780429281624-5

component specifies the sequence, and under what conditions, different sub-goals have to be achieved in order to satisfy the requirements of a superordinate goal. The plans are key in that they can specify the interrelations between goals, which is consistent with a systems lens.

Many applications of HTA undertaken in many different contexts can be found in the academic literature (see Stanton, 2006). Its status as the most popular task analysis method is largely down to its flexibility. HTA can be used in any domain and forms the starting point for many additional HFE analysis methods (Salmon et al., 2010; Stanton et al., 2013). For example, error identification methods such as the Systematic Human Error Reduction and Prediction Approach (Embrey, 1986) and more recently the Networked Hazard Analysis and Risk Management System (Net-HARMS; Dallat et al., 2018, see Chapter 2) risk assessment method require a HTA description as their primary data input. This has resulted in HTA being applied for all manner of system design and analysis purposes, including interface design and evaluation, job design, training program design and evaluation, error prediction and risk assessment, team task analysis, situation awareness requirements analysis, allocation of functions analysis, workload assessment and procedure design (Stanton, 2006; Stanton et al., 2013). Crucially for present purpose, it can be used as an input into further systems-based methods.

Another important aspect of the method's flexibility is the ability to support different levels of analysis within sociotechnical systems. For example, HTA can be used to analyse the behaviour of individuals, human-machine dyads, teams, organisations, multi-organisation tasks, and even entire sociotechnical systems.

DOMAIN OF APPLICATION

HTA is generic and can be applied in any domain for many different purposes. To date HTA has been applied in all manner of domains ranging from defence (Stanton et al., 2010), rail (Naweed et al., 2018), healthcare (Demirel et al., 2016); Yu et al., 2014), road transport (Walker et al., 2015) and maritime (Ramos et al., 2019) to more novel application areas such as crime scene investigation (Smith et al., 2008), air accident investigation (Nixon and Braithwaite, 2018), and vineyard cultivation (Fargnoli et al., 2019).

PROCEDURE AND ADVICE

A flowchart depicting the HTA procedure is presented in Figure 5.1. The process is relatively simple and involves collecting data about the task or system under analysis (through techniques such as observation, questionnaires, interviews with SMEs, walkthroughs, user trials, and documentation review). This data is then used to decompose and describe the goals and sub-goals involved in task or system operation. The standard HTA procedure is presented in Figure 5.1 (adapted from Stanton et al., 2013).

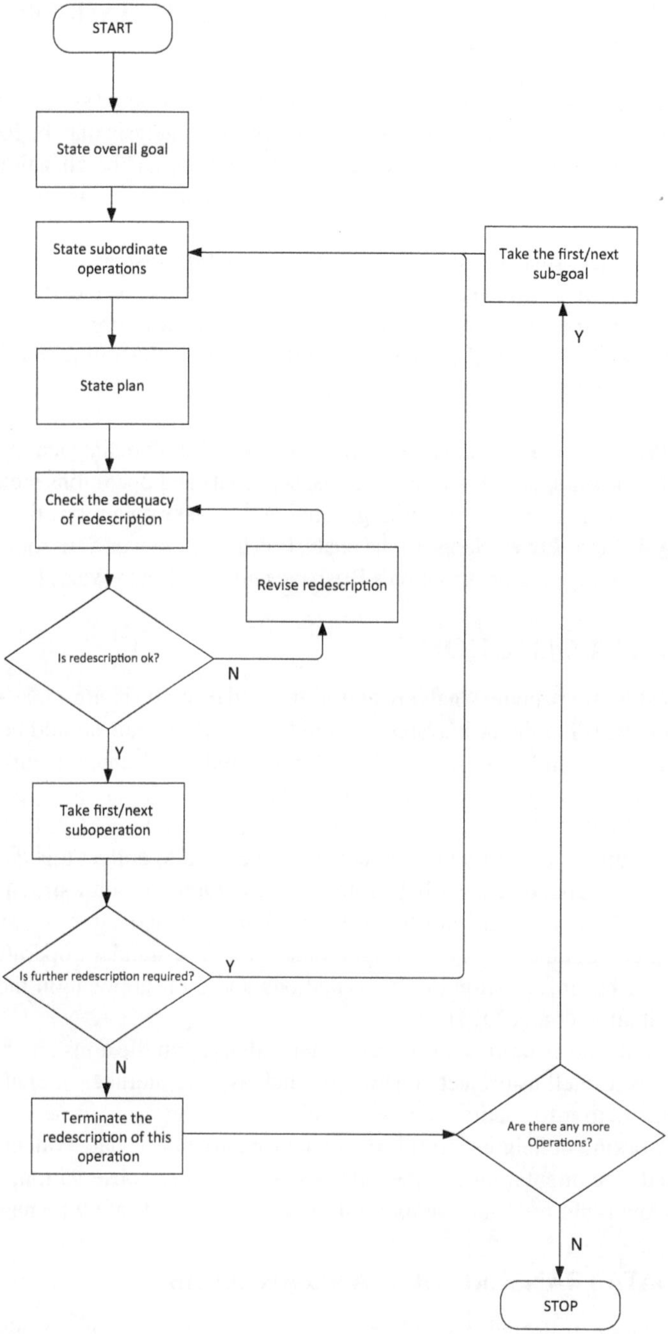

FIGURE 5.1 Flowchart depicting standard HTA procedure.

STEP 1: DEFINE AIMS OF THE ANALYSIS AND THE TASK OR SYSTEM UNDER ANALYSIS

The first step in applying HTA involves clearly defining the tasks or system under analysis along with any analysis boundaries. For example, the analysis may be focussed on individual behaviour in a given context (i.e. complete working at heights aircraft maintenance task), teamwork (i.e. in aircraft maintenance teams), organisational behaviour (i.e. airline company's attainment of organisational goals), or the behaviour of an entire sociotechnical system (i.e. operation of civil aviation system).

As part of this first step the aims of the analysis should also be clearly defined. For example, for the first HTA developed as part of the aviation organisation working from heights study described in the case study example the following was determined during step 1:

- **Task/System under analysis**: Working at heights aircraft maintenance system;
- **Aim of the analysis**: Describe the goals, sub-goals and operations required to deliver safe and compliant working from heights tasks on aircraft fleet; and
- **Analysis boundaries**: Scope of the analysis is limited to working from heights system employed in organisation's Brisbane and New South Wales branches.

STEP 2: DATA COLLECTION

Once the task/system under analysis and aims of the analysis are clearly defined, specific data regarding the task/system should be collected. Data should be collected regarding the goals and tasks involved, the human and non-human agents involved, the interactions between humans and non-human agents, the ordering of tasks and conditions that dictate task sequences, and information on the factors that influence behaviour. A number of different approaches can be used to collect this data, including observations, concurrent verbal protocols, structured or semi-structured interviews (e.g. the Critical Decision Method), questionnaires and surveys, walkthrough analysis and documentation review (e.g. incident reports, standard operating procedures). For further information on these methods and their application the reader is referred to Stanton et al. (2013).

The data collection approach selected is dependent upon the aims and boundaries of the analysis as well as project constraints, such as time, number of analysts available, and access to information and personnel.

For the working at heights aircraft maintenance case study, data collection activities included documentation review, site visits and direct observation, and SME workshops involving relevant workers, safety personnel, and safety managers.

STEP 3: DATA TRANSCRIPTION AND ANALYSIS

Once the data are collected it should be transcribed as appropriate. Analysis of the data should then proceed. This can take various forms depending on what data collection approaches are used. For example, when concurrent verbal protocols are used, the data analysis process might involve coding the data to identify goals, subgoals, tasks, operations, and performance shaping factors. Similarly, CDM interview transcripts are typically analysed thematically to identify goals, tasks, operations, the information used during task performance, and factors influencing behaviour (e.g. Mulvihill et al., 2016).

STEP 4: DETERMINE OVERALL GOAL

The analysis process begins by specifying an overall goal for the task or system under analysis. This should represent the overarching goal of the work or system under analysis and is placed centrally at the top of the hierarchy and is labelled 0. For example, in the aviation working at heights case study example, the overall goal was expressed as:

0. Deliver safe and compliant work at heights on aircraft

STEP 5: IDENTIFY AND RECORD SUB-GOALS

The next step involves decomposing the overall goal into a series of meaningful sub-goals (four or five is normally optimal however this is dependent on the task/system under analysis, and can extend beyond 10). These sub-goals should be numbered 1, 2, 3, 4, and so on. Where different sub-goals are undertaken by different actors, it is useful to make a note of this within the sub-goal description. For the working at heights system case study, the overall goal of 'Deliver safe and compliant work at heights on aircraft' was decomposed into the following sub-goals:

1. Design work tasks;
2. Identify and assess risks/hazards;
3. Develop and update procedures;
4. Communicate procedures;
5. Provide safe equipment;
6. Provide training;
7. Manage contractors;
8. Provide technical advice and assistance;
9. Conduct aircraft maintenance work at heights;
10. Supervise work at heights;
11. Report hazards;
12. Respond to emergency incidents;
13. Report and investigate incidents; and
14. Conduct assurance activities.

STEP 6: SUB-GOAL DECOMPOSITION

Next, the sub-goals identified in Step 5 should be broken down into further sub-goals and operations, according to the goals and tasks required. This process continues until an appropriate set of end operations is reached or the level of granularity required for the analysis is achieved. Note, the bottom level of any branch in a HTA should always be a series of operations. Whereas everything above an operation specifies goals, operations state what actually needs to be done. Thus, operations are the actions to be made by the operator (whether human or non-human). For the working at heights system, Sub-goal 6, 'Provide training', was decomposed as follows:

6. Provide training
   6.1. Identify training needs
   6.2. Design/update training program
   6.3. Deliver training
   6.4. Review training program
   6.5. Competency assessment and tracking
   6.6. Document training activities and outcomes

The decomposition process should continue until an appropriate set of end operations is reached or the level of granularity required for the analysis is achieved. For example, the sub-goal 'Identify training needs' was decomposed as follows:

6.1. Identify training needs
   6.1.1. Conduct training needs analysis for working at heights tasks
   6.1.2. Extract core competencies and knowledge, skills and attitudes for working at heights tasks
   6.1.3. Compile training needs for working at heights tasks
   6.1.4. Review and sign off on working at heights training needs

STEP 7: PLANS ANALYSIS

Once all sub-goals have been fully decomposed and described, the plans which dictate the sequencing of goals, sub-goals, and operations should be added. A simple linear plan would say "Do 1, then 2, and then 3 then EXIT". Once the plan is completed, the operator returns to the super-ordinate level and repeats the process for other decomposed sub-goals. Various forms of plan exist, and the method is flexible in that new types of plans can be added should the analysis require it. An overview of different plan types is presented in Table 5.1.

For the working at heights system example, the overall plan for 'Deliver safe and compliant work at heights on aircraft' was defined as:

Plan 0: Do 1, then 2 and 3 and 4, then do 5 continuously, then 6; when there is a contractor involved do 7, if clarification is required do 8, then do 9 and 10; if hazards are identified do 11, if emergency incident occurs do 12, if incident occurs do 13, if assurance is required do 14, then Exit

TABLE 5.1
HTA Plan Types

Plan	Example
Linear	Do 1, then do 2, then do 3, then EXIT
Non-linear	Do 1, 2, and 3 in any order, then EXIT
Simultaneous	Do 1, then 2 and 3 together, then EXIT
Branching	Do 1, if x is present, then do 2 and 3, then EXIT, if x is not present then do 4, then EXIT
Cyclical	Do 1, then do 2, then do 3, and repeat until X, then EXIT
Selection	Do 1, then do 2 or 3 as required

STEP 8: CONSTRUCT HTA DIAGRAM

Once all of the goals, sub-goals, operations, and plans are described, the next step involves creating the HTA diagram. This is typically a tree diagram (see Figure 5.2) or a tabular representation. Various HTA software tools are available to support this process; however, drawing packages such as Microsoft Visio can also be used.

STEP 9: SUBJECT MATTER EXPERT (SME) REVIEW

Once the initial draft is complete it is useful to have various SMEs review it. The HTA should then refined based on SME feedback. It is normal practice for the HTA to go through many iterations before it is finalised.

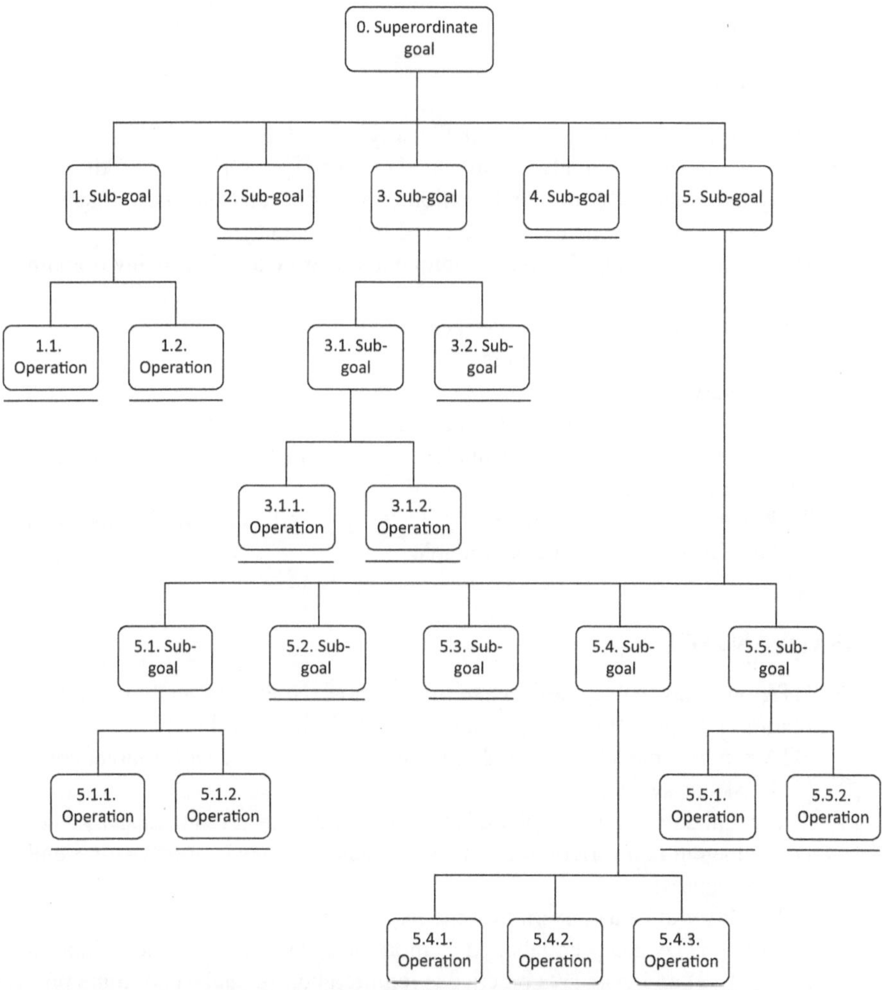

FIGURE 5.2 Example HTA tree diagram.

STEP 10: CONDUCT ADDITIONAL ANALYSES USING HTA EXTENSION METHODS

A key strength of HTA is that the output acts as the input to various additional analysis methods. For example, error identification methods such as the Systematic Human Error Reduction and Prediction Approach (Embrey, 1986) and more recently the Networked Hazard Analysis and Risk Management System (NET-HARMS; Dallat et al., 2018, see Chapter 2) risk assessment method require a HTA description as their primary data input. This enables HTA to be applied for various purposes, including interface design and evaluation, job design, training program design and evaluation, error prediction and risk assessment, team task analysis, situation awareness requirements analysis, allocation of functions analysis, workload assessment and procedure design (Stanton, 2006; Stanton et al., 2013).

Further information about HTA can be found in Stanton et al. (2013).

ADVANTAGES

- HTA requires minimal training and is easy to apply;
- The outputs are extremely useful and act as the input for various additional analyses, such as error identification, training needs analysis, interface design and evaluation, and allocation of function analysis;
- HTA is an extremely flexible technique and can be applied in any domain for a variety of purposes;
- HTA is quick to use in most instances;
- The output provides a comprehensive description of the task or system under analysis;
- HTA has been used extensively in a wide range of contexts;
- Conducting an HTA gives the analyst considerable insight into the task or system under analysis; and
- Tasks can be analysed to any required level of detail, depending on the aims of the analysis and resources available.

DISADVANTAGES

- HTA provides mainly descriptive information rather than analytical information. Interpretation is required to identify design solutions;
- HTA can become laborious and time-consuming to conduct for large, complex tasks or systems;
- The initial data collection phase is time-consuming and requires analysts to be competent in a variety of methods, such as interviews, observations, and questionnaires;
- Reliability can be a concern in some instances;
- Applying HTA has been labelled as more of an art than a science (Stanton et al., 2013) and extensive practice is required before analysts become proficient; and
- There are few prescriptive guidelines on how to apply it (Stanton, 2006).

RELATED METHODS

HTA is widely used and often forms the first step in various analyses, such as human error identification, human reliability analysis, training needs analysis, allocation of functions analysis, and workload assessment. Stanton (2006) outlines a range of additional analysis methods, including interface design, error prediction, workload analysis, team performance assessment, and training requirement identification. Mills (2007) argues that HTA is at its best when used alongside other methods.

APPROXIMATE TRAINING AND APPLICATION TIMES

The training time required for HTA can be substantial. Patrick et al. (2000) reported unsatisfactory results after providing a few hours training to students for a simple HTA task. Stanton and Young (2002) report that the training and application time for HTA is substantial. A survey by Ainsworth and Marshall (1998) found that more experienced practitioners produced more complete and acceptable analyses.

The application time associated with HTA is dependent upon the size and complexity of the task or system under analysis. For large, complex tasks, the application time for HTA will be high. Salmon et al. (2010) report that HTA application times are high, stating that the requirement for a detailed HTA can increase application times to almost double those of other methods such as CWA (see Chapter 6).

RELIABILITY AND VALIDITY

Whilst assessing the reliability and validity of HTA is difficult (Annett, 2004), studies have sought to do so. Stanton and Young (2002) reported that HTA achieved an acceptable level of validity but poor level of reliability. Stanton et al. (2013) suggest that analysts with differing levels of experience will likely produce different analyses of the same task (inter-rater reliability) and that the same analyst may produce different analyses on different occasions for the same task (intra-analyst reliability).

TOOLS NEEDED

HTA can be applied out using pen and paper only. The HTA output can be developed and presented in a number of software applications, such as Microsoft Visio, Microsoft Word, and Microsoft Excel. A number of dedicated HTA software tools also exist, such as the HTA tool (Salmon et al., 2010).

CASE STUDY EXAMPLE: IDENTIFYING THE RISKS ASSOCIATED WITH WORKING AT HEIGHTS IN AIRCRAFT MAINTENANCE

As part of a project aiming to identify the risks associated with aircraft maintenance working at heights tasks, the authors developed a HTA and task network to support a Net-HARMS risk assessment (see Chapter 2). The HTA was developed based on documentation review, site visit and observations, and an SME workshop. The HTA for the 'Deliver safe and compliant work at heights on aircraft' system is presented in Figure 5.3.

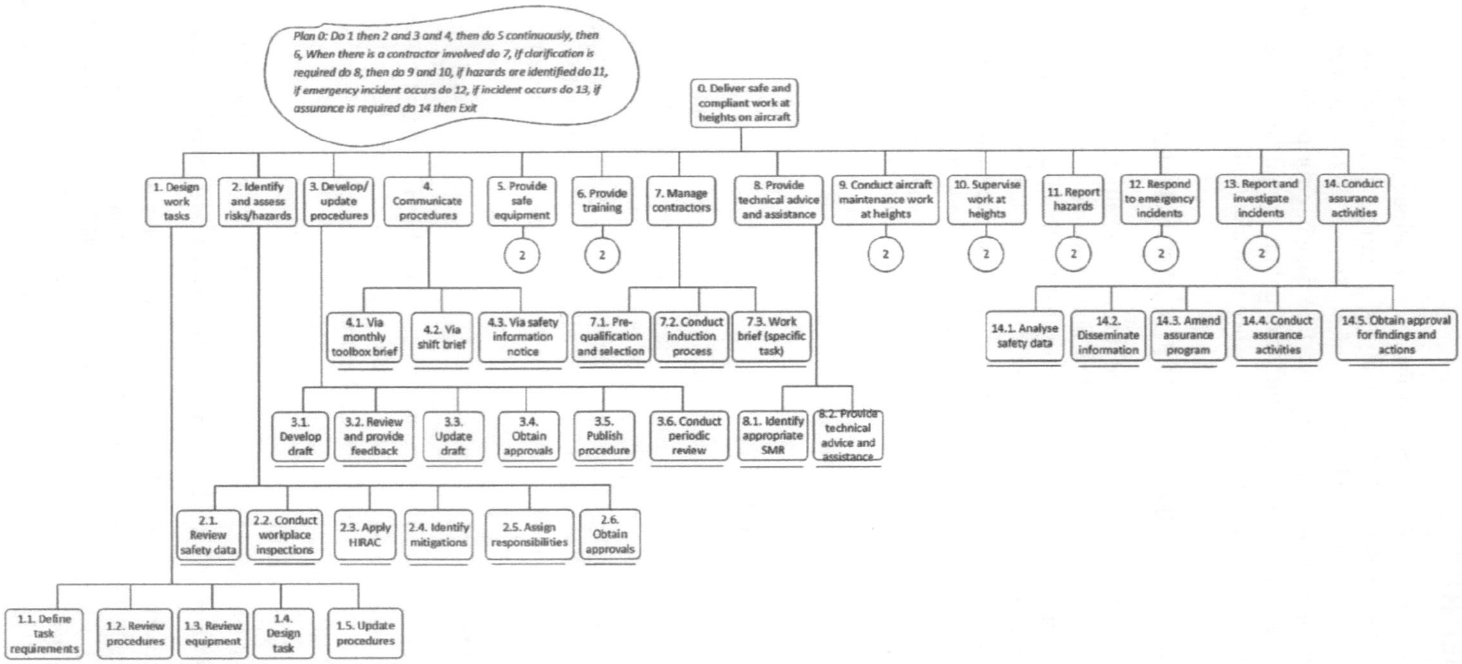

FIGURE 5.3 Deliver safe and compliant work at heights on aircraft system HTA.

RECOMMENDED READING

Annett, J. (2004). Hierarchical task analysis. In: D. Diaper, & N. A. Stanton (eds), *The Handbook of Task Analysis for Human-Computer Interaction*. Mahwah, NJ: Lawrence Erlbaum Associates, 67–82.

Annett, J., et al., 1971. *Task Analysis. Department of Employment Training Information Paper 6*. London: HMSO.

Salmon, P. M., Stanton, N. A., Jenkins, D. P., & Walker, G. H. (2010). Hierarchical task analysis versus cognitive work analysis: Comparison of theory, methodology, and contribution to system design. *Theoretical Issues in Ergonomics Science* 11(6), 504–531.

Stanton, N. A. (2006). Hierarchical task analysis: Developments, applications, and extensions. *Applied Ergonomics*, 37, (1), 55–79.

Stanton, N. A., Salmon, P. M., Rafferty, L., Walker, G., Baber, C., & Jenkins, D. P. (2013). *Human Factors Methods: A Practical Guide for Engineering and Design*. Second Edition, Aldershot, UK: Ashgate.

REFERENCES

Ainsworth, L., & Marshall, E. (1998). Issues of quality and practicability in task analysis: Preliminary results from two surveys. *Ergonomics*, 41(11), 1607–1617.

Annett, J. (2004). Hierarchical task analysis. In: D. Diaper and N. A. Stanton (eds), *The Handbook of Task Analysis for Human-Computer Interaction*. Mahwah, NJ: Lawrence Erlbaum Associates, 67–82.

Annett, J., et al., 1971. *Task Analysis: Department of Employment Training Information Paper 6*. London: HMSO.

Dallat, C., Salmon, P. M., & Goode, N. (2018). Identifying risks and emergent risks across sociotechnical systems: the NETworked hazard analysis and risk management system (NET-HARMS). *Theoretical Issues in Ergonomics Science*, 19(4), 456–482.

Demirel, D., Butler, K. L., Halic, T., Sankaranarayanan, G., Spindler, D., Cao, C., ... & deMoya, M. A. (2016). A hierarchical task analysis of cricothyroidotomy procedure for a virtual airway skills trainer simulator. *The American Journal of Surgery*, 212(3), 475–484.

Embrey, D. E. (1986). SHERPA: A systematic human error reduction and prediction approach. In *Proceedings of the International Topical Meeting on Advances in Human Factors in Nuclear Power Systems*.

Fargnoli, M., Lombardi, M., & Puri, D. (2019). Applying hierarchical task analysis to depict human safety errors during pesticide use in vineyard cultivation. *Agriculture*, 9(7), 158.

Kirwan, B., & Ainsworth, L. K. (1992). *A Guide to Task Analysis*. London, UK: Taylor & Francis Group.

Mills, S. (2007). Contextualising design: Aspects of using usability context analysis and hierarchical task analysis for software design. *Behaviour & Information Technology*, 26(6), 499–506.

Mulvihill, C. M., Salmon, P. M., Beanland, V., Lenné, M. G., Read, G. J., Walker, G. H., & Stanton, N. A. (2016). Using the decision ladder to understand road user decision making at actively controlled rail level crossings. *Applied Ergonomics*, 56, 1–10.

Naweed, A., Balakrishnan, G., & Dorrian, J. (2018). Going solo: hierarchical task analysis of the second driver in "two-up" (multi-person) freight rail operations. *Applied Ergonomics*, 70, 202–231.

Nixon, J., & Braithwaite, G. R. (2018). What do aircraft accident investigators do and what makes them good at it? Developing a competency framework for investigators using grounded theory. *Safety Science*, 103, 153–161.

Patrick, J., Gregov, A., & Halliday, P. (2000). Analysing and training task analysis. *Instructional Science*, 28(1), 51–79.

Ramos, M. A., Utne, I. B., & Mosleh, A. (2019). Collision avoidance on maritime autonomous surface ships: Operators' tasks and human failure events. *Safety Science*, 116, 33–44.

Salmon, P. M., Stanton, N. A., Jenkins, D. P., & Walker, G. H. (2010). Hierarchical task analysis versus cognitive work analysis: Comparison of theory, methodology, and contribution to system design. *Theoretical Issues in Ergonomics Science*. 11, 6, 504–531.

Smith, P. A., Baber, C., Hunter, J., & Butler, M. (2008). Measuring team skills in crime scene investigation: Exploring ad hoc teams. *Ergonomics*, 51(10), 1463–1488.

Stanton, N. A. (2006). Hierarchical task analysis: Developments, applications, and extensions. *Applied Ergonomics*, 37, 1, 55–79.

Stanton, N. A., Jenkins, D. P., Salmon, P. M., Walker, G. H., Rafferty, L., & Revell, K. (2010). *Digitising Command and Control: A Human Factors and Ergonomics Analysis of Mission Planning and Battlespace Management*. Aldershot, UK: Ashgate.

Stanton, N. A., Salmon, P. M., Rafferty, L., Walker, G., Baber, C., & Jenkins, D. P. (2013). *Human Factors Methods: A Practical Guide for Engineering and Design*. Second Edition, Aldershot, UK: Ashgate.

Stanton, N. A., & Young, M. S. (2002). *Guide to Methodology in Ergonomics: Designing for Human Use*. Boca Raton: CRC Press.

Walker, G. H., Stanton, N. A., & Salmon, P. M. (2015). *Human Factors in Automotive Engineering and Technology*. Aldershot, UK: Ashgate.

Yu, D., Minter, R. M., Armstrong, T. J., Frischknecht, A. C., Green, C., & Kasten, S. J. (2014). Identification of technique variations among microvascular surgeons and cases using hierarchical task analysis. *Ergonomics*, 57(2), 219–235.

6 Cognitive Work Analysis (CWA)

BACKGROUND

Cognitive Work Analysis (CWA; Rasmussen et al., 1994; Vicente, 1999) is a powerful framework which supports the analysis and design of complex sociotechnical systems. Developed in the nuclear power domain, it provides a means to identify and represent the constraints within a complex system, capturing the range of potential system functioning and the degrees of freedom for behaviour available to actors (human or non-human) operating within the system.

The framework is formative which means that it considers the possibilities for behaviour within constraints, as opposed to what behaviours should occur (as provided by normative analysis methods) or what behaviour actually occurs (as provided by descriptive analysis methods). This focus is what makes CWA unique in that it captures possibilities for flexible and adaptive behaviour, including under abnormal or emergency conditions, and in 'first-of-a-kind' future systems where we do not have data available to inform analysis.

The CWA framework comprises five phases, each modelling different constraints and behaviour from a different perspective: work domain analysis (WDA); control task analysis (ConTA); strategies analysis; social organisation and cooperation analysis (SOCA); and worker competencies analysis (WCA).

WDA is used to provide an event and actor independent description of the system under analysis. The aim is to describe the purposes of the system and the constraints imposed on the actions of those performing activities within it (Vicente, 1999).

ConTA is used to describe and analyse the tasks that are undertaken to achieve the purposes, priorities and functions of a particular work domain (Naikar et al., 2006). Two methods are available for this phase – Rasmussen's decision ladder (Rasmussen, 1986; cited in Vicente, 1999) and Naikar et al. (2006) Contextual Activity Template (CAT). The decision ladder is used to describe the decision-making process adopted for specific tasks along with the short cuts that can be made through this process by users with differing levels of expertise. The CAT is used to map functions and affordances across different contexts and locations in terms of where they are currently undertaken and where they could potentially be given the systems constraints.

Strategies analysis is used to identify the different strategies that can be used to achieve control tasks. SOCA is used to identify how the control tasks and strategies are distributed across human and non-human agents within the system. A formative element also enables analysts to determine how tasks and strategies could be allocated given modification of constraints. By examining who/what currently does what, and who/what could do what, SOCA aims to specify an optimum allocation of functions for the system under analysis.

DOI: 10.1201/9780429281624-6

WCA is used to identify the cognitive skills required when performing different control tasks and undertaken different strategies. This phase uses Rasmussen's (1983) Skills, Rules and Knowledge (SRK) framework to identify the cognitive behaviour levels required for control tasks. This is then used to evaluate how well the system supports cognition during different tasks.

To date, CWA has been applied in many diverse domains including aviation (Borst et al., 2006), rail transport (Jansson et al., 2006; Salmon et al., 2016), road transport (e.g. Birrell et al., 2012; Cornelissen et al., 2013a; Salmon et al., 2019), sport (Hulme et al., 2019; McLean et al., 2017), military systems (e.g. Jenkins et al., 2011), maritime (Stanton and Bessell, 2014), public health (Salmon et al., 2021) and social networking (e.g. Whitworth and de Moor, 2009) to name only a selection.

PROCEDURE AND ADVICE

While it is beyond the scope of this chapter to fully describe the CWA procedure, the following section provides general guidance for each of the phases. Additional descriptions can be found in other texts (e.g. Vicente, 1999; Jenkins et al., 2009; Naikar; 2013). Further, it should be noted that CWA is a framework that is subject to continuing evolution and refinement. Several extensions have been proposed to the framework over time. For example, Ashoori and Burns (2013) developed an approach called Team Cognitive Work Analysis, which provides a set of methods for analysing the work of teams using CWA. In addition, Cornelissen et al. (2013) developed the Strategies Analysis Diagram to provide a more structured approach to the Strategies Analysis (StrA) phase of CWA. Finally, more recently, Naikar and Elix (2016) proposed extensions to improve the capacity of the framework to support integrated system design which enables adaptations in both actor behaviour as well as in the structure and organisation of work. Due to space limitations, this chapter refers to the commonly applied phases and methods outlined by Vicente (1999) and Jenkins et al. (2009).

STEP 1: DETERMINE THE AIMS AND OBJECTIVES OF THE ANALYSIS

The first step in a CWA is to clearly define the purpose of the analysis. The exact aims of the analysis and the boundaries of the system under consideration should be clearly specified to guide the CWA application. Consideration of scope is important to avoid being too narrow and thus not taking an appropriate systems-thinking lens, but also not having such a broad scope that the analysis becomes unwieldly. Given that CWA is an iterative framework, the scope and boundary of analysis may be re-considered after the analysis has commenced.

STEP 2: SELECT APPROPRIATE CWA PHASES AND METHODS

Once the nature and desired outputs of the analysis are initially defined, the analysis team carefully select the most appropriate CWA phases and methods to be employed during the analysis. For example, when using the framework for the design of a novel interface, it may be that only the first phase, Work Domain Analysis (WDA), is required (e.g. McLean et al., 2017; Salmon et al., 2021). Alternatively, when seeking to specify an optimal allocation of functions for a particular system, WDA and SOCA will be useful together. In general, it is recommended that WDA be applied as a starting point as it provides a holistic view of the system. Based on the selection of phases and methods, Steps 3–8 are then conducted as appropriate.

STEP 3: DATA COLLECTION

Once the aims of the analysis are clearly defined and the appropriate phases are chosen, the next step involves collecting targeted data about the system and its behaviour. The specific data collected is dependent on the phases being applied; however, data collection for CWA typically involves combinations of direct observation, concurrent verbal protocols, structured or semi-structured interviews (e.g. the Critical Decision Method), questionnaires and surveys, walkthrough analysis and documentation review (e.g. incident reports, standard operating procedures). Recommended data sources for each phase are presented in Table 6.1.

STEP 4: WORK DOMAIN ANALYSIS (WDA)

The initial phase within the CWA framework, WDA, provides a description of the constraints that govern the purpose and the function of the system under analysis. The abstraction hierarchy (Rasmussen, 1986; Vicente, 1999) is used to provide a description of the domain that is independent of context and of actors. The top three levels of the abstraction hierarchy identify the overall objectives of the domain, and what it can achieve, whereas the bottom two levels concentrate on the physical components and their affordances. Through a series of 'means-ends' links, it is possible to analyse how individual components can impact on the overall purpose of the system.

The abstraction hierarchy is constructed by considering the system's objectives (top-down) and the system's capabilities (bottom-up). The diagram can be constructed based upon a range of data. The exact data collection procedure is dependent on the system in question and the availability of data. In most cases, the procedure commences with document analysis including a review of documents such as strategic plans, organisational policies, standard operating procedures, laws and

TABLE 6.1
Example Data Sources for Each CWA Phase

	WDA	ConTa	StrA	SOCA	WCA
Documentation review	✓		✓	✓	
SME workshops	✓	✓	✓	✓	✓
Direct observation	✓	✓	✓	✓	
Cognitive task analysis interview		✓	✓	✓	✓
Task walkthrough	✓	✓	✓	✓	
Concurrent verbal protocols		✓	✓	✓	✓
Questionnaire/survey	✓			✓	

regulations, user guides, and maintenance manuals. Other data collection activities might involve engaging with domain experts via semi-structured interviews, workshops, or surveys. Wherever possible, observation of the activity in context is highly recommended.

The abstraction hierarchy consists of five levels of abstraction, ranging from the most abstract level of purposes to the most concrete level of physical form (Vicente, 1999). The labels used for each of the levels of the hierarchy can differ, depending on the aims of the analysis.

The following are generic labels for the hierarchical levels:

- **Functional purpose:** the domain purpose (or purposes if multiple exist), displayed at the very top of the diagram, describes the reason why the system exists. This purpose is independent of any specific situation and is also independent of time;
- **Values and priority measures:** this level captures the key values that can be used to assess how well the system is performing its domain purpose(s);
- **Purpose-related functions:** the middle layer of hierarchy identifies the functions that are performed within the system in order for it to meet its purposes;
- **Object-related processes:** this level identifies the physical processes or 'affordances' that the physical objects provide. These should be listed generically such that multiple objects could be linked to each process via means-ends links; and
- **Physical objects:** the physical objects within the system are identified at the base of the hierarchy. Analyst judgement is required to determine the level of detail of specification of the physical objects, based on the aims, objectives and scope of the analysis, and to limit the object list to a manageable size.

Nodes at each level of the abstraction hierarchy are connected to nodes at the levels above and below via means-ends links. These are used to show the relationships between objects, processes, functions, values, and functional purposes: the linked nodes at the level above a node describe why that node is required, and the linked nodes at the level below describe how the node is achieved. For example, for a particular function at the purpose-related functions level the nodes linked above represent the values and priorities that the function aims to support, and the nodes linked below represent the object-related processes that are undertaken to achieve the function (see Figure 6.1).

A set of prompts to aid development of the abstraction hierarchy are presented in Table 6.2 (adapted from Naikar, 2013).

The structure of the abstraction hierarchy framework acts as a guide to acquiring the knowledge necessary to understand the domain. The framework helps to direct the search for deep knowledge about the system, providing structure to the document analysis process, particularly for the domain novice. While the output may initially appear somewhat overwhelming, its value to the analysis cannot be overstated. The abstraction hierarchy defines the systemic constraints at the highest level.

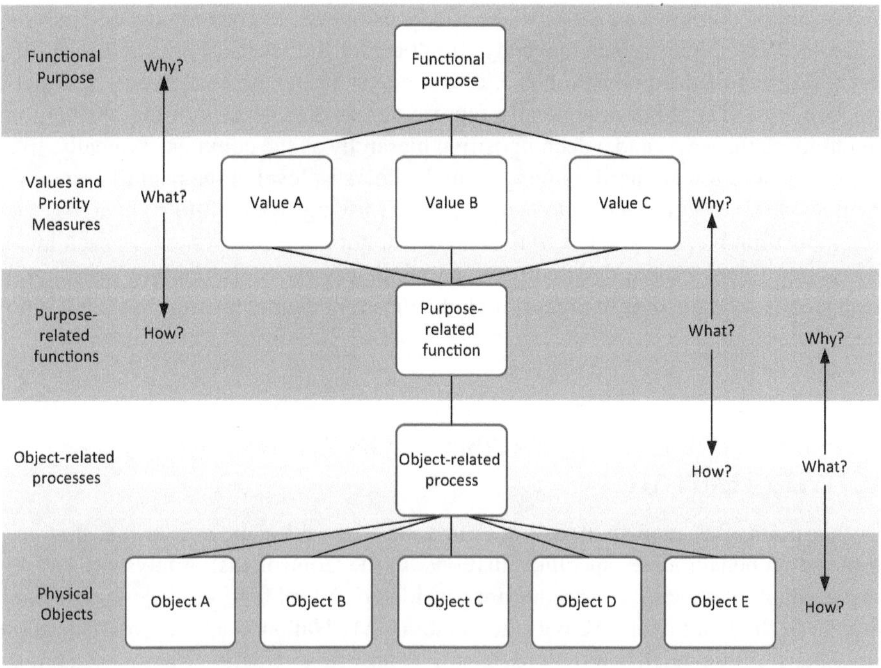

FIGURE 6.1 Means-ends links example.

TABLE 6.2
Prompts to Aid Abstraction Hierarchy Development

Abstraction Hierarchy Level	Description	Abstraction Hierarchy Development Prompts
Functional purpose	The overall purposes of the system and the external constraints imposed on its operation	Why does the system exist?
Values and priority measures	The criteria that stakeholders use for measuring progress towards the functional purposes	How can stakeholders tell whether the system is achieving its purposes? What is currently measured to enable assessment of whether the system is achieving its purposes?
Purpose-related functions	The general functions of the system that are necessary for achieving the functional purposes	What functions must be performed for the system to achieve its purposes?
Object-related processes	The functional capabilities and limitations of the physical objects within the system that enable the purpose-related functions	What processes are physical objects used to achieve within the system?
Physical objects	The physical objects within the system that are used to undertake the purpose-related functions	What physical objects are used within the return from lockdown system?

Source: Adapted from Naikar, 2013.

The work domain can also be described using an abstraction–decomposition space (ADS), which enables the analyst to consider the levels of abstraction at different degrees of decomposition (i.e. a whole system perspective, sub-system perspective, etc). The ADS is generally represented as a matrix, with the abstraction hierarchy as the rows, and a decomposition hierarchy as the columns. Typically, five levels of resolution are used, ranging from the coarsest level of the overall system to the finest level of components (Vicente, 1999). Moving from left to right across the decomposition hierarchy is the equivalent of zooming into the system, as each level across represents a more detailed representation. The ADS also employs structural means-ends relationships in order to link the different elements of the system within the ADS.

STEP 5: CONTROL TASK ANALYSIS – CONTEXTUAL ACTIVITY TEMPLATE

To this point, the analysis does not deliberately consider the constraints that are imposed on behaviour by specific situations. In the Control Task Analysis (ConTA) phase, situation-specific constraints are considered. A tool for considering such constraints is the Contextual Activity Template (CAT; Naikar et al., 2006). This tool plots the purpose-related functions identified in the abstraction hierarchy against a number of specific 'situations'. Within the ConTA phase of CWA, the analysis remains independent of the actor. That is, the focus is on functions specified generally, not on tasks performed by any individual human or technological actor. The first stage of the process is to define the situations. Situations can be characterised by either time or location, or a combination of the two. In many cases, it is appropriate to explore more than one set of situations using multiple CAT representations to meet a range of analytic goals.

The CAT is generally presented as a matrix with the functions represented in rows and the situations as columns (see Figure 6.2). The matrix identifies the situations in which it is possible for functions to be performed (indicated by the dashed boxes) and those in which the function is typically performed (indicated by circles and whiskers). If a cell in the matrix is blank this indicates that the function cannot be performed within that situation.

STEP 6: CONTROL TASK ANALYSIS – DECISION LADDERS

Continuing with the theme of describing additional constraints, key function-situation cells within the CAT can be explored in terms of decision-making. The decision ladder (Rasmussen et al. 1994) is the tool most commonly used within CWA to describe decision-making activity. Its focus is on the entire decision-making activity rather than the moment of selection between options. It is not specific to any single actor; instead, it represents the decision-making process of the combined system. In many cases, the decision-making process may be collaborative, distributed between a range of human and technical decision-makers.

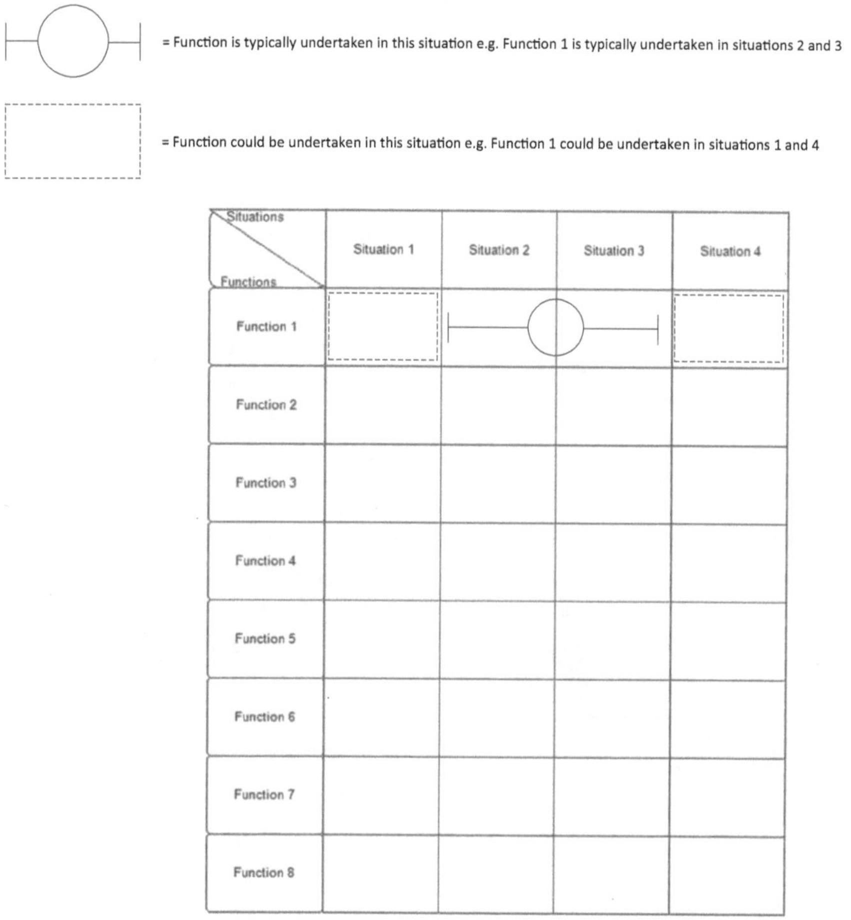

FIGURE 6.2 CAT template and example.

A decision ladder template is presented in Figure 6.3. The decision ladder contains two different types of node: rectangular boxes that represent data-processing activities, and circles that represent resultant states of knowledge. The left side of the decision ladder represents the observation of the current system state, whereas the right side represents the planning and execution of tasks and procedures to achieve a target system state. Actors who are novices (within that situation) are expected to follow the decision ladder in a linear fashion, whereas experts are expected to take 'shortcuts' through the ladder.

Decision ladders are generally populated based on interview data from users. Critical decision method interviews (Klein and Armstrong, 2005; Klein et al., 1989) have been used for this purpose. Ostensibly, in its raw form, decision ladder models provide a list of the information requirements for making a decision. At this stage of the analysis, the relative importance of these information elements is not considered.

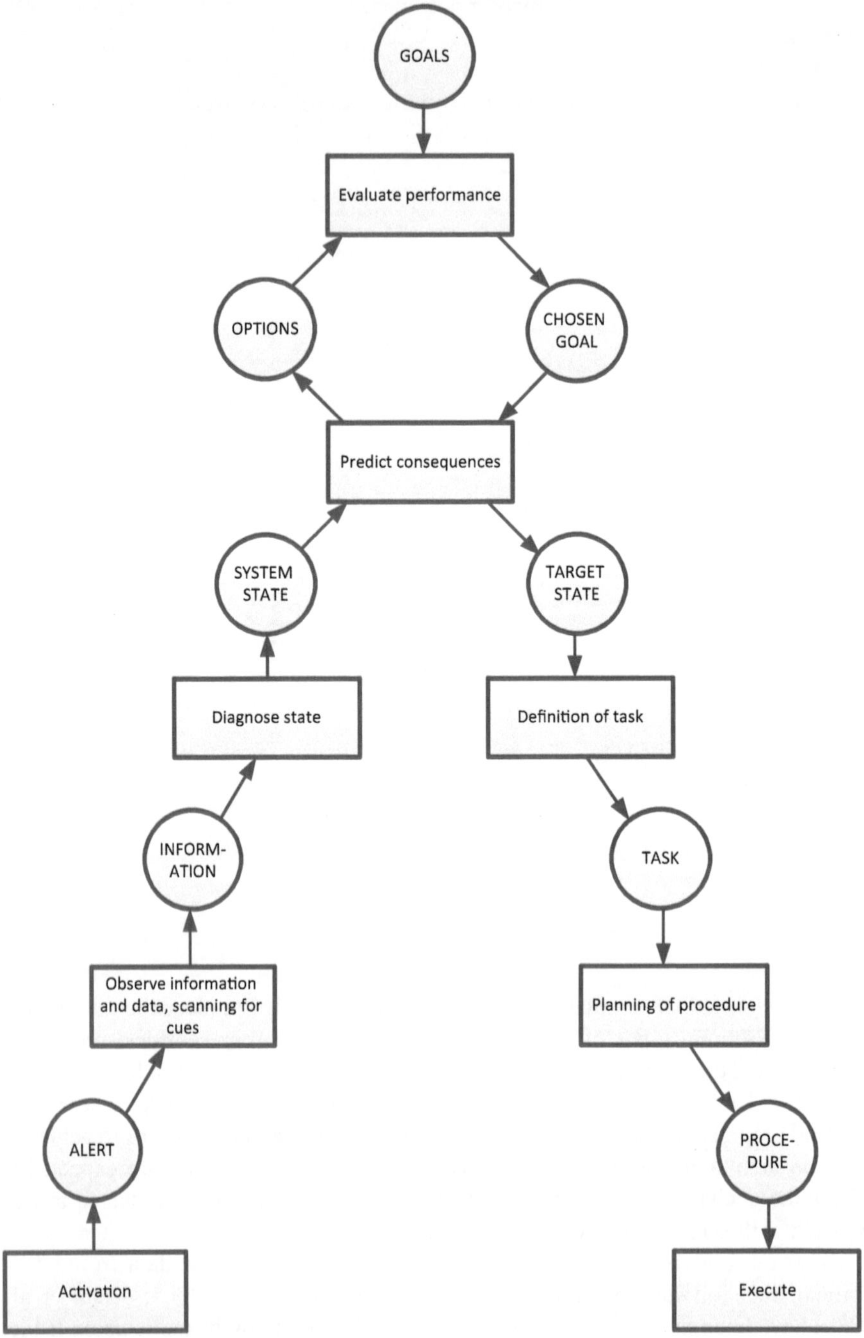

FIGURE 6.3 Decision ladder template.

STEP 7: STRATEGIES ANALYSIS

The strategies analysis phase of CWA addresses the constraints influencing the way in which activity can be conducted. In keeping with the remainder of the framework, it introduces additional detail to the analyses developed in the previous phases. The aim of the strategies analysis phase is to describe the constraints that dictate how a system can be (rather than how it should be or currently is) moved from one state to another. This phase can be particularly useful for exploring flexibility within a system (Jenkins et al., 2009). Information flow maps are typically used (see Figure 6.4. for an information flow map template). The first step in developing the information flow map is to identify the relevant start state and end state, based on the aims and objectives of the analysis. Next, the analyst identifies all of the strategies that can move the system from the start state to the end state. Again, at this point of the analysis, it remains actor independent.

Although information flow maps are typically used for the Strategies Analysis component of CWA, other tools, such as the Strategies Analysis Diagram (SAD; Cornelissen et al., 2013) can also be applied. SAD builds on the abstraction hierarchy developed in the WDA phase and involves the addition of two levels to the diagram: verbs and criteria. The verbs are used to specify how the physical objects can be used. The criteria are then used to specify the circumstances under which different strategies might be chosen.

STEP 8: SOCIAL ORGANISATION AND COOPERATION ANALYSIS

The SOCA phase of CWA involves identifying how activities and strategies are and can be distributed between human and technological actors within the system. This phase of CWA builds upon outputs such as the abstraction hierarchy, decision ladders, and information flow maps developed during the preceding phases. Conducting the SOCA involves first identifying relevant human and technological actors and then analysing the abstraction hierarchy (SOCA-WDA), CAT (SOCA-CAT), decision

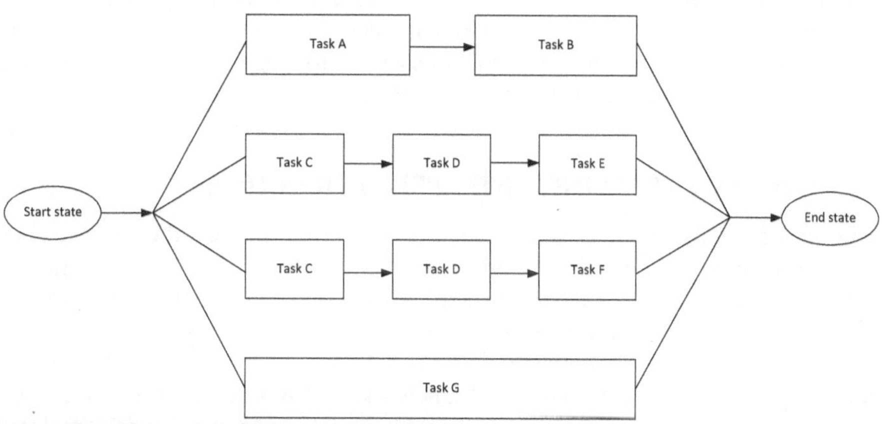

FIGURE 6.4 Information flow map template.

ladders, and/or strategies analysis outputs to show what functions, tasks, and strategies the different actors current perform and could perform given design modifications. This involves shading nodes in the abstraction hierarchy, CAT, decision ladder, and information flow map to show who or what can undertake functions, provide affordances, engage in parts of the decision making process, and complete different components of strategies. This provides a systemic description that can be used to explore optimal allocation of functions across the system.

STEP 9: WORKER COMPETENCIES ANALYSIS

The final phase of a CWA involves identifying the cognitive competencies required for task performance. Worker Competencies Analysis (WCA) uses Rasmussen's (1983) skills, rules, and knowledge framework to classify the cognitive activities employed by agents during control task performance. The skills, rules, and knowledge taxonomy (Rasmussen, 1983) enables the analyst to map, for each critical strategy, how the system supports each level of cognitive processing.

In the skills, rules, and knowledge taxonomy, skill-based behaviour is associated with sensory-motor performance which occurs in skilled activity without the requirement for conscious control. Rule-based behaviour refers to the application of stored rules, based on past experience, to determine behaviour. Finally, knowledge-based behaviour is engaged in unfamiliar situations where it is not possible to draw upon past experience, and the actor must engage in reasoning to understand the situation and select an appropriate course of action.

RELATED METHODS

As it is a framework there are various methods which can be applied across the five phases. These include the abstraction hierarchy and abstraction–decomposition space (WDA), decision ladders and the CAT (ConTa), information flow maps and the SAD (Strategies analysis) and the SRK framework (WCA). In addition, various methods can be used to gather the data required for each phase including cognitive task analysis interviews such as the Critical Decision Method (CDM; Klein et al., 1989), concurrent verbal protocols, direct observations, cognitive walkthroughs, and Hierarchical Task Analysis (HTA; see Chapter 5).

APPROXIMATE TRAINING AND APPLICATION TIMES

Due to the number of phases and methods within the phases, the training time associated with the CWA framework is high. Similarly, the application time can also be high, if all five phases are applied. However, in many cases, analysts choose to use a selection of phases that will be sufficient to answer their aims and objectives. A number of applications involve the WDA phase only, and assuming the analyst has experience with the framework, and has access to the necessary information or domain experts to input to the model, this phase can be completed relatively quickly.

RELIABILITY AND VALIDITY

The formative nature of CWA makes the assessment of its the reliability and validity somewhat difficult. However, some testing has been conducted. For example, Burns et al. (2004) conducted a qualitative comparison of two independently created WDA models which described similar military command and control environments. Although the models were developed with different scopes and purposes, they found a reasonable level of consistency in the models, suggesting a level of reliability within the method. In a later study, Cornelissen and colleagues (2014) tested the criterion-referenced validity and test-retest reliability of the Strategies Analysis Diagram method, using novice analysts. They reported that the method did not show high levels of reliability and validity. However, when individual results were pooled, the validity was much improved, suggesting that group modelling processes may be superior to individual model development processes. A comparison of the findings of this study, using novice analysts, to those of Burns et al. (2004) suggests that, as analyst expertise grows more reliable outcomes are obtained.

TOOLS REQUIRED

No specific tools are required to conduct CWA; however, a software tool has been developed to support users to develop the graphical representations required in CWA, including user guidance for all five phases (Jenkins et al., 2009). If a specialised software tool is not used, the representations can be drawn in any illustration software such as Microsoft Visio™ or they could be hand drawn with key aspects drawn formally in illustration software for presentation purposes.

CASE STUDY EXAMPLE: AUTONOMOUS VEHICLES

Advanced Autonomous Vehicles (AVs) are currently being introduced into road systems internationally and represent a growing area of interest within HFE. As with the introduction of many automated technologies, AVs are promoted with the promise to improve safety by reducing crashes caused by 'human error'. However, while automation will create safety and efficiency benefits, it also poses a range of risks (Hancock, 2019; Stanton and Marsden, 1996). Importantly, little is known about how other road users (e.g. human drivers, cyclists, pedestrians) will respond and adapt to sharing the road with AVs. The openness and complexity of the road system, and its use by all members of society, makes this an important area for research given the potential for new types of risks to safety.

Governments worldwide have begun to respond to some of the challenges posed by the imminent introduction of advanced AVs onto public roads. In Australia, approaches such as risk-based regulation are currently being explored in relation to the introduction of autonomous vehicles (e.g. National Transport Commission, 2017). Policy development, however, is difficult given that the exact nature of the technologies remains unknown. Instead, policy and regulation will need to be agile; to respond to risks that emerge as technology progresses. Of particular concern is the transition period, where a mix of manual, semi-automated and fully automated vehicles will share the road (van Lint et al., 2016). This application of CWA aimed to identify the potential safety risks and other issues associated with the introduction of AVs.

The aim of the analysis was to model the future intersection system, with a focus on the next 5 to 10 years', to identify potential risks associated with the introduction of AVs and associated insights for system design. As an illustrative case study, this analysis drew on document review, pre-existing datasets and the authors' knowledge of the road transport domain through extensive previous research (e.g. Salmon et al., 2018; Salmon et al., 2019; Read et al., 2018).

The scope of the analysis was road user traversal of signalised intersections in Australia. Road users in this case study were considered to be both human and technological actors, including drivers, motorcyclists, cyclists, and AVs. The case study analysis is based on a right-angle intersection of two dual-lane roads in an urban area of Australia. The roads have 50km/h speed limits and medium traffic density. Traffic lights are present on all approaches, including pedestrian walk/don't walk displays (in the form of red man/green man symbols). No turning arrows are presented on the traffic light assemblies. The road rules require drivers wishing to turn right to give way to oncoming traffic and to any pedestrians crossing the road.

For this analysis, we chose to use all five CWA phases: WDA, ConTA, StrA, SOCA, and WCA. This was done to illustrate the use of each phase in providing a unique perspective on the system. The CWA software tool was used to conduct the analysis, except for the decision ladders which were drawn using Microsoft Visio.

RESULTS

Work Domain Analysis

The abstraction hierarchy method was used for the WDA phase. Given the novelty of advanced AVs, especially in Australia where this research was conducted, the abstraction hierarchy was developed through document review and pre-existing datasets. A key input was the review of two previously developed abstraction hierarchies for road transport. The first had been developed to understand interactions between different types of road users when conducting a right-hand turn at a signalised intersection (Cornelissen et al., 2013b). The other was developed to describe the entire road transport system (Salmon et al., 2019). These existing abstraction hierarchies were reviewed, updated, and refined to describe the broader intersection system, and also to represent what would be expected to comprise the future intersection system, with a timescale of 5 to 10 years, including the incorporation of automated vehicles.

Figure 6.5 shows the abstraction hierarchy for the future intersection system. The abstraction hierarchy has one functional purpose, or 'reason for being'; to support the negotiation of road users through the intersection. Eight value and priority measures were identified. These were focused on measures of safety, efficiency, compliance, reaching the desired destination, and user experience. A specific value and priority measure for two-wheelers was to maintain balance and keep upright, and a specific measure relating to AVs was that of trust. At the purpose-related functions level of the abstraction hierarchy 12 nodes were identified. These can generally be

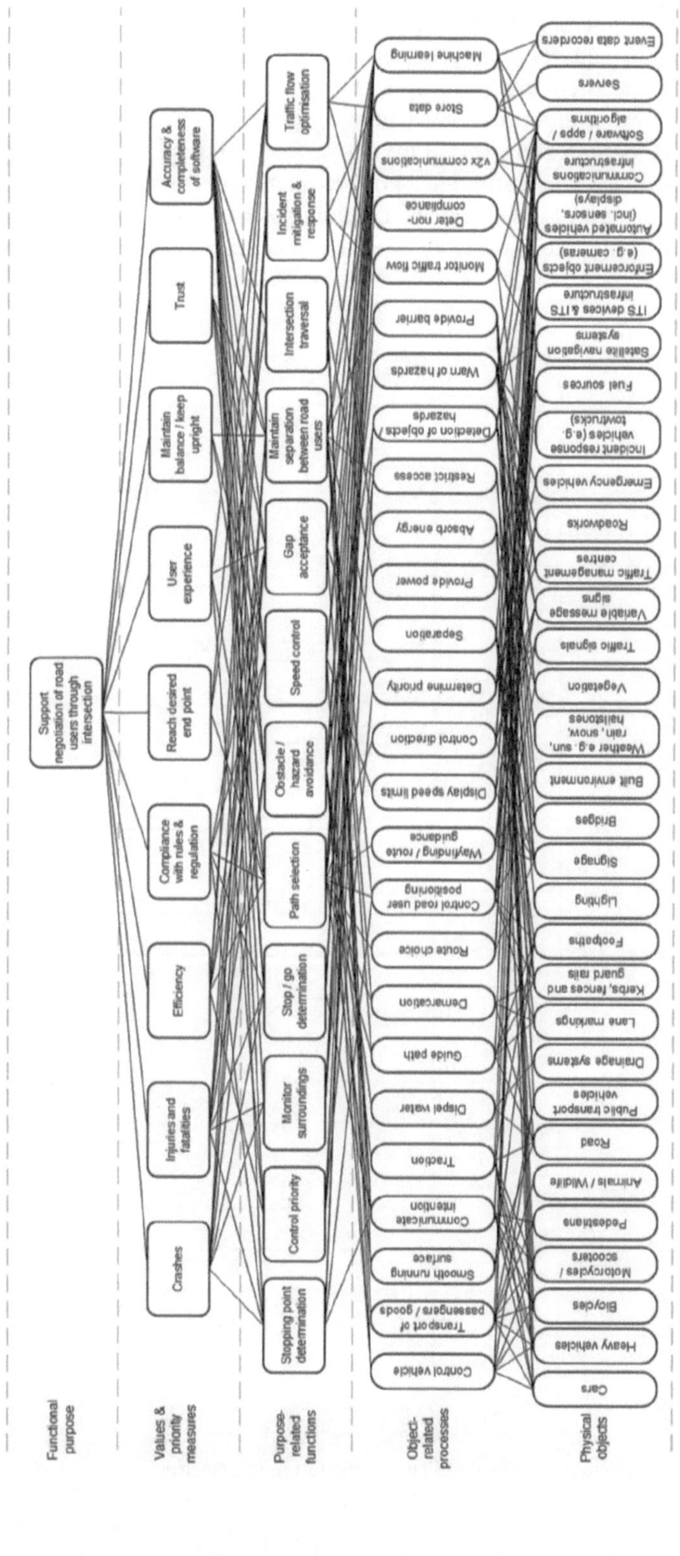

FIGURE 6.5 Abstraction hierarchy of the future road intersection. Control Task Analysis – Contextual Activity Template.

categorised into two types: 1) Road user functions (e.g. stopping point determination, monitor surroundings, stop/go determination, path selection, obstacle/hazard avoidance, and speed control); and 2) Traffic management functions (e.g. control priority, traffic flow optimisation, incident mitigation, and response).

Finally, at the bottom two levels of the abstraction hierarchy, there were 33 physical objects identified, which afforded 24 object-related processes.

The CAT is presented in Figure 6.6. The columns of the CAT were based on six different road user situations which represented stages of the road users' approach to

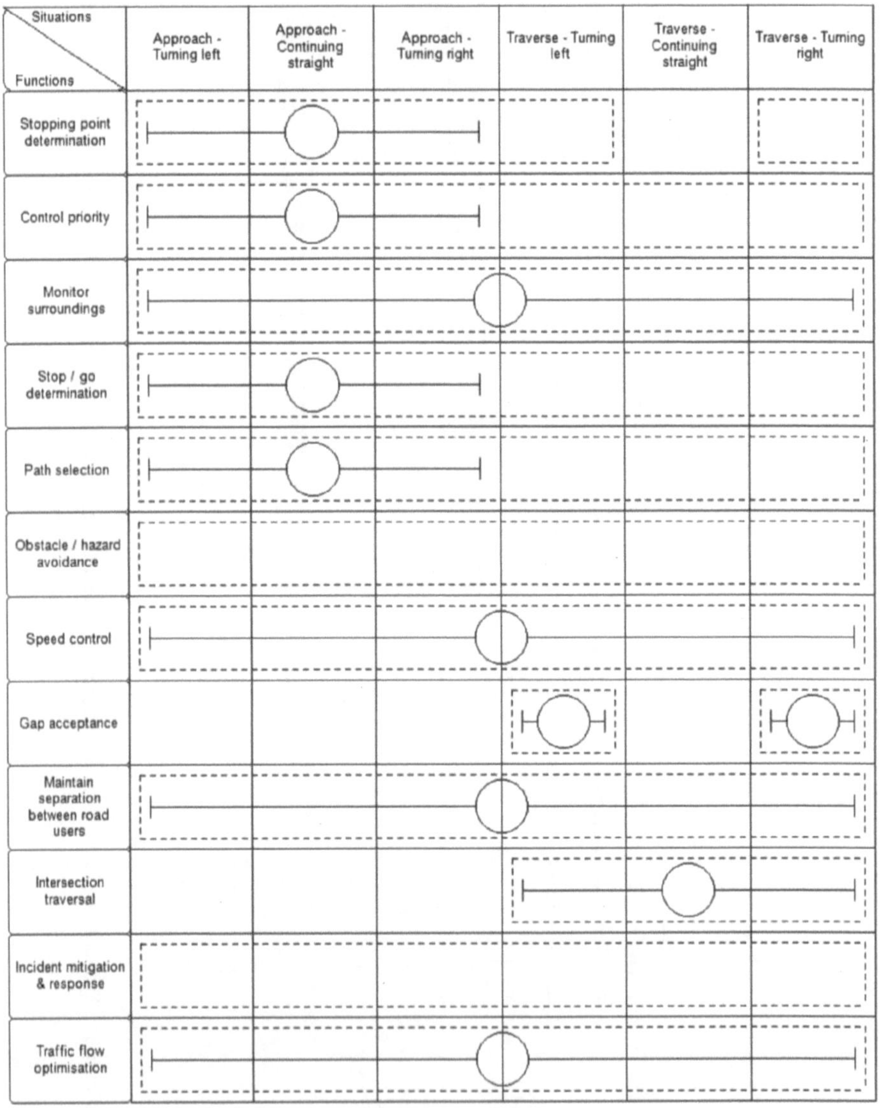

FIGURE 6.6 Contextual activity template for the future intersection.

the intersection. Three situations were defined as being when the road user is on approach and three when the road user is within the boundary of the intersection.

The CAT shows a group of functions are performed continuous within the system, regardless of the situation. These functions include: Monitor surroundings, Speed control, and Traffic flow optimisation. Another group of functions (Stopping point determination, Control priority, Stop/go determination, and Path selection) are typically performed when the road user is approaching the intersection, and can continue to be performed when they have entered the intersection, but are not typically undertaken at this point. The Gap acceptance function is only relevant in two situations: where a road user is within the intersection and is either turning left or right. Finally, the functions of Obstacle/hazard avoidance and Incident mitigation and response can be undertaken in any situation, but are not typically undertaken.

Given that in Australia, the right-hand turn scenario is the riskier situation as it involves turning against oncoming traffic, we chose to focus further analyses on two key scenarios within the CAT. These were:

- Stop/go determination when approaching and turning right.
- Stop/go determination when traversing the intersection and turning right.

CONTROL TASK ANALYSIS – DECISION LADDERS

The decision ladders for the two key scenarios are shown in Figures 6.7 and 6.8.

Considering the approach situation, and reading up from the bottom left-hand side of the ladder in Figure 6.7, it can be seen that a number of alerts might indicate to a road user that they need to make a stop/go determination (e.g. seeing the intersection, seeing the traffic lights). Then, once the road user is alerted, they gather information (e.g. what is the status of the traffic light?) to understand the system state (e.g. is it safe to enter the intersection?) and then evaluate their options (e.g. proceed or stop), taking into account their chosen goal (e.g. safety, efficiency, compliance with laws). Moving down the right-hand side of the ladder, the road user defines how they will enact their chosen option, and execute the action.

Two shortcuts through the decision ladder are shown in Figure 6.7. The first, from Information to Execute, represents occasions whereby a simple rule-based decision can be made. For example, the road user sees the traffic lights and is thus alerted to the need to make a decision, they then gather the information that the traffic light is showing green, and without the need for further cognitive processing, execute the task of proceeding into the intersection to prepare to make a right-hand turn. The second shortcut is shown from System state from Execute. This also represents rule-based decision-making. For example, the road user may be alerted to the need to make a decision on seeing the traffic lights. They may gather information that the traffic light has just changed from green to amber. The road user may consider the system state in two ways. Firstly, whether, by the time they reach the intersection, they might be fined if they enter the intersection once the light has changed to red. Secondly, whether it would be safe to stop, particularly where there is traffic following behind. The user, in making this 'dilemma zone' decision, is likely to rely on

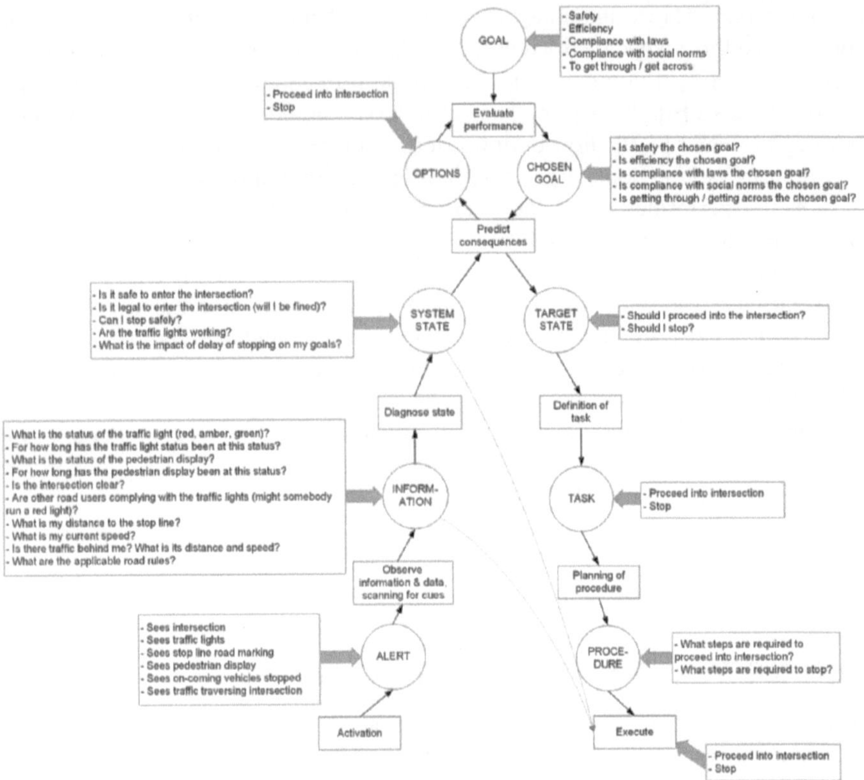

FIGURE 6.7 Decision ladder for the stop/go determination when approaching and turning right.

experience integrating these considerations and other information (e.g. their speed and distance to the crossing) to make a decision. The shortcuts demonstrate how experienced road users are unlikely to need to refer to higher level considerations such as their options and goals.

Considering the situation when the road user is within the intersection (Figure 6.8), it can be seen that different alerts and information apply. Key aspects of information are related to the need to understand the status of oncoming traffic, whether pedestrians are crossing the pedestrian crossing, and whether the road that the driver will turn into is clear of traffic (so that if the turn is executed the road user won't be blocking the intersection). Much of the remainder of the decision ladder remains the same as in Figure 6.8, with the exception of another option – to 'Change path'. This might be used where the road user initially intended to turn, but due to circumstances once they enter the intersection, they decide to continue ahead instead. For example, in heavy congestion, the road user may block the path of other following road users who are intending to continue straight on the road. If the road user is sensitive to social norms, impatient road users behind may influence them to select this option.

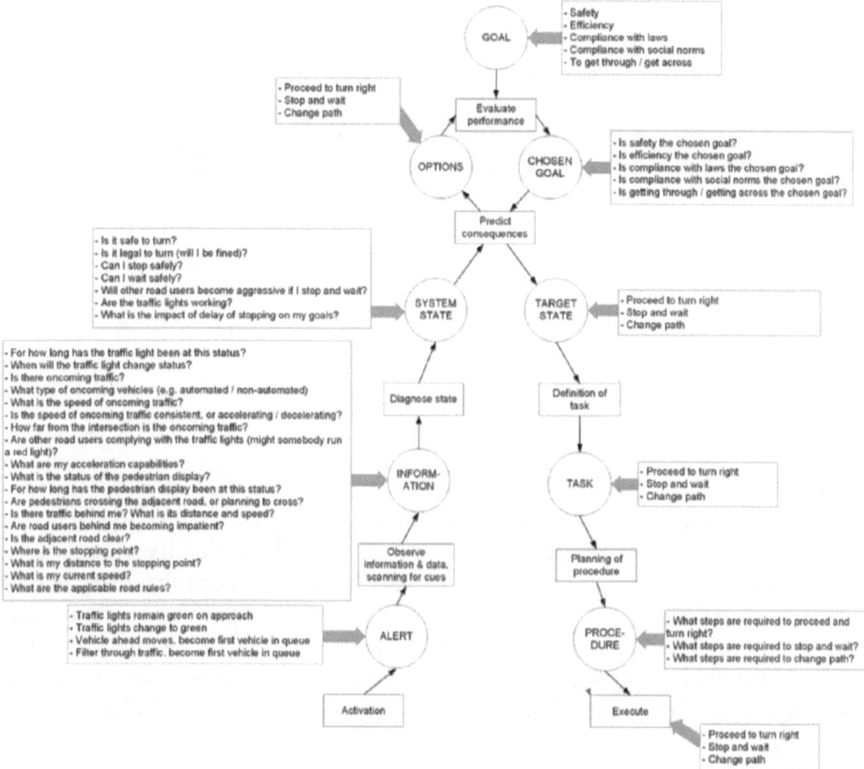

FIGURE 6.8 Decision ladder for the stop/go determination when traversing the intersection and turning right.

For this ladder, one key shortcut was identified, between Information and Execute. This rule-based decision could occur where the road user is alerted to the need to make a decision once the traffic lights change to green and they proceed into the intersection, there is no oncoming traffic, nor pedestrians using the crossing, and the road they are proceeding into is clear. In this instance they would then proceed to execute the right-hand turn.

STRATEGIES ANALYSIS

The information flow map was used to identify the strategies that could be undertaken by road users at a signalised intersection for the two key scenarios identified in the CAT. Figure 6.9 shows the results of the information flow map which identifies that strategies that can be taken to move from the state of 'intention to turn right' through to 'turning right'. A total of six strategies were identified. The first strategy would be used in the case that the road user is initially in the left-hand side lane, and requires them to move to the right lane first. Then, in Strategies 1 and 2, the road user identifies and moves to a safe stopping/holding point within the intersection, past the

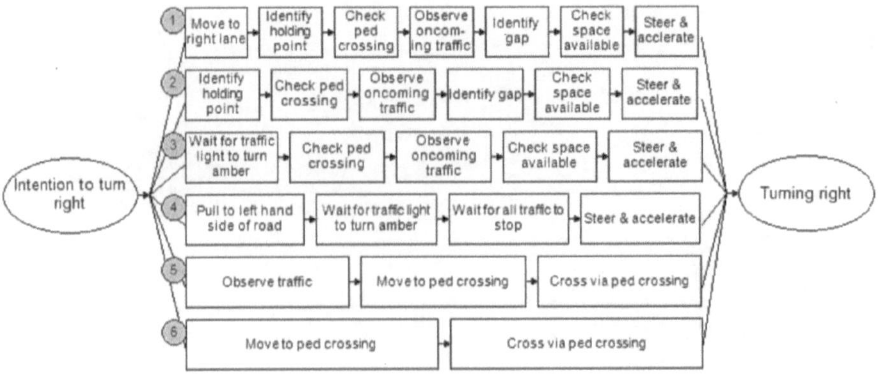

FIGURE 6.9 Information flow map for turning right at the future intersection.

stop line. They then check the pedestrian crossing as they are required to give way to pedestrians on the crossing, and observe oncoming traffic to identify an appropriate gap in the traffic to turn. At this stage road users will also check that there is space available in the road they are turning into to ensure they will not block the intersection if they turn. Finally, they will accelerate and steer to perform the right-hand turn. These strategies are interesting for AVs, as there are no line markings to indicate where a safe holding point within the intersection is for these manoeuvres. Further, sensors need to be able to judge pedestrian crossing behaviour, and whether there is space available on the road into which they are turning. This may lead to conservative decisions made by the AV such as where pedestrians are waiting near the kerb, or have stepped off the kerb onto the road but are giving way to traffic.

The third strategy may be undertaken if the oncoming traffic is particularly dense, or if the road user is less confident. Depending on its programming, it may also be used by AVs. This strategy involves waiting at or just past the stop line for the traffic lights to turn amber. The road user then checks the pedestrian crossing for any pedestrians who may be crossing late, and observes the oncoming traffic to ensure it has come to a stop on the amber light. They also need to ensure that space is available in the road onto which they are turning, and then they accelerate and steer to perform the turn. This strategy may lead to frustration from vehicles waiting behind the stopped road user. It is less likely to be undertaken by cyclists or motorcyclists as they may experience vehicles overtaking them or experience a sense of pressure from other road users waiting behind. However, interestingly, it is not expected that AVs would be sensitive to social pressures and thus may be more likely to take a conservative approach.

The final three strategies might be used by cyclists, particularly under conditions where the oncoming traffic is particularly dense, or if the cyclist is less confident travelling in traffic. Strategy 4 involves using a hook turn, where the cyclist pulls to the left-hand side of the traffic stream and waits for the light to turn to amber. They then wait for all traffic to stop before accelerating and turning across the whole intersection. Strategies 5 and 6 are similar in that they involve the cyclist choosing to leave the road and use the pedestrian crossing instead. In Strategy 5, the cyclist observes the traffic density and decides to move off the road because they feel unsafe turning in the traffic stream. In Strategy 6, the cyclist takes this option without prior consideration of the traffic conditions (i.e. this is their regular strategy either at this intersection or at all intersections).

SOCIAL ORGANISATION AND COOPERATION ANALYSIS

Figure 6.10 shows the output of the SOCA phase. SOCA was applied to the information flow map with the following actors identified: drivers, AVs, motorcyclists, cyclists, as well as communications systems. These communication system technologies could include messages provided traffic lights to vehicles (e.g. assisting with the strategy of waiting for the lights to turn amber), or include messages between vehicles (e.g. assisting with the strategy of observing oncoming traffic), or messages communicated to pedestrians via smartphones (e.g. assisting with the strategy of checking the pedestrian crossing). It was assumed that cyclists will not access connected vehicle technology. Thus, Strategies 4, 5, and 6 involve the cyclist only.

WORKER COMPETENCIES ANALYSIS

Table 6.3 shows the outputs of the WCA phase, using the SRK taxonomy to identify competencies at each level of cognitive processing. The WCA analysis is based on five key strategies identified from the Information flow map. We chose to focus the analysis on the strategies that involve future technologies (i.e. AVs and communications systems).

The table identifies a range of skill-based competencies required by road users to successfully turn right at the future intersection. These competencies generally involve perceiving elements in the environment. In the current road transport system, this generally requires visual perception of elements in the road environment. However, this is likely to change in the future. For example, AVs will use lidar and

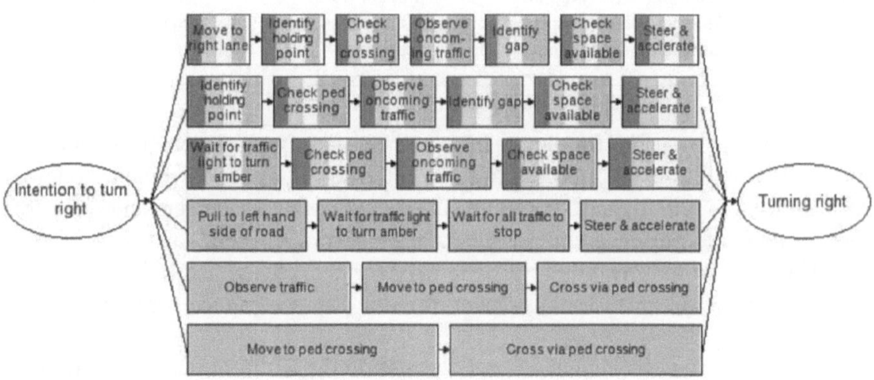

FIGURE 6.10 SOCA information flow map. Note, purple = driver, green = AV, yellow = motorcyclist, pink = cyclist, blue = communications systems.

TABLE 6.3
WCA for Turning Right at the Future Intersection

Strategy	Skill-Based Competency	Rule-Based Competency	Knowledge-Based Competency
Identify holding point	– Ability to judge space so as not to block oncoming traffic	– Ability to use experience to judge appropriate holding point	– Knowledge that it is acceptable to move beyond the stop line and wait within the intersection
Check pedestrian crossing	– Ability to perceive location of pedestrian crossing – Ability to perceive/interpret pedestrian collision warnings – Ability to perceive pedestrians (including children, pedestrians using wheelchairs, pedestrians on scooters, etc.)	– Ability to anticipate pedestrian behaviour (i.e. identify intention to cross) based on experience	– Knowledge that must give way to pedestrians on right-hand turn – Knowledge that pedestrians regularly do not comply with road signals
Observe oncoming traffic	– Ability to perceive oncoming traffic (including its direction of travel)	– Ability to anticipate traffic from side roads/driveways based on experience	– Knowledge that must give way to oncoming traffic
Identify gap	– Ability to perceive time to arrival (e.g. speed and distance) – Ability to perceive/interpret gap acceptance support systems. – Ability to perceive cues regarding intention of other road users (e.g. indicators, slowing/speeding up, visual focus)	– Ability to anticipate time to arrival of other road users based on experience	– Knowledge that other road users may act in unexpected ways (e.g. use indicator but fail to slow down and turn). – Knowledge that other road users may not comply with road signals (i.e. run a red light)
Check space available	– Ability to perceive traffic in road that are turning into – Ability to perceive speed of traffic and predict time to clear	– Ability to anticipate time for traffic to clear based on experience	– Knowledge that the intersection needs to remain clear

radar to 'see' visual information such as other road users and environmental elements. This will give them access to different information than available to human road users. AVs will also use communications with other road users and infrastructure to judge the presence and possibly the behaviour of other road users. Drivers are also likely to have access to decision support systems which collect data from sensors and communications systems to provide gap acceptance support and collision warnings.

The rule-based competencies generally refer to the act of drawing on experience to anticipate or judge conditions. While for human road users this reflects their schema for intersections, for AVs this is likely to relate to the stimuli used to train AV algorithms.

The knowledge-based competencies represent a mix of knowledge of formal road rules (i.e. knowledge that must give way to pedestrians when turning) and knowledge of actual behaviour on the roads (i.e. that pedestrians regularly do not comply with road signals). This demonstrates that AVs cannot be designed to simply follow the road rules, expecting other road users to do the same.

The aim of this analysis was to model the future intersection system, with a focus on the next 5 to 10 years, to identify insights for system design. The five phases of CWA were applied, beginning with a focus on the entire road transport system, and concluding with the identification of competencies of the key strategies undertaken to turn right at a signalised intersection. The case study demonstrates how the phases of CWA are interlinked and how they allow the analyst to focus on particular aspects of the system, while maintaining a whole system perspective through the AH.

The analysis highlights that while the future road system will likely incorporate new advanced technologies, these will not be relevant to all road users. For example, as demonstrated in the SOCA analysis, cyclists are not expected to use communications system technologies, thus they will not have the benefit of decision support systems, and they will also be reliant upon other road users to visually perceive their presence, unless they become required to fit sensors to their vehicles.

Although the analysis presented here was a case study only, it has identified a range of practical insights to support the design of future road systems. These include that:

- The expanding introduction of connected technologies, with additional data storage and sharing identified in the AH highlights the need for the design of new management systems around data including in relation to storage capacity, cybersecurity, and privacy;
- Trust, identified as a new value and priority measure in the future road system, needs to be measured and influence through design initiatives. For example, 'explainable AI' where the user is presented with information about how the AV has made certain decisions, could be adopted within AVs;
- Consideration is needed as to the different goals of road users identified in the decision ladder analysis (e.g. safety, efficiency, compliance with laws, compliance with social norms), and how these should be balanced in the future road system. For example, should AVs have settings that their user can adjust as to their own goals for the trip?
- AVs will not be influenced by social pressures in the same way as human road users. They may therefore not follow established social norms. AV designers should consider whether desirable social norms should be programmed into vehicle behaviour. Where AVs do not follow social norms, education of human road users may be required to avoid undesirable interactions due to mis-matched expectations;
- The outputs of the decision ladder and WCA could assist to inform future decision support systems for human drivers for aspects of the task of turning right such as gap acceptance and collision avoidance;

- AVs need to be designed to meet the competencies identified in the WCA. For example, machine learning training sets should incorporate examples of other road users acting in unexpected ways (e.g. using their indicator but fail to slow down and turn); and
- Individual human drivers must experience various traffic situations to build up their schema of the road system, particularly for unexpected or emergency situations (i.e. edge cases). AVs, on the other hand, have the advantage of access to the collective experience of a network of vehicles who are experiencing and learning from situations. While on-going learning is currently occurring within companies (e.g. Tesla vehicles upload data to the cloud daily; Stilgoe, 2017) such learning should be shared across companies to improve safety.

Beyond insights for design, the CWA outputs could also be useful from a practical perspective if they can structure safety assurance processes for AV manufacturers and road safety regulators. For example, testing could assess the competencies identified within the WDA. Further, more exhaustive analyses could identify design requirements for future driver decision support systems that integrate the range of data that will be available in a manner that best supports driver performance.

RECOMMENDED READING

Bisantz, A. M., & Burns, C. M. (2008). *Applications of Cognitive Work Analysis*. Boca Raton, FL: CRC Press, Taylor & Francis Group.

Jenkins, D. P., Stanton, N. A., Salmon, P. M., & Walker, G. H. (2009). *Cognitive Work Analysis: Coping with Complexity*. Aldershot, UK: Ashgate.

Read, G. J. M., Salmon, P. M., Goode, N., & Lenné, M. G. (2018). A sociotechnical design toolkit for bridging the gap between systems-based analysis and system design. *Human Factors in the Manufacturing and Services Industries* 28(6), 327–341.

Stanton, N. A., Salmon, P. M., Walker, G. H., & Jenkins, D. P. (2017). *Cognitive Work Analysis: Applications, Extensions and Future*. Boca Raton, FL: CRC Press.

Vicente, K. J. (1999). *Cognitive Work Analysis: Toward Safe, Productive, and Healthy Computer-Based Work*. Mahwah, NJ: Lawrence Erlbaum Associates.

REFERENCES

Ashoori, M., & Burns, C. (2013). Team cognitive work analysis: Structure and control tasks. *Journal of Cognitive Engineering and Decision Making* 7, 123–140.

Birrell, S. A., Young, M. S., Jenkins, D. P., & Stanton, N. A. (2012). Cognitive Work Analysis for safe and efficient driving. *Theoretical Issues in Ergonomics Science* 13, 430–449.

Borst, C., Suijkerbuijk, H. C. H., Mulder, M., & Van Paassen, M. M. (2006). Ecological interface design for terrain awareness. *The International Journal of Aviation Psychology*, 16, 375–400.

Burns, B. & Roth, E. M., Burns, C. M., Bisantz, A. M., & Roth, E. M. (2004). Lessons from a comparison of work domain models: Representational choices and their implications. *Human Factors* 46, 711–727.

Cornelissen, M., Salmon, P. M., McClure, R., & Stanton, N. A. (2013a). Using cognitive work analysis and the strategies analysis diagram to understand variability in road user behaviour at intersections. *Ergonomics*, 56(5), 764–780.

Cornelissen, M., Salmon, P. M., McClure, R., & Stanton, N. A. (2013c). Using cognitive work analysis and the strategies analysis diagram to understand variability in road user behaviour at intersections. *Ergonomics* 56(5), 764–780.

Cornelissen, M., Salmon, P. M., & Young, K. L. (2013). Same but different? Understanding road user behaviour at intersections using cognitive work analysis. *Theoretical Issues in Ergonomics Science*, 14(6), 592–615.

Hancock, P. (2019). Some pitfalls in the promises of automated and autonomous vehicles. *Ergonomics* 62(4), 479–495.

Hulme, A., McLean, S., Read, G., Dallat, C., Bedford, A., & Salmon, P. M. (2019). Sports organisations as complex systems: Using Cognitive Work Analysis to identify the factors influencing performance in an elite Netball organisation. *Front. Sports Act Living – Sports Management and Marketing* 1, 1–56.

Jansson, A., Olsson, E., & Erlandsson, M. (2006). Bridging the gap between analysis and design: Improving existing driver interfaces with tools from the framework of cognitive work analysis. *Cognition, Technology & Work* 8, 41–49.

Jenkins, D. P., Stanton, N. A., Salmon, P. M., & Walker, G. H. (2009). *Cognitive Work Analysis: Coping with Complexity*. Surrey: Ashgate.

Jenkins, D. P., Stanton, N. A., Salmon, P. M., & Walker, G. H. (2011). A formative approach to developing synthetic environment fidelity requirements for decision-making training. *Applied Ergonomics* 42, 757–769.

Klein, G. & Armstrong, A. A. 2005. Critical decision method. In *Handbook of Human Factors and Ergonomics Methods*, edited by N. Stanton, A. Hedge, K. Brookhuis, E. Salas, & H. Hendrick, 347–356. Boca Raton, FL: CRC Press.

Klein, G. A., Calderwood, R., & MacGregor, D. (1989). Critical decision method for eliciting knowledge. *IEEE Transactions on Systems, Man and Cybernetics* 19 (3), 462–472.

van Lint, H., Schakel, W., Tamminga, G., Knoppers, P., & Verbraeck, A. (2016). Getting the human factor into traffic flow models: new open-source design to simulate next generation of traffic operations. *Transportation Research Record*, 2561(1), 25–33.

McLean, S., Soloman, C., Gorman, A., & Salmon, P. M. (2017). What's in a game? A systems approach to enhancing performance analysis in football. *Plos One*, 1–15.

Naikar, N. (2013). *Work Domain Analysis: Concepts, Guidelines and Cases*. Boca Raton, FL: Taylor & Francis Group.

Naikar, N. & Elix, B. (2016). Integrated system design: Promoting the capacity of sociotechnical systems for adaptation through extensions of cognitive work analysis. *Frontiers in Psychology* 7, 962.

Naikar, N., Moylan, A., & Pearce, B. (2006). Analysing activity in complex systems with cognitive work analysis: Concepts, guidelines and case study for control task analysis. *Theoretical Issues in Ergonomics Science*, 7(4), 371–394.

National Transport Commission. (2017). *Assuring the Safety of Automated Vehicles: Policy Paper*. Melbourne: National Transport Commission. https://www.ntc.gov.au/sites/default/files/assets/files/NTC-policy-paper-assuring-the-safety-of-automated-vehicles_0.pdf

Rasmussen, J. (1983). Skills, rules and knowledge: Signals, signs and symbols, and other distinctions in human performance models. *IEEE Transactions on Systems, Man and Cybernetics* 13, 257–266.

Rasmussen, J. (1986). *Information Processing and Human-Machine Interaction: An Approach to Cognitive Engineering*.

Rasmussen, J., Pejtersen, A. M., & Goodstein, L. P. (1994). *Cognitive Systems Engineering*. New York: Wiley-Interscience.

Read, G. J. M., Stevens, E., Lenné, M. G., Stanton, N. A., Walker, G. H., & Salmon, P. M. (2018). Walking the talk: Comparing pedestrian 'activity as imagined' with 'activity as done'. *Accident Analysis and Prevention* 113, 74–84.

Salmon, P. M., Lenné, M. G., Mulvihill, C., Young, K., Cornelissen, M., Walker, G. H., & Stanton, N. A. (2016). More than meets the eye: Using cognitive work analysis to identify design requirements for safer rail level crossing systems. *Applied Ergonomics* 53, Part B, March 312–322.

Salmon, P. M., Read, G. J. M., Stevens, N. A., Walker, G. H., Beanland, V., McClure, R., Hughes, B., Johnston, I., & Stanton, N. A. (2019). Using the abstraction hierarchy to identify how the purpose and structure of road transport systems contributes to road trauma. *Transportation Research Interdisciplinary Perspectives* 3, 100067.

Salmon, P. M., Read, G. J. M., Walker, G. H., Lenné, M. G., & Stanton, N. A. (2018). *Distributed Situation Awareness in Road Transport: Theory, Measurement, and Application to Intersection Design*. Boca Raton, FL: CRC Press.

Salmon, P. M., Stevens, N. J., McLean, S., Hulme, A., & Read, G. J. M. (2021). Human Factors and Ergonomics and the management of existential threats: a work domain analysis of a COVID-19 return from lockdown restrictions system. *Human Factors and Ergonomics in the Manufacturing and Service Industries* 31(4), 412–424.

Stanton, N. A. & Bessell, K. (2014). How a submarine returns to periscope depth: Analysing complex socio-technical systems using Cognitive Work Analysis. *Applied Ergonomics* 45(1), 110–125.

Stanton, N. A., & Marsden, P. (1996). From fly-by-wire to drive-by-wire: Safety implications of automation in vehicles. *Safety Science*, 24(1), 35–49.

Stilgoe, J. (2017). Machine learning, social learning and the governance of self-driving cars. *Social Studies of Science*, 48(1), 25–56.

Vicente, K. J. (1999). *Cognitive Work Analysis: Toward Safe, Productive, and Healthy Computer-Based Work*. Mahwah, NJ: Lawrence Erlbaum Associates.

Whitworth, B., & de Moor, A. (Eds.). (2009). *Handbook of Research on Socio-Technical Design and Social Networking Systems*. Hershey, New York: IGI Global.

7 The Event Analysis of Systemic Teamwork (EAST)

BACKGROUND

The Event Analysis of Systemic Teamwork (EAST; Stanton et al., 2013) provides an integrated suite of methods for analysing the behaviour of teams, organisations, and sociotechnical systems. EAST has the following propositions:

- Systems can be explained in terms of a set of interrelated networks;
- EAST is a model-fit-data approach (rather than a data-fit-model approach);
- Some networks perform better than others;
- EAST can model distributed cognition in systems;
- All accidents result from a failure to communicate information between agents and tasks;
- EAST can be used to investigate accidents and forecast risks; and
- EAST models can be used to inform the design of future systems.

The framework provides methods to describe, analyse and integrate three network-based representations of activity: task, social, and information networks. Task networks are used to provide a representation of the interrelated tasks which are undertaken within a system. Social networks are used to describe the human and non-human agents performing tasks within the system and the communications that take place between them during task performance. Information networks describe the information that agents use in support of task performance and how this information is distributed across different tasks and agents. The three network representations are subsequently analysed using a range of network analysis metrics to examine network structure and the importance of different nodes within the networks. The three networks can also be brought together to form composite networks showing the relationships between tasks, agents, and information. The EAST 'network of networks' approach is presented in Figure 7.1.

DOMAIN OF APPLICATION

To date, EAST has been applied in many domains, including land and naval warfare (Stanton et al., 2006; Stanton, 2014; Stanton and Roberts, 2020), aviation (Stanton et al., 2019; Stewart et al., 2008), air traffic control (Walker et al., 2010), road and rail transport (Salmon et al., 2014; 2018), cybersecurity (Lane et al., 2019), the emergency services (Houghton et al., 2006), and elite cycling (Salmon et al., 2017).

DOI: 10.1201/9780429281624-7

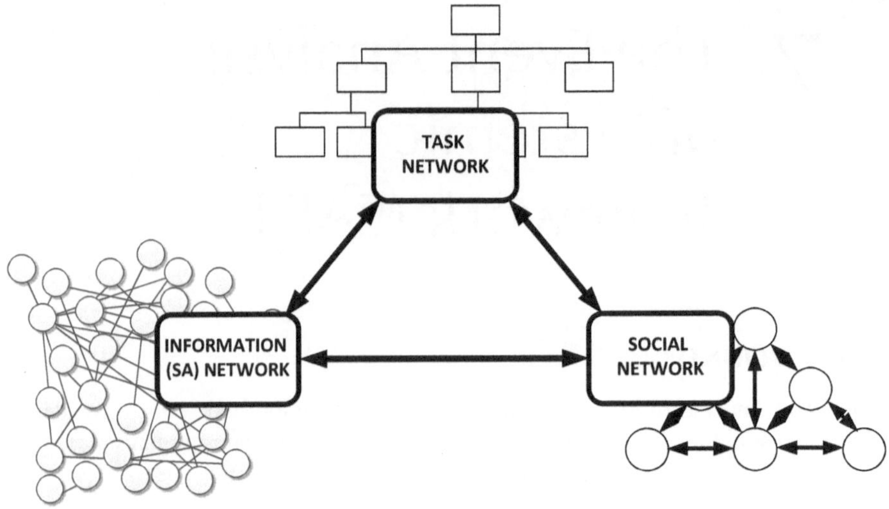

FIGURE 7.1 EAST network of networks approach.

PROCEDURE AND ADVICE

A flowchart depicting the EAST procedure is presented in Figure 7.2. The following step-by-step procedure is intended to act as a broad set of guidelines for applying EAST.

STEP 1: DEFINE ANALYSIS AIMS

First, the aims of the analysis should be clearly defined so that appropriate scenarios are used, and relevant data is collected. In addition, not all components of the EAST framework may be required, so it is important to clearly define the aims at this point to ensure that the appropriate EAST methods are applied.

STEP 2: DEFINE THE TASK/SCENARIO/SYSTEM UNDER ANALYSIS

Next, the task(s)/scenario(s)/system under analysis should be clearly defined. EAST is a flexible framework which can be used to describe and analyse the behaviour of individuals (e.g. Salmon et al., 2014), teams (e.g. Roberts et al., 2017), organisations (e.g. Walker et al., 2010) and entire sociotechnical systems (e.g. Stanton et al., 2019). It is therefore important to clearly to clearly state the analysis focus and boundaries so that all analysts understand them and what sits within and outside the scope of the analysis. Consideration of scope is important to avoid being too narrow and thus not taking an appropriate systems thinking lens, but also not having such a broad scope that the analysis becomes unwieldly. The scope and boundary of analysis may be re-considered after the analysis has commenced based on identification of new data or areas that relate to the analysis aims.

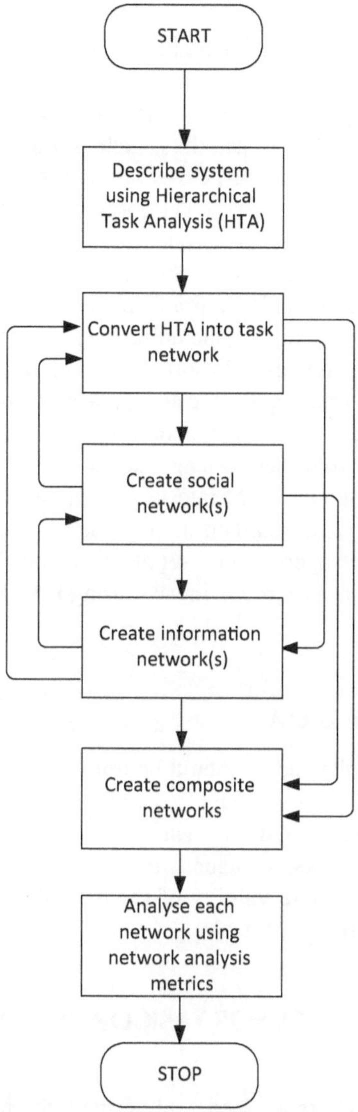

FIGURE 7.2 EAST procedure.

STEP 3: DATA COLLECTION

Once the aims of the analysis are clearly defined, the next step involves collecting targeted data about the system and the activities of interest. Data collection for EAST typically involves a combination of documentation review (e.g. standard operating procedures), direct observations of the activities of interest, gathering recordings of communications that occur between agents, eliciting concurrent verbal protocols from the agents performing tasks, conducting post hoc structured or semi-structured interviews (e.g. the Critical Decision Method), and task walkthrough analyses. For further information on these methods and their application the reader is referred to Stanton et al. (2013).

The data collection approach selected is dependent upon the aims and boundaries of the analysis as well as project constraints, such as time and the number of analysts available and opportunity and access. Observational data is often a critical requirement. Typically, several analysts are used to observe the activities under analysis. All activities should be recorded, and an activity timeline should be created, including a description of the activity undertaken, the agents involved, any communications made between agents and the tools, equipment, and technology involved. Additional notes should be made where required, including the purpose of the activities observed, any documents or instructions used to support each activity, activity outcomes, any errors made, and any information that the agents involved feel is relevant. It is normal practice, where permitted, to make video recordings of the activities and audio recordings of communications between agents.

It is often useful to conduct CDM interviews with agents who play an important role in the activities under analysis. This involves dividing the activity into key phases and then interviewing each agent using a set of pre-defined CDM probes (e.g. O'Hare et al., 2000). A set of probes used previously during EAST applications is presented in Table 7.1 below.

STEP 4: TRANSCRIBE DATA

Once all of the data is collected, it should be transcribed in order to make it usable during the EAST analysis phase. When transcribing the data it is recommended that an event transcript be constructed. This should describe the activity over a timeline, including descriptions of tasks, the agents involved, any communications made, and the technology used. To ensure validity of the data, the event transcript should be reviewed by as many SMEs as possible.

STEP 5: CONSTRUCT HTA FOR TASK OR SCENARIO UNDER ANALYSIS

Once the data is collected, transcribed, and verified, the next step involves creating a Hierarchical Task Analysis (HTA; Stanton, 2006) for the activities under analysis (see Chapter 5). HTA works by decomposing systems and behaviour into a hierarchy of goals, subordinate goals, operations, and plans. In doing so, it focuses on "what an operator … is required to do, in terms of actions and/or cognitive processes to achieve a system goal" (Kirwan and Ainsworth, 1992, p. 1). HTA outputs specify the overall goal of a particular system, the sub-goals to be undertaken to achieve this goal, the operations required to achieve each of the sub-goals specified and the plans that trigger engagement and achievement of the different goals and operations. The plans component specifies the sequence, and under what conditions, different sub-goals have to be achieved in order to satisfy the requirements of a super-ordinate goal. Once the HTA is complete it is important to have it reviewed by appropriate

TABLE 7.1
CDM Probes

Goal Specification	What Were You Aiming to Accomplish Through This Activity?
Assessment	Suppose you were to describe the situation at this point to someone else. How would you summarise the situation?
Cue identification	What features were you looking for when you formulated your decision? How did you know that you needed to make the decision? How did you know when to make the decision?
Expectancy	Were you expecting to make this sort of decision during the course of the event? Describe how this affected your decision-making process.
Options	What courses of action were available to you? Were there any other alternatives available to you other than the decision you made? How/why was the chosen option selected? Why were the other options rejected? Was there a rule that you were following at this point?
Influencing factors	What factors influenced your decision-making at this point? What was the most influential factor/piece of information that influenced your decision-making at this point?
Situation awareness	What information did you have available to you at the time of the decision?
Situation assessment	Did you use all of the information available to you when formulating the decision? Was there any additional information that you might have used to assist in the formulation of the decision?
Experience	What specific training or experience was necessary or helpful in making this decision? Do you think further training is required to support decision-making for this task?
Mental models	Did you imagine the possible consequences of this action? Did you create some sort of picture in your head? Did you imagine the events and how they would unfold?
Decision-making	How much time pressure was involved in making the decision? How long did it actually take to make this decision?
Conceptual	Are there any situations in which your decision would have turned out differently?
Guidance	Did you seek any guidance at this point in the task/incident? Was guidance available?
Basis of choice	Do you think that you could develop a rule, based on your experience, which could assist another person to make the same decision successfully?

Source: Adapted from O'Hare et al., 2000 and Crandall et al., 2006.

SMEs. Following this, the HTA is often used to identify activity phases for which task, social, and information networks will be constructed.

STEP 6: DEFINE ACTIVITY PHASES

It is useful at this stage to review the HTA and determine whether the analysis should focus on an overall activity or a set of discrete activity phases. For example, when examining driving the EAST analysis could focus on the overall task of 'Drive car' or instead on discrete car driving activities or phases, such as 'negotiate intersection', 'negotiate roundabouts', 'drive on motorway', 'drive on urban roads' and so on. The specific focus is dependent on the analysis aims and boundaries and the quality of the

TABLE 7.2
Analysis Rules Regarding the Relationships Between Nodes Within EAST Networks

Network	Nodes	Example Nodes	Relationships	Example Relationships
Task network	Represent high level tasks that are undertaken during the activity under analysis. The high-level tasks are typically extracted from the subordinate goals level of the HTA	– Control vehicle speed – Control vehicle direction – Monitor road ahead – Monitor area behind vehicle	Represent instances where the conduct of one task (i.e. task network node) influences, is undertaken in combination with, or is dependent on, another task	Control vehicle speed and control vehicle direction are linked in the task network as both tasks are undertaken together
Social network	Represent human, technological, or organisational agents who undertake one or more of the tasks involved in the activity under analysis (as identified in the HTA and task network)	– Driver – Speedometer – Pedestrian – Traffic lights	Represent instances where agents within the social network interact with one another during the activity under analysis	The driver and speedometer are linked as the driver checks the speedometer when monitoring speed
Information network	Represent pieces of information that is required by agents when undertaking the activity under analysis (as identified in the task and social network)	– Own vehicle – Speed – Direction – Traffic light status	Represent instances where information influences other information or is used in combination with other information during the activity under analysis	Own vehicle is linked to speed and direction as the vehicle has a current speed and is travelling in a particular direction

data collected. If the analysis is to be an overall analysis of the activity in question, then an overall task, social, and information network should be constructed during Steps 7 to 9. If there are a set of activity phases, then task, social, and information networks should be constructed for each activity phase.

STEP 7: CONSTRUCT TASK NETWORK

Steps 7–9 involve developing task, social, and information networks for the activities under analysis. A set of network construction rules are presented in Table 7.2. It is recommended that each network is created first in Microsoft Excel in the form of a matrix and is then drawn visually using drawing software packages such as Microsoft Visio. It is important to note that not all networks may be required. For example, when analysing situation awareness failures during the Air France 447

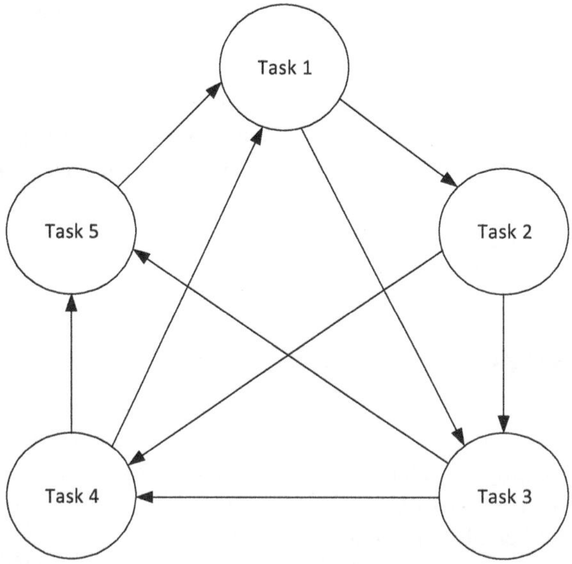

FIGURE 7.3 Task network example showing the relationships between five tasks.

disaster, Salmon et al. (2016) only developed information networks as the task and social networks were not required.

The first analysis step involves constructing a task network for the activity under analysis. Task networks (see Figure 7.3) are used to represent HTA outputs in the form of a network which shows key tasks and the relationships between them (Stanton et al., 2013; 2018). This enables analysts to understand the interactions and coupling that exists between tasks across the work system. Within Figure 7.3 the circular nodes represent tasks and the arrows linking the tasks represent relationships between tasks. Relationships between tasks are included in the task network when:

1. Tasks are undertaken sequentially e.g. task three is undertaken after completion of task two;
2. Task are undertaken together (e.g. tasks one and two are undertaken together);
3. The outcomes of one task influence the conduct of another (e.g. the outcomes of task four influence how task five is undertaken); or
4. The conduct of one task is dependent on completion of the other e.g. task five cannot be undertaken until task four is completed.

Task networks are constructed by taking the first layer of sub-goals from the HTA and identifying which of them are related with one another based on the relationship types described above. Once the initial draft task network is complete, it is useful to have SMEs review it. The task network should then be refined based on the SMEs feedback. Changes at this stage typically involve the addition and/or removal of relationships between tasks as the tasks or the modification of task descriptions.

Like HTA, it is normal practice for the task network to go through various iterations before it is finalised.

It should also be noted that whilst we recommend that the development of a task network should be based on HTA, it is possible to develop a task network based directly on raw data collection activities, SME input and verification, and document reviews (Stanton, 2014). This is often the case where project constraints do not enable development of an initial HTA. If the HTA step is indeed not utilised, then we recommend that analysts use SME review to ensure that the core tasks of interest are included in the task network.

STEP 8: CONSTRUCT SOCIAL NETWORK

Social Network Analysis (SNA; Driskell and Mullen, 2004) is used to analyse the relationships between the agents involved in the activities under analysis. This involves first creating a social network matrix showing the relationships between agents in terms of interactions (e.g. communications) and the frequency of interactions (e.g. number of communications between agents). Following this a social network diagram is constructed which provides a visual representation of the social network (Figure 7.4). Typically the direction (i.e. from actor A to actor B) and frequency of communications is depicted; however, the type and content of communications can also be included, as well as the communications media used (e.g. Walker et al., 2010).

STEP 9: CONSTRUCT INFORMATION NETWORK

The third network to be constructed is the information network, also previously referred to as propositional networks (e.g. Stanton et al., 2005) or situation awareness networks (Salmon et al., 2018). The information network includes all of the 'information elements' that are used when conducting the tasks described within the task network. Information networks can be constructed either manually or via the Leximancer software tool using data derived from communications logs or concurrent verbal protocols provided by participants as they perform the activities under analysis. If this data is not available, however, information network scan also be developed based on cognitive task analysis interview data, procedures, or accident investigation reports (e.g. Salmon et al., 2016).

Information networks depict the information or concepts underlying situation awareness and the relationships between them. For example, the verbal protocol transcript extract "the red car behind is slowing down" from a driver would produce the linked concepts 'Car', 'Red', 'Behind', and 'Slowing' as in the car 'is' red, it is with 'located' behind the participant's vehicle and it 'is' slowing down. This is shown below in the form of an information network in Figure 7.5.

Manual construction of the networks involves performing a content analysis on the data to identify concepts and the relationships between them. It is often useful to have a second analyst also construct an information network (or at least a portion of it) from the same data for the purposes of reliability testing.

From \ To	Agent A	Agent B	Agent C	Agent D
Agent A	-	20	2	10
Agent B	17	-	4	8
Agent C	1	5	-	8
Agent D	10	10	1	-

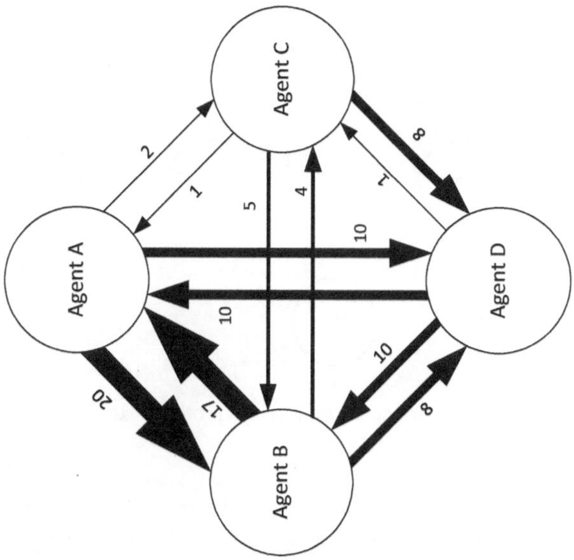

FIGURE 7.4 Simple social network matrix and associated social network diagram.

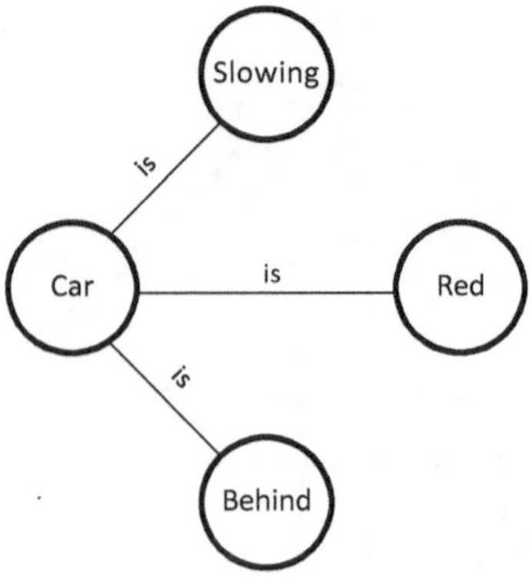

FIGURE 7.5 Example SA network for concurrent verbal protocol extract "the red car behind is slowing down".

The Leximancer thematic analysis software can be used to auto-construct information networks. Leximancer uses text representations of natural language to interrogate verbal transcripts and identify themes, concepts, and the relationships between them. The software does this by using algorithms linked to an in-built thesaurus and by focusing on features within the verbal transcripts such as word proximity, quantity, and salience. Leximancer thus automates the content analysis procedure by processing verbal transcript data through five stages: conversion of raw text data, concept identification, thesaurus learning, concept location, and mapping (i.e. creation of network). The output is a concept map, similar to an information network, which shows concepts derived from the verbal transcript and the relationships between them reflected within the verbalisations.

STEP 10: CONSTRUCT COMPOSITE NETWORKS

Composite networks are used to explore the relationships between tasks, agents, and information and are constructed by combining the different networks. For example, a task by agent network can be constructed by combining the task and social network to show which tasks are undertaken by which agents. This involves assigning a colour to the different agents within the social network and shading each node within the task network to show which agent performs that task. Useful composite networks to construct include:

- Task by agent network (task-social network showing which tasks are performed by different agents);
- Information by agent network (social-information network showing which agents use what information);
- Task and associated information network (task-information network showing what information is used when performing each task); and
- Fully composite network showing information by agents and related to tasks (combined task, social, and information network).

STEP 11: ANALYSE NETWORKS USING NETWORK ANALYSIS METRICS

An important component of EAST analyses involved using network analysis metrics to analyse the task, social, information, and composite networks. Both whole of network and nodal metrics can be used, enabling analysis of the structure of the networks (e.g. density) and identification of key nodes (e.g. tasks, agents, information) within the networks. The size of the network determines the number of possible relations, and the number of possible relations grows exponentially with the size of the network. This defines the network's complexity. The first set of metrics are used to analyse the whole network:

- The **density** of a network is defined by the number of social relations that are actually observed and can be represented as some fraction of the total possible;
- **Cohesion** is defined as the number of reciprocal connections in the network divided by the maximum number of possible connections; and
- The largest geodesic distance within a network defines its **diameter**, which can be thought of as another metric of the network's size (i.e. the number of hops to get from one side of the network to the other).

The second set of metrics are used to analyse the individual nodes within each network:

- **Emission** and **reception** degree are the number of ties emanating from, and going to, each agent in the network;
- **Eccentricity** is defined by the largest number of hops an agent has to make to get from one side of the network to another;
- The **sociometric status** of each agent refers to the number of communications received and emitted, relative to the number of nodes in the network;
- Agent **centrality** is calculated in order to determine the central or key agent(s) within the network. There are a number of different centrality calculations that can be made. For example, agent centrality can be calculated using Bavelas-Leavitt's index;
- **Closeness** is the inverse of the sum of the shortest distances between each individual and every other person in the network. It reflects the ability to access information through the 'grapevine' of network members;
- **Farness** is the index of centrality for each node in the network, computed as the sum of each node to all other nodes in the network by the shortest path; and
- **Betweenness** is defined by the presence of an agent between two other agents, which may be able to exert power through its role as an information broker.

For sociometric status and centrality, nodes that achieve values above the mean + standard deviation value for the network are classified as 'key' nodes within the network (e.g. key tasks, key agents, key information). This approach has previously been used to identify key nodes within EAST networks (e.g. Houghton et al., 2006; Salmon et al., 2018; Stanton, 2014; Stanton et al., 2017).

ADVANTAGES

- As the analysis covers tasks, agents, and information (and the relationships between all three) it provides a comprehensive overview of the activities under analysis;
- The analysis is both qualitative (network content) and quantitative (network analysis) in nature and useful insights can be drawn from both forms of analysis;
- EAST is highly flexible and can be used to study work as imagined, work as is done, and accidents. The EAST-BL method can also be used proactively to identify risks (see Chapter 3);
- Composite networks enable analysts to explore the direct relationships between tasks, agents, and information;
- The use of network analysis metrics is a useful feature of the method that enables analysts to identify key tasks, agents, and information and/or to compare activities in different contexts;
- The framework approach allows data collection and analysis methods to be chosen based on analysis requirements;
- EAST has been applied in a wide range of different domains for various purposes. The approach is generic and can be used in any domain;
- The networks enable analysis of various concepts, including distributed situation awareness, distributed cognition, teamwork, decision-making, workload, and human and non-human agent teaming;
- EAST can be applied at an individual, team, organisational, or systems level; and
- Software support is available for developing and analysing all three networks.

DISADVANTAGES

- EAST analyses can become large, complex, and unwieldy, especially when applied to overall sociotechnical systems;
- Complete analyses involving the construction of all three networks and composite networks are time consuming to undertake;
- There is no reliability and validity evidence associated with EAST;
- The use of various data collection and analysis methods ensures that the training time required can be higher than other systems analysis methods; and
- A high level of access to the domain, activities, and relevant SMEs is often required.

APPROXIMATE TRAINING AND APPLICATION TIMES

Due to the number of different methods involved, the training time associated with the EAST framework is high. Similarly, application time is typically high, although this is dependent upon the task under analysis and the scope of the analysis.

RELIABILITY AND VALIDITY

The reliability and validity of the EAST methods is difficult to assess and at the time of writing there have been no formal EAST reliability and validity studies.

TOOLS NEEDED

Normally, video and audio recording devices are used to record the activities under analysis. There is no EAST software tool; however, there are various tools which can be used to help build and analyse the networks. Network matrices are typically created in Microsoft Excel. Drawing software packages such as Microsoft Visio can be used to draw the networks, however, for larger more complex networks the use of SNA tools such as SocNET V or Gephi is recommended. The Leximancer thematic analysis tool is often used to construct information networks directly from verbal or communications transcripts (e.g. Salmon et al., 2014). Social network analysis tools are also used to quantitatively analyse the networks.

CASE STUDY EXAMPLE: SUBMARINE CONTROL ROOM DESIGN

For the purposes of this case study example, a Return To Periscope Depth (RTPD) scenario is presented. The data were collected over a series of observations in a control room and sound room layout in the Talisman Command Team Trainer. There is a strong communication link between the Officer of the Watch (OOW), Chief Petty Officer for Tactical Systems (CHOPS(TS): sometimes called the OpsO) and the Sound Room Controller (SRC, sometimes called the Sonar Controller or CHOPS(S)). The layout of the control room and sound room is fixed by the equipment layout, but has been optimised over the decades to the current design. The photograph in Figure 7.6 shows a view into the control room with the periscope on the right hand side.

All of the internal communications were recorded by plugging into the system and recording on a laptop computer using Audacity (version 1.3.14 software). The ambient communications were recorded using an external Yoga BM-26D boundary microphone.

TASK NETWORK ANALYSIS

The task network was constructed from the main activities involved in returning the submarine to periscope depth. Rather than represent the phases as a linear process, the task network portrays the relationships between the tasks which are non-sequential. The decision to RTPD is continually being assessed

FIGURE 7.6 View into the control room with periscope on right hand side.

and reassessed. If an unexpected contact (i.e. another vessel) appears in the area (or going towards the area) where the submarine is heading, then the decision to RTPD will be cancelled and the submarine will head down to the safe depth (below that of the deepest ship's hull; i.e. greater than 30 metres). A depiction of the task network is shown in Figure 7.7. As Figure 7.7 shows, the focus of the task is on identifying all of the contacts surrounding the area where the submarine is intending to RTPD. This focus continues all the way up and when the 'look' (i.e. viewing the surface using the periscope) is established. At any point in the manoeuvre the ship will return to a safe depth if a contact is detected. The decision to RTPD begins with the OOW calling an outstations briefing to ensure all people and systems on the submarine are in order and ready for the manoeuvre. This is followed by ballasting (to ensure the submarine is in trim) and clearing of stern arcs (so that the sound room can check for vessels behind the submarine). Then the sound room and control room are engaged in the activities to range all contacts within the local area to find a safe area of sea to RTPD. If no safe area is found, then the submarine will need to change area and continue to range contacts. When all contacts have been ranged and an area of sea has been identified, the OOW reports to the Captain for permission to RTPD. If permission is forthcoming, the OOW will request final reports from outstations to check that the people and systems on the submarine are in order and ready for the manoeuvre. A decision will be made by the OOW whether to conduct the standard routine (where range and bearing of all contacts are called out as the submarine returns to periscope depth) or a silent routine (where the submarine returns to periscope depth as stealthily as possible). At periscope depth (called PD in Figure 7.7) the OOW calls "breaking" and conducts two sweeps with the periscope to check that the submarine is clear from potential collision with contacts, and then will range the contacts manually to update the Submarine Maritime Command System (SMCS) operators. Then the

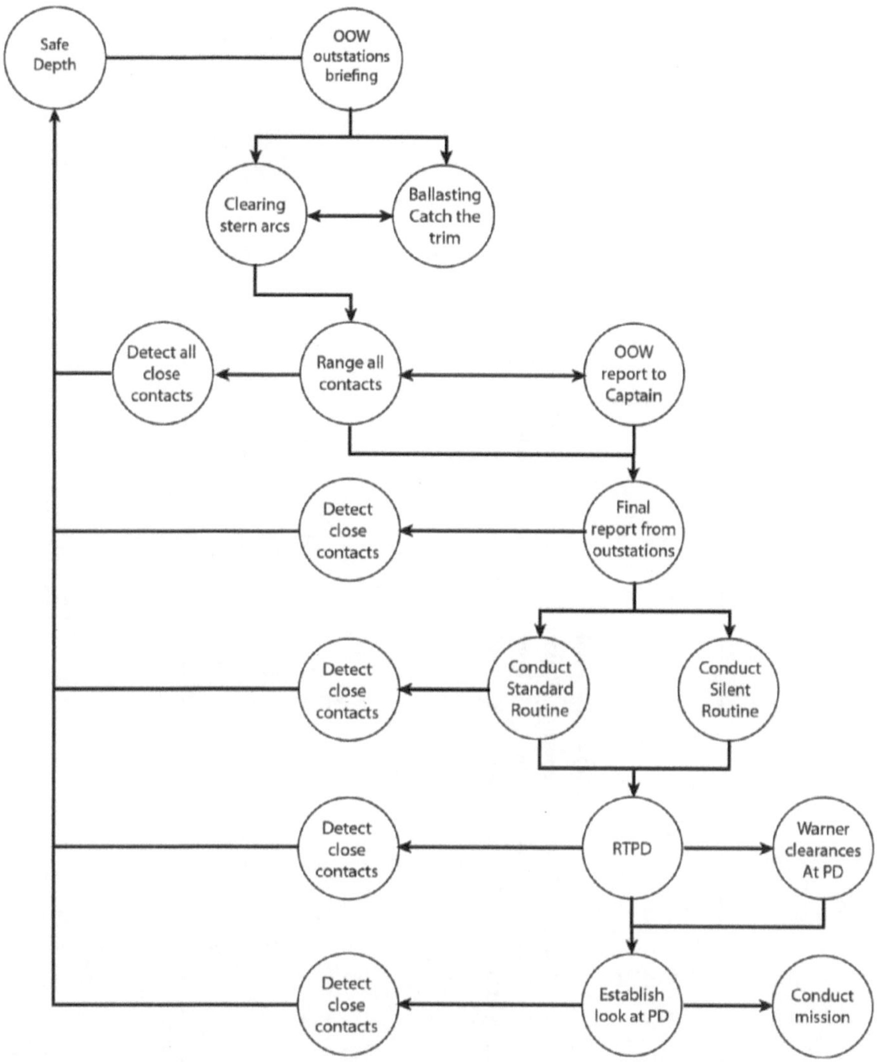

FIGURE 7.7 Task network for RTPD.

mission intentions can be carried out. If the submarine is on a potential collision course with any of the contacts, the OOW will give the order to dive to the safe depth. As stated previously, the task network helps to identify interdependencies and relationships between tasks.

The network in Figure 7.7 was used to construct an association matrix to indicate the presence or absence of links between tasks. The number of links into and out of each task are shown in the 'reception' row and 'emission' rows respectively of Table 7.3. The network statistics (as described in the section on data collection and analysis) were then computed using AGNA. Obviously, the task 'detect close contacts' is critical in RTPD, which is reflected in most of the metrics. Some of the other

TABLE 7.3
Task Network Analysis Outputs

Task Network Analysis	Safe Depth	OOW Outstations Briefing	Clear Stern Arcs	Ballasting	Range All Contacts	Detect Close Contacts	OOW Report to Captain	Final Report from Outstations	Conduct Standard Routine	Conduct Silent Routine	Return to Periscope Depth	Warner Clearances at Periscope Depth	Establish Look at Periscope Depth	Conduct Mission
Reception	**2**	**1**	**2**	**2**	**1**	**7**	**1**	**2**	**1**	**1**	**2**	**1**	**2**	**1**
Emission	**1**	**3**	**2**	**1**	**3**	**1**	**1**	**3**	**2**	**2**	**3**	**2**	**2**	**0**
Eccentricity	8	7	6	7	5	9	6	6	7	7	7	8	9	0
Sociometric Status	0.2	0.3	0.3	0.2	0.3	0.6	0.2	0.4	0.2	0.2	0.4	0.2	0.3	0.1
Centrality (B-L)	7.7	7.8	7.9	6.9	7.7	7.5	6.5	7.3	6.4	6.4	6.3	5.7	5.8	10.4
Closeness	0.2	0.3	0.3	0.2	0.4	0.2	0.3	0.3	0.3	0.3	0.3	0.2	0.2	0
Farness	58	46	41	53	36	67	46	39	46	46	49	56	63	0
Betweenness	67	68	70	0	70	65	0	59	17	17	31	0	12	0

tasks are less obvious. The two tasks of 'clear stern arcs' and 'range all contacts' score quite low on sociometric status but quite high on Centrality and Betweenness.

The network density was 0.14 (i.e. a low distribution network) and cohesion was 0.02 (i.e. a very low level of reciprocal links – suggesting very strict dependences in the task network, as shown in the task network diagram) with a diameter of 9 (i.e. nine hops from one side of the network to the other). The network can also be described as weighted (i.e. non-uniform) and asymmetric (i.e. unbalanced).

SOCIAL NETWORK ANALYSIS

The social network analysed the relationships (e.g. communications and activity) between the agents during the scenario, i.e. the Captain, OOW, Ships Control, WECDIS, Weapons Engineering Officer (WEO), OpsO, Sound Control and SMCS. The matrix shows the frequency of communications between each agent in the network. The association matrix for the RTPD scenario is presented in Table 7.4. This matrix shows whether or not an agent within the system can be associated with any other agent, specifically through frequency of communications.

Once the matrix of association is completed, the social network diagram can be created. The social network depicts each agent in the network and the communications between them. The communications are represented by directional arrows, and the frequency of communications is also presented, as shown in Figure 7.8. The thicker the line, the more communications have occurred, and vice versa.

Table 7.5 shows the metrics for the social network in the RTPD scenario. This highlights the Office of the Watch (OOW) and Operations Officer (OpsO) as the highest on sociometric status and betweenness metrics whereas the SMCS operator is highest on Centrality. The Captain, who has the most executive power in the ship is modestly placed on the sociometric status metric for the RTPD scenario. The Captain's betweenness score hints at his executive power. The higher the value means the higher each task is scored on that metric.

Network density is equal to the total number of links between the agents in the network divided by the total number of possible links. Low network density figures

TABLE 7.4
Association Matrix for the RTPD Scenario

FROM/TO	Captain	OOW	OpsO	Sound	SMCS	WEO	Ships Control	Warner	WECDIS
Captain		2	1	1	1	2	1	1	1
OOW	9		9	4	4	4	21	4	4
OpsO	3	3		38	5	3	3	8	3
Sound Control	0	0	28		0	0	0	0	0
SMCS	0	0	0	0		0	0	0	0
WEO	2	0	0	0	0		0	0	0
Ships Control	0	31	0	0	0	0		0	0
Warner	0	0	4	0	0	0	0		0
WECDIS	0	1	0	0	0	0	0	0	

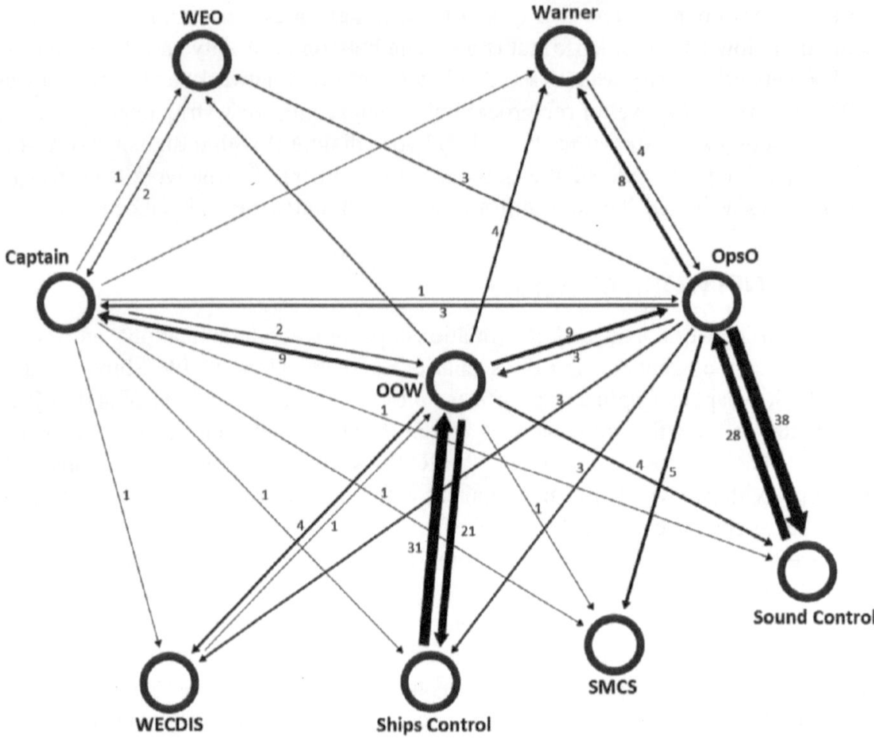

FIGURE 7.8 Social network diagram for the RTPD scenario.

TABLE 7.5
Social Network Metrics for the Network for the RTPD Scenario

Social Network Analysis	Captain	OOW	OpsO	Sound	SMCS	WEO	Ship Cont	Warner	WEC DIS
Reception	**14**	**37**	**37**	**43**	**10**	**9**	**25**	**13**	**8**
Emission	**10**	**54**	**66**	**28**	**0**	**2**	**31**	**4**	**1**
Eccentricity	1	1	1	2	0	2	2	2	2
Sociometric Status	3	11	12	9	1	1	7	2	1
Centrality (B-L)	5.2	5.5	5.5	3.8	7.6	3.8	3.8	3.8	3.8
Closeness	1	1	1	0.5	0	0.5	0.5	0.5	0.5
Farness	8	8	8	15	0	15	15	15	15
Betweenness	7	14	14	0	0	0	0	0	0

are indicative of a broadly spread network with few links. High density figures indicate a tight network that has many links. In this case the network density was 0.4 (i.e. a medium distributed network – which is likely to have good resilience against failure of the network) and cohesion was 0.2 (i.e. low level of reciprocal links) with a diameter of 2 (i.e. two hops from one side of the network to the other), and 29 edges (i.e. when an in-link to an agent had a corresponding out-link to another agent) for a network of nine nodes. The network can be described as weighted (i.e. non-uniform) and asymmetric (i.e. unbalanced). As a baseline study, this characterised the network for returning the submarine to periscope depth.

INFORMATION NETWORK ANALYSIS

The information networks were created by analysing the communications transcripts. Each 'concept' in the transcript was identified and paired with its nearest relation (i.e. concepts from the same sentence) to build up a network of information concepts in the control room related to the activities of returning the submarine to periscope depth. The networks were then presented to the SMEs for verification. The information may be related temporally or spatially to agent or tasks, or both. In this way it is possible to build up a picture of who knows what, when, and how this relates to task performance. This notion takes the activities of the sound room and control room into the realm of distributed cognition. In this view, cognition is not an individual phenomenon but rather a systemic endeavour performed by the whole team working together. It is argued that cognition transcends the boundaries of individual actors and becomes a function that is achieved by coordination between the human and technological agents working within the collaborative system. Systems thinking argues that the cognitive system needs to be analysed as a whole, rather than its constituent parts. To this end, the entire transcript is presented in the information network as the unit of analysis is not the individual person but the entire system under investigation. The information network for returning the submarine to periscope depth is presented in Figure 7.9 (note that the node 'cuts' at the bottom right of Figure 7.9 is duplicated merely as a convenience to disentangle the lines). By viewing the system as a whole, it does not matter if humans or technology own this information, just that the right information is activated and passed to the right agent at the right time. Individual human agents are not required to know everything, provided that the system has the information in some form or other.

As Figure 7.9 shows, the information network has 68 nodes. Central nodes in the network (informally defined on the basis of the centrality in the network) appear to be 'cuts', 'tracks', 'contacts', 'classification', 'reports', 'manoeuvre', 'depth' and 'trim'. The informational nodes coming off these more central key nodes provide the detail that would pertain to the situational specifics, such as the 'bearing and 'range' for the 'fire control solution' or the type and nature of the vessel classification. The purpose of the information network was to identify the type of information required to return the submarine to periscope depth. A simplified version of

FIGURE 7.9 Information network for returning the submarine to periscope depth.

the network was formed for practical convenience of presenting the SNA metrics in this chapter, as is shown in Figure 7.10.

The metrics in Table 7.6 suggest three classes of information in the network. ‘Contacts’ has the highest sociometric status and is probably the most important piece of information in the control room. Next are two clusters of information. One refers to information about the activities of the submarine (i.e. depth, manoeuvre and course) and one refers to the activities of other vessels (i.e. tracks, picture, and reports). The network statistics are useful in highlighting the important underlying features of the relationships between information elements from the control room. The higher the value means the higher each task is scored on that metric.

The network density was 0.18 (i.e. a low distribution network) and cohesion was 0.17 (i.e. a very low level of reciprocal links – suggesting very strict dependences in the information network, as shown in the task network diagram) with a diameter of 7 (i.e. seven hops from one side of the network to the other). The network can also be described as weighted (i.e. non-uniform) and asymmetric (i.e. unbalanced).

COMBINING NETWORK MODELS

As stated earlier the purpose of EAST is not only to consider the task, social, and information networks in isolation of each other, but also to consider the relationships between the three networks. These relationships are explored in the following

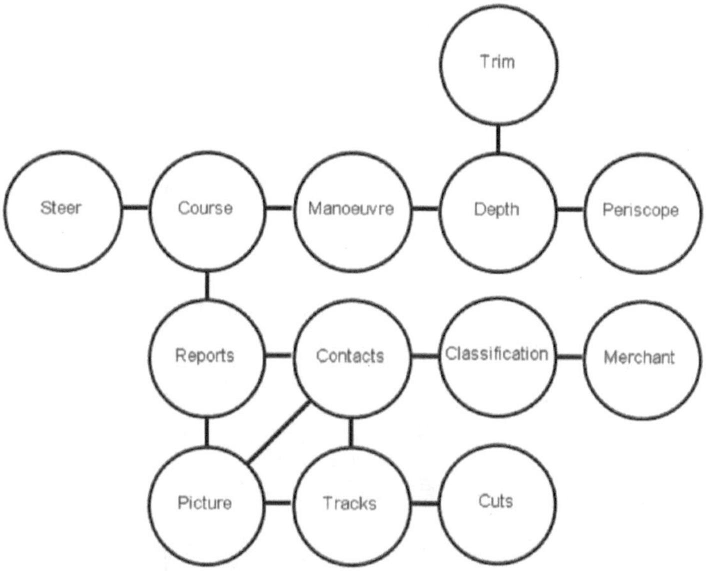

FIGURE 7.10 Simplified information network.

sections. First, the combined task and social networks are presented, where the agents are mapped onto the task network. Next, the information and social networks are presented where the agents are mapped onto the information network. Finally, the three networks are integrated, where the information network is coded by the agents and tasks.

TASK AND SOCIAL NETWORKS

The relationships between the task and social networks show which agent is primarily involved in each aspect of the task network. As Figure 7.11 shows, the role specialists are involved in different tasks throughout the RTPD. The OOW calls the outstation briefing. The Ships Control Officer (SCO) and Planesman clear the stern arcs and ballast the submarine. The Sound Room Controller (SRC), Sound Room Operators (SRO), Tactical Picture Supervisor (also called the Ops Officer, or OpsO) and SMCS are responsible for ranging contacts and reporting to the OOW. The OOW reports to the Captain that the submarine is ready to RTPD. The OOW receives final reports from the outstations. All are involved in the silent or standard routine. The OOW, SCO and Planesman are responsible for getting the submarine to periscope depth whilst the SRC and SRO are listening for close contacts (which if detected would lead the OOW to order the submarine back to a safe depth). The Warner is responsible for giving clearances at periscope depth. Then the OOW and Periscope Watch Keeper (PWK) are responsible for establishing the look at periscope depth. The first sweep is to search for close contacts and the second sweep to contacts at greater distance. Any contacts on a potential collision course would lead the OOW to

TABLE 7.6
Analysis of Simplified Information Network

Information Network Analysis	Periscope	Trim	Depth	Manoeuvre	Course	Steer	Report	Contacts	Classification	Merchant	Picture	Tracks	Cuts
Reception	**1**	**1**	**3**	**2**	**3**	**1**	**3**	**3**	**3**	**1**	**3**	**3**	**1**
Emission	**1**	**1**	**3**	**2**	**3**	**1**	**3**	**4**	**2**	**1**	**3**	**3**	**1**
Eccentricity	7	7	6	5	4	5	4	5	6	7	5	6	7
Sociometric Status	0.2	0.2	0.5	0.3	0.5	0.2	0.5	0.6	0.4	0.2	0.5	0.5	0.2
Centrality (B-L)	4.9	4.9	6.2	7.5	8.8	6.4	9.1	8.2	6.7	5.2	8.0	6.8	5.3
Closeness	0.2	0.2	0.3	0.3	0.4	0.3	0.4	0.4	0.3	0.,2	0.4	0.3	0.2
Farness	52	52	41	34	29	40	28	30	39	50	32	37	48
Betweenness	0	0	42	54	78	0	72	42	23	0	22	23	0

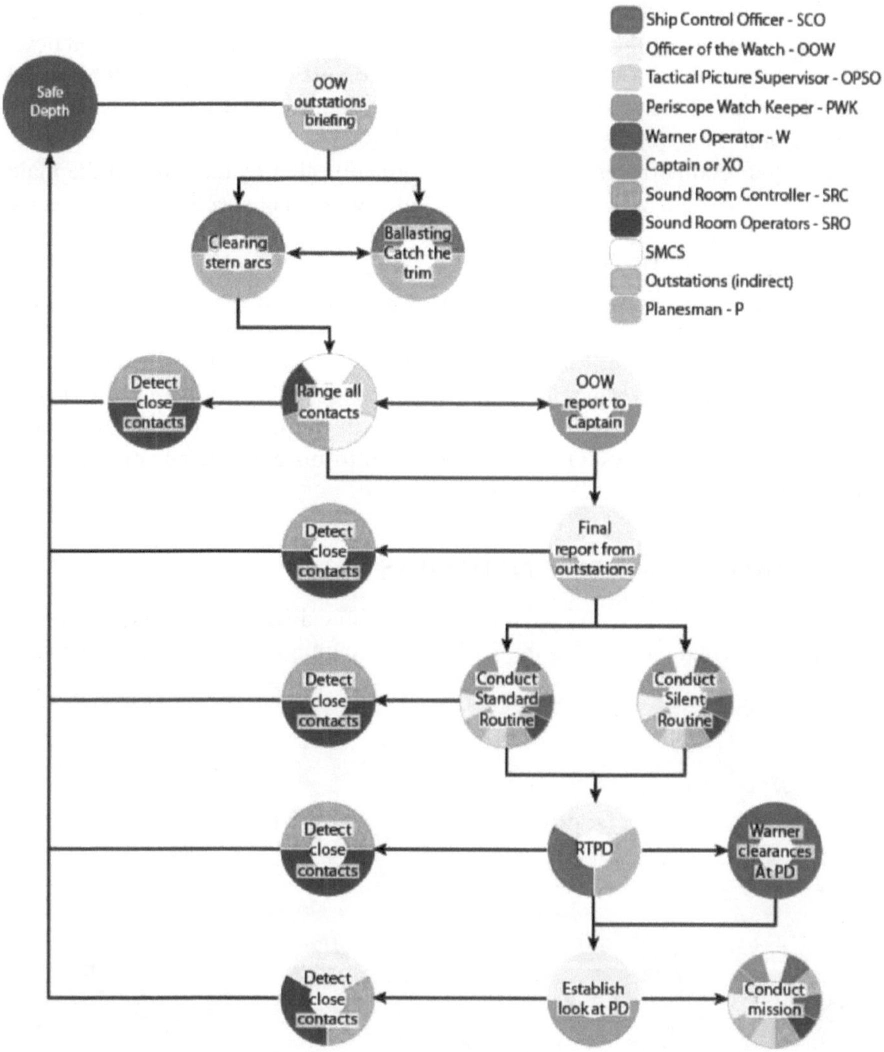

FIGURE 7.11 Task network coded by social agents.

order the submarine back to a safe depth. If the submarine is safe at periscope depth, then the mission intentions can be carried out, in which all of the specialist roles will be performing their respective tasks.

INFORMATION AND SOCIAL NETWORKS

The dynamic nature of information communication means it is subject to change moment by moment, in light of changes in the task, environment and interactions (both social and technological). These changes need to be tracked if the phenomena are to be properly understood. As Figure 7.12 shows, the sound room and control room function as a distributed cognition system, with ownership of information flowing around the network (a coding of the information network in Figure 7.9 by social agent). No one individual owns all of the information; rather the system works as a functional unit.

The colour coding reveals clusters of information that feature around the main information nodes identified in the section 'information networks' (i.e. cuts, tracks, picture, merchant, classification, contacts, reports, steer, course, manoeuvre, depth, trim, and periscope). These are presented in Table 7.7.

The relationships between the key concepts and the roles in Table 7.7 show the information exchanges that occur in the course of the submarine returning to periscope depth. These information exchanges are called transactions, as the information flows both ways. In a simple exchange, this may be a request, followed by an information transfer and ending with a confirmation (as revealed in the communication analyses). The transaction informs both the person requested and the requestor.

INFORMATION AND TASK NETWORKS

The grouping of information by the tasks shows how different tasks use, distribute and share information (see Figure 7.13). The groupings show the links between

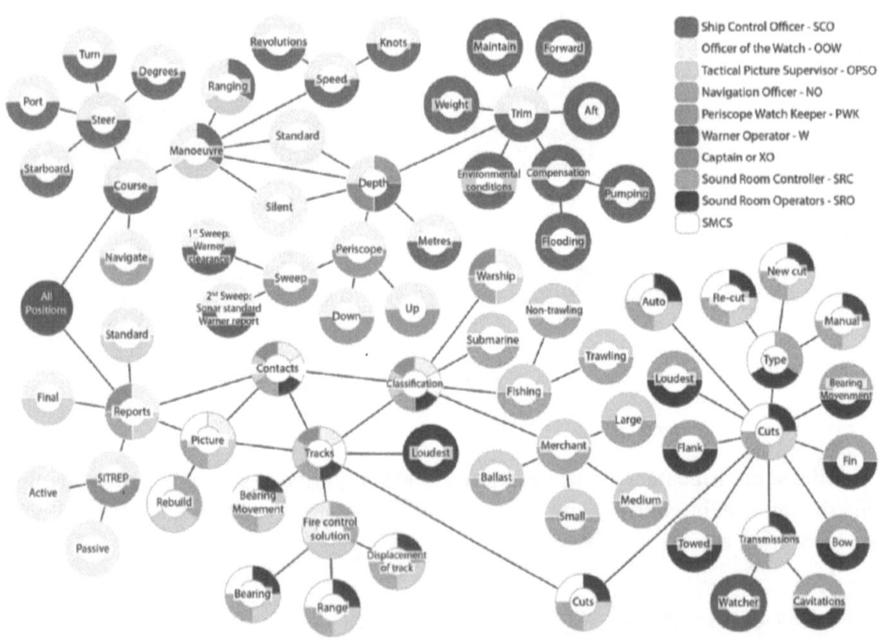

FIGURE 7.12 Information network coded by social agents.

TABLE 7.7
Analysis of Concepts by Social Agents

Key Concept	XO	OOW	SCO	OPSO	SMCS	SRC	SRO	PWK
Cuts								
Tracks								
Picture								
Merchant								
Classification								
Contacts								
Reports								
Steer								
Course								
Manoeuvre								
Depth								
Trim								
Periscope								

information and tasks, such that 'cuts' are linked to 'tracks', 'tracks' are linked to 'contacts', 'contacts' are linked to the 'picture' and 'reports' which are linked to the 'course' (via 'all positions' reports), the 'course' is linked to 'manoeuvre', which is linked to 'speed' and 'depth' and 'depth' is linked to 'trim' and 'periscope'.

COMBINED TASK, SOCIAL, AND INFORMATION NETWORKS

The final combined picture puts the task, social, and information networks into one network as presented in Figure 7.14. The dependences between tasks cannot be represented in this view (the reader is referred to Figures 7.7 and Figure 7.11 for those views). The integrated network model reveals the multi-modal nature of the work in the sound and control rooms in returning the submarine to periscope depth, i.e. how people utilise information to conduct tasks.

As Figure 7.14 shows, no one individual possessed all of the information to perform the RTPD task. The tasks were distributed amongst the crew and artefacts they use.

In summary, this case study analysis has described the application of the EAST to the analysis of the RTPD task in the control room of the Trafalgar class submarine. EAST acknowledges that systems are inherently complex and multiple perspectives on the problem are required to better appreciate the relationships between the social and technical aspects of the system. EAST accepts that systems are intertwined and analyse the system as a whole, rather than constituent parts. This case study has shown that the network models are able to characterise the domain in different, but complementary, ways. The seven outputs, namely the individual task networks, social, and information networks (and associated metrics), and the combined networks (i.e. task and social network, information and social network, information and

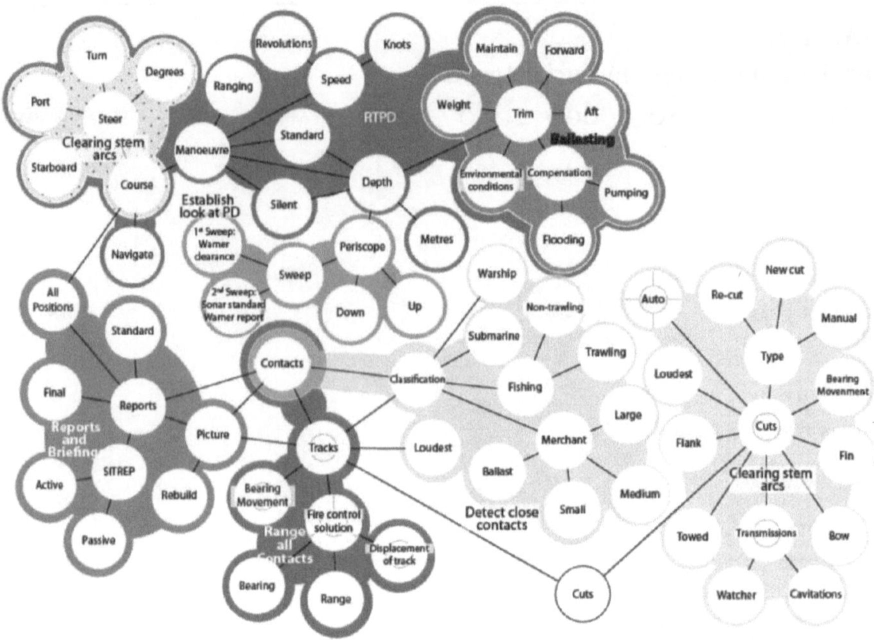

FIGURE 7.13 Combined information and task networks.

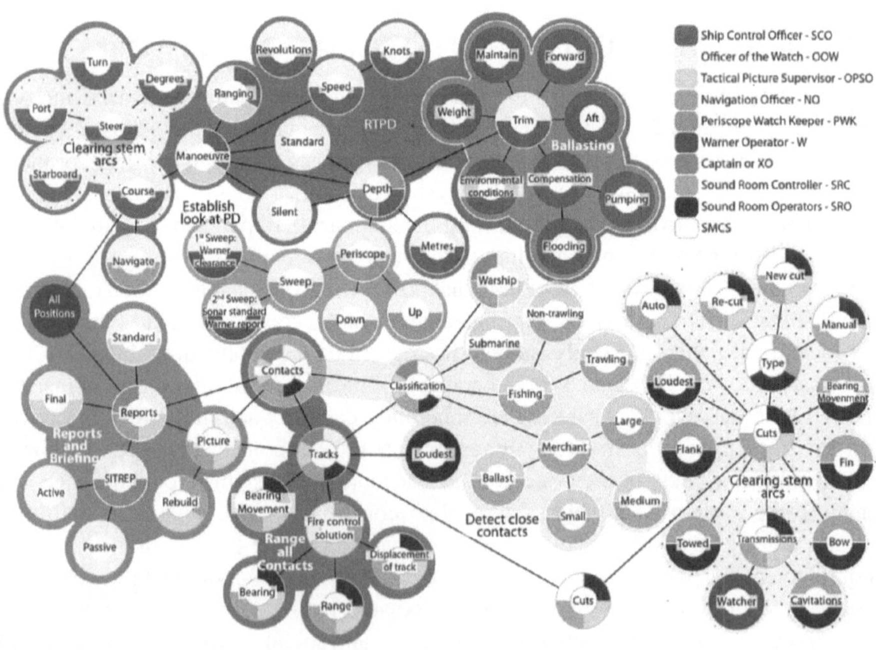

FIGURE 7.14 Integrated networks model.

task network, and task, social, and information network) offer a graphical representation of distributed cognition (e.g. the sound and control room) from different perspectives. The different perspectives offered by the representations are an attempt to characterise the activities between the sound and control rooms in returning the submarine to periscope depth.

RECOMMENDED READING

Stanton, N. A., Salmon, N. A., & Walker, G. H. (2018). *Systems Thinking in Practice: The Event Analysis of Systemic Teamwork*. Boca Raton, FL: CRC Press.

Stanton, N. A., Salmon, P. M., Rafferty, L., Walker, G. H., Jenkins, D. P., & Baber, C. (2013). *Human Factors Methods: A Practical Guide for Engineering and Design*. 2nd Edition, Aldershot, UK: Ashgate.

REFERENCES

Crandall, B., Klein, G. A., & Hoffman, R. R. (2006). *Working Minds: A Practitioner's Guide to Cognitive Task Analysis*. Mit Press.

Driskell, J. E., & Mullen, B. (2004). Social network analysis. In *Handbook of Human Factors and Ergonomics Methods* (pp. 565–571). CRC Press.

Houghton, R. J., Baber, C., McMaster, R., Stanton, N. A., Salmon, P., Stewart, R., & Walker, G. (2006). Command and control in emergency services operations: a social network analysis. *Ergonomics*, 49(12–13), 1204–1225.

Kirwan, B., & Ainsworth, L. K. (Eds.). (1992). *A Guide to Task Analysis: The Task Analysis Working Group*. CRC Press.

Lane, B. R., Salmon, P. M., Cherney, A., Lacey, D., & Stanton, N. A. (2019) Using the Event Analysis of Systemic Teamwork (EAST) broken-links approach to understand vulnerabilities to disruption in a darknet market. *Ergonomics*, 62(9), 1134–1149.

O'Hare, D., Wiggins, M., Williams, A., & Wong, W. (2000). Cognitive task analyses for decision centred design and training. In *Task Analysis* (pp. 176–196). CRC Press.

Salmon, P., Lenne, M., Walker, G. H., Stanton, N. A., & Filtness, A. (2014). Exploring schema-driven differences in situation awareness across road users: an on-road study of driver, cyclist and motorcyclist situation awareness. *Ergonomics*, 57(2), 191–209.

Salmon, P. M., Clacy, A., & Dallat, C. (2017). It's not all about the bike: Distributed situation awareness and teamwork in elite women's cycling teams. *Contemporary Ergonomics*, 2017, 240–248.

Salmon, P. M., Read, G. J. M., Walker, G. H., Goode, N., Grant, E., Dallat, C., Carden, T., Naweed, A., & Stanton, N. A. (2018). STAMP goes EAST: Integrating systems ergonomics methods for the analysis of railway level crossing safety management. *Safety Science*, 110(B), 31–46.

Salmon, P. M., Walker, G. H., & Stanton, N. A. (2016). Pilot error versus sociotechnical systems failure? A distributed situation awareness analysis of Air France 447. *Theoretical Issues in Ergonomics Science*, 17(1), 64–79.

Stanton, N. A. (2006). Hierarchical task analysis: Developments, applications, and extensions. *Applied Ergonomics*, 37(1), 55–79.

Stanton, N. A. (2014). Representing Distributed Cognition in Complex Systems: How a submarine returns to periscope depth. *Ergonomics*, 57(3), 403–418.

Stanton, N. A., Plant, K. L., Revell, K. M. A., Griffin, T. G. C., Moffat, S., & Stanton, M. J. (2019) Distributed Cognition in Aviation Operations: A gate-to-gate study with implications for distributed crewing. *Ergonomics*, 62(2), 138–155.

Stanton, N. A., & Roberts, A. P. J. (2017). Examining social, information, and task networks in submarine command and control. *IEEE Transactions on Human-Machine Systems*, 48(3), 252–265.

Stanton, N. A. & Roberts, A. P. (2020). Better together? Investigating new control room configurations and reduced crew size in submarine command and control. *Ergonomics*, *63*(3), 307–323.

Stanton, N. A., Roberts, A. P. J., & Fay, D. (2017). Up periscope: Understanding submarine command and control teamwork during a simulated return to periscope depth. *Cognition, Technology and Work*, 19(2–3), 399–417.

Stanton, N. A., Salmon, P. M., Baber, C., & Walker, G. (2005). *Human Factors Methods: A Practical Guide for Engineering and Design*. Aldershot, UK: Ashgate.

Stanton, N. A., Salmon, P. M., Rafferty, L. A., Walker, G. H., Baber, C., & Jenkins, D. P. (2013). *Human Factors Methods: A Practical Guide for Engineering and Design*. CRC Press.

Stanton, N. A., Stewart, R., Harris, D., Houghton, R. J., Baber, C., McMaster, R., … & Green, D. (2006). Distributed situation awareness in dynamic systems: theoretical development and application of an ergonomics methodology. *Ergonomics*, 49(12–13), 1288–1311.

Stewart, R., Stanton, N. A., Harris, D., Baber, C., Salmon, P. M., Mock, M., Tatlock, K., Wells, L., & Kay, A. (2008). Distributed situation awareness in an airborne warning and control system: Application of novel ergonomics methodology. *Cognition Technology and Work*, 10(3), 221–229.

Walker, G. H., Stanton, N. A., Baber, C., Wells, L., Gibson, H., Salmon, P. M., & Jenkins, D. P. (2010) From ethnography to the EAST method: A tractable approach for representing distributed cognition in air traffic control. *Ergonomics*, 53(2), 184–197.

8 Causal Loop Diagrams (CLDs)

BACKGROUND

All systems comprise interacting networks of positive and negative feedback loops that influence behaviour (Sterman, 2000). CLDs is a method that is used to elicit and represent stakeholder views on what these positive and negative feedback loops comprise, what relationships exist between them, and how they interact dynamically to influence behaviour. CLDs are useful for capturing hypotheses about causal dynamics, for eliciting and capturing mental models, and for communicating the feedback loops that may be responsible for a particular problem (Sterman, 2000). CLDs are typically used to support the development of system dynamics models (ibid.); however, they can also be used independently as a systems analysis method (e.g. Allender et al., 2015). Their flexibility is such that they have previously been used to model a diverse set of issues ranging from the transmission of COVID-19 (Kumar et al., 2021; Sahin et al., 2020), major accidents (e.g. Lin and Chien, 2019), and obesity (Allender et al., 2015) to issues impacting research quality and translation (McLean et al., 2021), racism (Burrell et al., 2021), transport safety (Shepherd, 2014) and climate change (Richards et al., 2021).

CLDs comprise relevant variables connected by arrows which depict the causal influences between them (Sterman, 2000). Two forms of causal loop are modelled: positive, or reinforcing loops, and negative, or balancing loops. Positive feedback loops are self-reinforcing; for example, as shown by the left-hand loop in Figure 8.1, as a population of chickens increases in number, more eggs are laid, which in turn increases the chicken population, which in turn leads to more eggs, and so on (Sterman, 2000). Negative loops are self-correcting and act to balance or limit growth. As shown by the right-hand loop in Figure 8.1, as the population of chickens increases, more chickens will attempt road crossings, leading to more fatal collisions with vehicles, and a reduction in the overall chicken population (Sterman, 2000). The road crossings loop therefore acts as balance on the increasing chicken population.

PROCEDURE AND ADVICE

A flowchart depicting the CLD procedure is presented in Figure 8.2. The main elements of the CLD are presented in Table 8.1. Step-by-step guidance is presented below. CLDs are typically developed using a group model building process (Berard, 2010; Sterman, 2000). Berard (2010) describes two forms of group model building. The first involves the modellers themselves constructing the CLD using data derived from sources such as documentation review, interviews, and research study findings. The second form involves engaging relevant subject matter experts (SMEs)

DOI: 10.1201/9780429281624-8

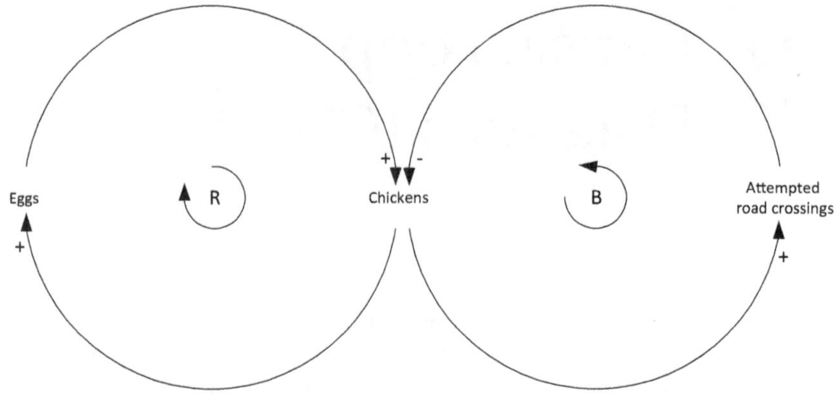

FIGURE 8.1 Conceptual causal loop diagram showing reinforcing and balancing loops (adapted from Sterman, 2000). Positive links (+) refer to links where, if the cause increases, the effect increases also. Negative links (–) refer to links where, if the cause increases, the effect decreases, and if the cause decreases, the effect increases.

to assist in building the models. This may be a superior approach, as studies have demonstrated the importance of involving multiple participants in the model building process (Berard, 2010). It is often useful to combine these approaches, with initial development of the model based on data, followed by SMEs being involved in the review and refinement of the model.

STEP 1: DEFINE AIMS AND FOCUS OF THE ANALYSIS

The first step in an applying CLDs involves clearly defining the aims, focus, and boundaries of the analysis. The CLDs presented in the case study example in this chapter were developed to support the design of interventions aiming to prevent road crashes caused by the fatal five behaviours of speeding, driving while distracted, driving while under the influence of alcohol or drugs, driving while fatigued, and driving without a seatbelt. The following aims, focus, and boundaries were defined during Step 1:

- **Issue(s) under analysis**: road crashes involving the fatal five road safety behaviours;
- **Aim(s) of the analysis**: to describe the societal and road transport system variables which interact to create road crashes involving the fatal five road safety behaviours; and
- **Analysis boundaries**: the scope was limited to fatal five-related road crashes occurring in Queensland, Australia.

STEP 2: PRE-GROUP MODELLING DATA COLLECTION

Once the issue under analysis and aims of the analysis are clearly defined, specific data regarding the issue and system should be collected. The data collection activities undertaken are dependent upon the aims and boundaries of the analysis as well as

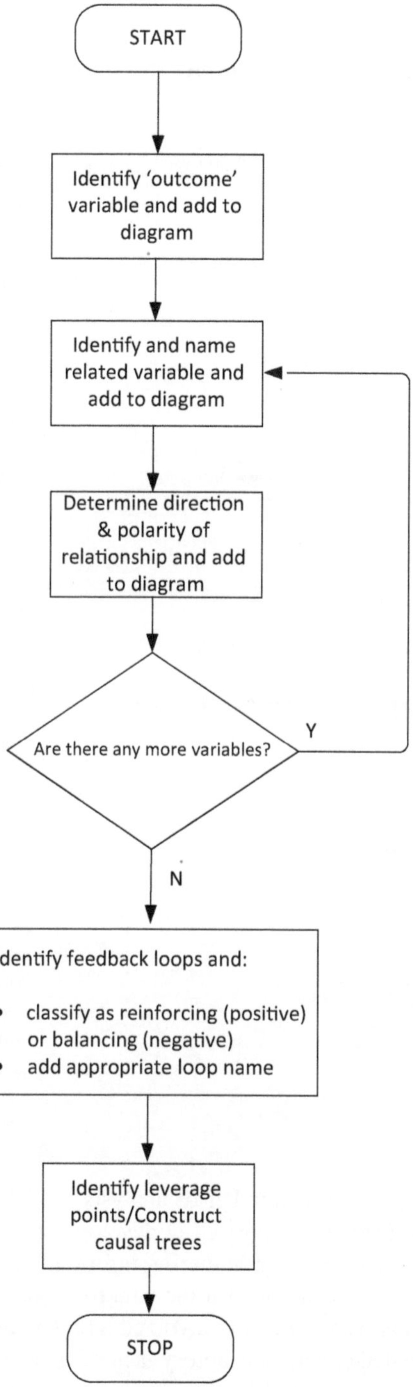

FIGURE 8.2 Causal loop diagram procedure following definition of aims and collection of data.

TABLE 8.1
Main CLD Elements (Kim, 1995; Lannon, 2012)

CLD Diagram Element	Explanation
	Arrows in the CLD represent causality. In the example variable A causes a change in variable B
	A plus sign in the CLD indicates that variable A causes a change in variable B, with the change in the same direction as the cause. As variable A increases, variable B increases, or, as variable A decreases, variable B decreases
	A minus sign in the CLD indicates that variable A causes a change in variable B, with the change in the opposite direction to the cause. As variable A increases, variable B decreases, or, as variable A decreases, variable B increases
	Two lines crossing an arrow with the CLD are used to represent a delay whereby the change in variable B is not immediate
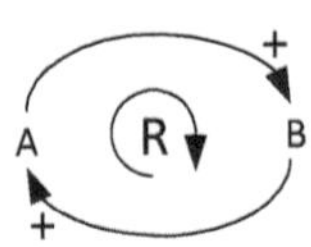	An R within the centre of a loop within the CLD represents a positive or reinforcing loop which acts to create growth or decline within the system
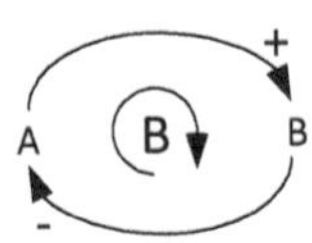	A B within the centre of a loop within the CLD represents a balancing loop which acts to move the variables towards equilibrium

project constraints, such as time, number of analysts available, and access to information and personnel. Data to support CLD development can be collected from a variety of sources, with the focus on gathering information on the known variables that influence behaviour. For example, for the fatal five road safety case study example, pre-group modelling data collection involved a review of the research literature on known fatal five-related crash contributory factors, an analysis of fatal road crash data from Queensland, Australia, driver surveys, and discussions with road safety experts on the variables that influence driver engagement in the fatal five behaviours (Salmon et al., 2019).

STEP 3: IDENTIFY AND NAME VARIABLES

Steps 3–6 can be either be undertaken by the analyst team, or directly with SMEs as part of a workshop process. It is recommended, however, that Steps 3–6 be undertaken initially by the analyst team and then SMEs are used during Steps 7 and 8 to review and refine the CLD and identify leverage points.

Step 3 involves identifying and naming the different variables that interact to influence the behaviour of interest. It can be useful at this stage to begin with the outcome variable. For example, for the fatal five road safety case study example, construction of the CLDs began with the fatal five behaviour in question (e.g. 'Drivers under the influence of alcohol'). Next, the group should identify other related variables, such as 'Alcohol addiction' and 'Societal alcohol consumption levels'. De Pinho & Larsen (2015) provides the following advice when identifying and naming variables:

- Use nouns for variable names and try to avoid the use of verbs (e.g. "Crashes" is more appropriate than "Drivers having crashes");
- Variable names should be neutral in terms of direction (e.g. "Crashes" should be used and not "Increasing crashes" or "Decreasing crashes");
- Variables should be measurable and observable and can increase or decrease over time.
- Where possible the positive instantiation of the variable should be used rather than the negative (e.g. "Growth" rather than "Reduction"); and
- Try to distinguish between perceived and actual states (e.g. "Perceptions of societal alcohol consumption levels" compared to "actual societal alcohol consumption levels").

It is often useful when identifying variables to use scenarios, case studies, or stories to help the group think about the range of different influences on behaviour.

STEP 4: DETERMINE AND RECORD RELATIONSHIPS BETWEEN THE VARIABLES

Step 4 can occur in conjunction with Step 3 or following Step 3. Once the key variables are identified, the next step involves identifying how the variables affect one another and marking this on the CLD. This involves the group discussing each variable and its relationships with other variables in the CLD. Where a relationship is specified, a directional link should be added to the CLD. For example, for the fatal five road safety case study example, a relationship was identified between 'Drivers under the influence of alcohol' and 'Crashes', as the group determined that an increase in the number of drivers driving under the influence of alcohol would lead to an increase in the number of crashes (See Figure 8.3).

Delays should also be added to relationships where the effect of the cause is not immediate and takes time to occur. For example, for the fatal five road safety case study example, a delay was added to the relationship between 'Gap between goal state and actual state' and 'Political will' as the increase in political will is not immediate and can take some time to occur. This is shown in Figure 8.4.

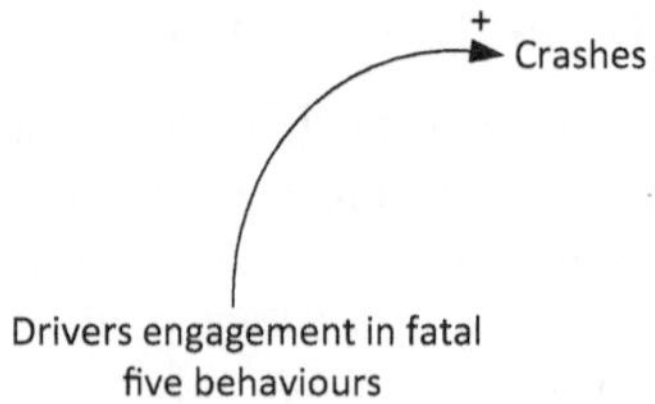

FIGURE 8.3 Adding relationships example. The + indicates that crashes increase as more drivers engage in fatal five behaviours (e.g. speeding, driving while distracted, under the influence of alcohol or drugs, or fatigued). Alternatively, the + also indicates that a reduction in crashes will result from a reduction in the number of drivers engaging in fatal five behaviours (i.e. the + indicates an effect in the same direction of the cause).

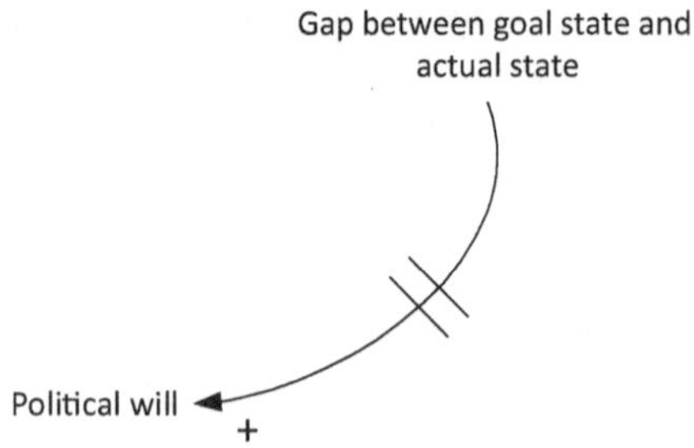

FIGURE 8.4 Delay demarcation. As the gap between the goal state in terms of acceptable levels of road trauma and actual levels of road trauma increases, the amount of political will increases also. The delay demarcation is used to indicate that the increase in political will is not immediate.

New variables are also likely to be found during this step when the group identify relationships with variables not already identified. In this case, the new variables and identified relationships should also be added to the CLD. In addition to identifying new variables, undertaking this step also prompts the group to refine and validate variable names or combine variables, where possible, to create a more parsimonious model.

Steps 3 and 4 should continue until there is group consensus that all relevant variables and relationships have been identified and added to the draft CLD.

STEP 5. CLASSIFY AND NAME LOOPS

Once all relevant variables and relationships have been added to the CLD, the next step involves identifying key causal loops and classifying them as either reinforcing (positive) loops or balancing (negative) loops. Each identified loop should also be named at this point. For example, for the fatal five road safety case study example, the 'Road safety interventions' loop presented in Figure 8.5 is a balancing loop that shows how the increase in crashes and fatalities caused by more drivers driving

FIGURE 8.5 Conceptual drink driving causal loop diagram (Adapted from Salmon et al., 2020).

under the influence of alcohol is checked by a concomitant increase in road safety interventions. According to the loop, as drink driving-related crashes increase, crash-related injuries and fatalities increase, which in turn leads to the implementation of more road safety interventions such as increased enforcement and driver education campaigns. As the interventions take effect, the number of drivers driving under the influence of alcohol decreases as they are either prevented from driving under the influence in the first place (e.g. through education or via an alcohol interlock device) or are caught engaging in the behaviours and removed from the road (e.g. through random breath testing). This reduction in drivers driving under the influence of alcohol leads to a decrease in the number of crashes, injuries, and fatalities (Salmon et al., 2020).

The identification and naming of pertinent causal loops can also be supported by the creation of behaviour over time graphs or reference modes. These do not need to be data driven at this stage and can be developed based on the group's perceptions regarding the behaviour of key variables over time. For the fatal five road safety case study example, a behaviour over time graph showing oscillations in road fatalities alongside oscillations in the level of road safety effort was created to assist the group when developing the CLD (see Figure 8.6.).

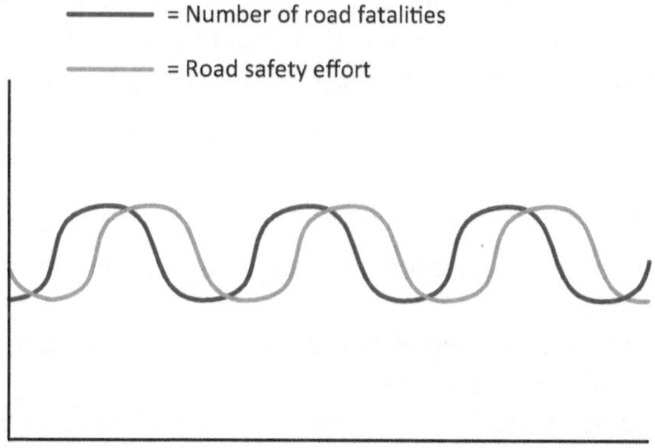

FIGURE 8.6 Behaviour over time graph showing oscillations in road fatality and road safety effort levels.

STEP 6: BUILD CLD IN SOFTWARE TOOL

Once the draft CLD is complete, the next step involves building the CLD within a suitable software package such as VenSim or Stella Architect (assuming it was not developed initially within a software program). Further refinement can also occur during this process. It is recommended that the CLD diagram is formatted in a manner that makes it easy to interpret. For example, the analyst should attempt to avoid arrows crossing over one another, to provide sufficient space between variable names, and ensure all polarity markings are clearly presented.

STEP 7: REVIEW AND REFINE ANALYSIS

SME review is a critical component of CLD development. Once the initial draft CLD is complete it is useful to have various SMEs review it, including those involved in the development of CLD and others external to the process. SME review can be undertaken in various ways, including workshops, focus groups, Delphi study, or one-on-one sessions. It is recommended that the SMEs are first given a description of the CLD, including all variables, relationships, and positive and negative feedback loops. Next the SMEs should be asked to review each variable and relationship and comment on variables names, relationships and their direction, feedback loop names, and also whether any variables or relationships are missing from the CLD. It is useful during this stage to have a data dictionary which provides a clear definition of each variable and relationship within the CLD. The CLD should then be refined based on the SMEs feedback. This typically involves adding new variables and relationships or modifying variable and/or loop names.

STEP 8: IDENTIFY LEVERAGE POINTS

CLDs are useful in that they support the identification of leverage points that reflect places in the system where small interventions could have significant impacts on system behaviour (Meadows, 2008). If required, the analyst team and SMEs should review the CLD and seek to identify leverage points. This can be supported via software tools' construction of causal trees which show the incoming and outgoing relationships from individual variables. Variables that are highly connected to other variables can often be classified as leverage points as they have a strong influence on system behaviour.

ADVANTAGES

- The output provides an intuitive model of the causal structure underlying system behaviour;
- The output provides a comprehensive overview of the variables that interact to create the issue under analysis;
- Polarity and delay markings showing cause directionality and temporality add an additional layer to the analysis;
- Causal trees and network analysis metrics can help in the interpretation of the CLD;

- CLDs are a generic method and can be applied to any issue in any domain;
- CLDs are highly flexible and can be used with varying degrees of granularity and scope;
- Initial development can be quick, noting that further work is required as part of a group modelling process;
- CLDs are straightforward to apply and require minimal training;
- Various software programs are available to support the development of CLDs, including VenSim and Stella Architect; and
- CLDs can be used to inform the development of quantitative system dynamics models.

DISADVANTAGES

- The quality of the CLD is highly dependent on the involvement of multiple SMEs during development;
- Initial data collection activities and group modelling processes can be time consuming;
- CLDs can become overly complex, unwieldy, and difficult to interpret;
- It is possible to create erroneous or 'faulty' models with incorrect assumptions about causality and system dynamics (Mirchi et al., 2012);
- Unless converted into system dynamics models, CLDs are difficult to quantitively test, and
- There is no reliability and validity evidence associated with CLDs.

RELATED METHODS

CLDs are often developed to inform the development of quantitative system dynamics models. Specifically, the CLD is used to identify the positive and negative feedback loops which define the functional architecture of the system (Sterman, 2000). The key loops within the CLD are then articulated as formal dynamic hypotheses, and the overall architecture of the system is captured in a quantitative system dynamics model. The system dynamics model comprises stocks, flows, and information (auxiliary) variables and can be used to simulate system behaviour over time. Important in this conversion, however, is an appreciation that units of measurement within CLDs are ill-defined. By contrast, units of measurement in system dynamics models must be explicitly defined and conserved throughout the operation of the model.

APPROXIMATE TRAINING AND APPLICATION TIMES

CLDs are a relatively simple method that requires little training. Various courses are available on-line. Application time depends on the system under analysis; however, CLD analyses typically require significant time as they require initial data collection activities and multiple iterations of the modelling. This may require multiple group modelling sessions with SMEs to refine the CLD. They can, however, also be entirely conceptual, requiring little to no 'real-world' data to either create or qualitatively validate among model builders and stakeholders.

TOOLS NEEDED

CLDs can be created using pen and paper, or a whiteboard and marker; however, once a draft model is developed it useful to use a software program such as VenSim or Stella Architect to create the final CLD. Further, tools such as VenSim have additional features such as causal trees that can be used to interpret the model.

CASE STUDY EXAMPLE: THE FATAL FIVE CAUSES OF ROAD TRAUMA

Road transport systems worldwide continue to create unacceptable levels of trauma. Whilst there are many systemic and societal factors that contribute to road collisions and trauma, there are a set of adverse driver behaviours that are known to play a direct causal role in road trauma. Known in Queensland (Qld), Australia as the fatal five, these behaviours are: 1) driving under the influence of alcohol and drugs; 2) driving while distracted; 3) driving while fatigued; 4) speeding; and 5) driving with occupants not wearing restraints (i.e. seatbelts and child restraints).

Historically, road safety interventions such as education, enforcement, or engineering have been used to prevent these behaviours. Whilst this approach has proven successful, in the past decade, reductions in road trauma levels have plateaued, and in some jurisdictions road crash fatalities have begun to increase again. There are various variables interacting to create increases in crash, injury, and trauma levels; however, it has been argued that the slowing of road safety progress relates, in part, to the driver-focussed road safety intervention approach. Critics have argued, for example, that directing interventions at drivers alone does not deal with the complex and dynamic set of variables across road systems and society that influence drivers' engagement in adverse behaviours (Salmon et al., 2019).

As part of a research program which aimed to produce novel interventions for preventing fatal five-related road crashes, we developed a CLD for each fatal five behaviour showing how variables interact to influence drivers' engagement in the fatal five behaviours. The aim was to identify the set of variables that interact to influence drivers' engagement in each behaviour. CLDs were developed for each behaviour and were subsequently reviewed and refined via a group model building process involving road safety SMEs. The boundary for the analysis was established as relating to fatal five-related road crashes and trauma occurring within the Australian road transport system; however, the analysis was intended to be applicable to other countries in the developed world.

BEHAVIOURS ANALYSED

The behaviours of interest were categorised into eight distinct items. Specifically, drink and drug driving was decomposed into 'drink driving', 'drug driving (illegal drugs)' and 'drug driving (legal drugs)', and driving with occupants not wearing restraints was decomposed into 'driving with occupants not wearing seat belts' and 'driving without children wearing child restraints'.

This yielded the following six behaviours that were included in the study:

1. Driving under the influence of alcohol;
2. Driving under the influence of illegal drugs;
3. Driving under the influence of legal drugs (i.e. prescription or non-prescription medications);
4. Distraction and inattention;
5. Driving whilst fatigued;
6. Speeding;
7. Driving with occupants not wearing seatbelts; and
8. Driving with children not wearing child restraints.

Note, the CLDs for driving with occupants not wearing seatbelts and driving with children not wearing child restraints are not included in this chapter.

Participants

Twelve SMEs (Age: 41.9 ± 10.5 years) participated in two separate group model building workshops. SMEs described their roles as policy officer (n = 5), research officer (n = 2), road safety research academic (n = 2), cycling, pedestrian and motorcycle technologist (n = 1), public servant (n = 1), and customer designer (n = 1). The SMEs had self-reported expertise across the following six areas of road safety: 1) road infrastructure design, construction or maintenance; 2) behaviour and human factors; 3) road rules and regulations; 4) Crash analysis and investigations; 5) risk and and/or safety management; and 6) road safety policy development and implementation.

CLD Development

Two group modelling processes were adopted, with the first undertaken initially by the research team, followed by a subject matter expert group process to review and refine the CLDs.

Initially, the research team held a workshop to develop a generic control loop structure of road crashes and trauma based on existing research literature, publicly available data on road crashes, and existing systems analyses undertaken by the authors (e.g. McClure et al., 2015; Salmon et al., 2016; 2019). Subsequently, draft CLDs, based on the generic CLD structure, were developed for each of the six behaviours. This involved the use of existing research literature, publicly available data on road crashes, and existing systems analyses as the primary data inputs. Once finalised, the draft CLDs were drawn in Microsoft Visio.

The draft CLDs were subsequently reviewed and refined via two SME workshops.

Subject Matter Expert Workshop 1

In the first workshop CLDs for each behaviour were presented and explained to the 12 SMEs. This included an explanation of each of the feedback loops and variables within each CLD and a description of how they interact with one another to create road crashes and trauma. The participants were then asked the following questions:

1. Are the variables included in the CLD appropriate? If not, which ones should be modified or removed?
2. Are there any relevant variables not currently included in the CLD? If yes, what are they?
3. Are the relationships between variables, and the direction of the relationships included in the CLD appropriate? If not, which ones should be modified or removed?
4. Are there any relevant relationships between variables not currently included in the CLD? If yes, what are they?

The CLDs were revised based on the workshop participants' responses. Specifically, variables and relationships were either confirmed, modified, removed, or added. Variables and relationships were confirmed if all participants agreed that they were correct and should be included. Variables were modified if all participants agreed that the variable should be included, but that the terminology used was not correct. Relationships were modified if all participants agreed that the relationship should be included, but that the direction of the relationship was incorrect. Variables and relationships were removed if all participants agreed that they did not influence behaviour as stated and should not be included. Where new variables or relationships were identified and agreed upon by the group, they were added to the CLDs.

Subject Matter Expert Workshop 2

During the second workshop, the revised CLDs were presented to seven of the SMEs who attended the first workshop. The SMEs were asked whether each CLD provided a valid representation of the causal loops underpinning road crashes in relation to each of the six behaviours. No further changes were proposed during the second workshop as all participants agreed that the CLDs provided a valid representation.

Following the workshop, the final CLDs were drawn using the systems dynamics software program VenSim.

RESULTS

Generic Causal Loop Structure

The group model building process identified a generic set of central causal loops that apply to road crashes regardless of the driver behaviour that is seen to be the primary cause. The CLD representing this generic set of causal loops is presented in Figure 8.7. According to the CLD, road crashes are primarily caused by drivers engaging in the fatal five behaviours. Crashes lead to injuries and fatalities, and as the number of injuries and fatalities rises, the gap between the goal state (i.e. road safety targets relating to crashes, injuries and fatalities) and actual state increases, which in turn heightens recognition of the problem and acts to increase the political will to intervene. Both generate policy action, which in turn creates an impetus and funding for road safety interventions that aim to prevent drivers from engaging in the crash-causing behaviours. These interventions typically encompass the "Three Es" of Education, Enforcement,

FIGURE 8.7 Generic causal loop that describes the influences on driver behaviour and road trauma. (Adapted from Salmon et al., 2020).

and Engineering and are designed to improve road user knowledge and manage behaviour so that engagement in unsafe behaviours is reduced.

An additional set of interventions included in Figure 8.7 are public health initiatives that focus more on the societal issues that may be influencing adverse behaviours, such as drug and alcohol addiction, time-poor lifestyles, and a desire to be connected through mobile phones and social media. It should be noted that these do not tend to be considered by policy makers to be road safety interventions, yet the analysis suggests that they can have an impact on road user behaviour. A key feature of the generic CLD is the 'resistance' variable, which is a common phenomenon in complex systems (Sterman, 2000). In this case, policy resistance acts to inhibit policy action and includes resistance from the general public that is created through the increased level of road safety interventions such as enforcement and education. In addition, resistance in government and stakeholders exists and increases if the gap between the goal state in terms of acceptable levels of road injuries and fatalities and the actual state becomes too small and road safety targets are being achieved. This reflects a dynamic whereby a reduction in accidents and injuries is followed by complacency and a related reduction in the accident prevention activities that led to the improvements in the first place (Marais et al., 2006).

CLDS FOR EACH FATAL FIVE BEHAVIOUR

The driving under the influence of alcohol CLD is presented in Figure 8.8. The model shows that a key issue influencing the number of drink driving-related road trauma is societal alcohol consumption levels which create a population of addicted alcohol users and a population of incidental alcohol users. An output of this is a set of drivers who either intentionally or unintentionally drive under the influence of alcohol at any given time, which in turn leads to crashes and injuries and fatalities (NHTSA, 2013). As the number of injuries and fatalities rises, the gap between the actual road safety state and goal road safety state increases, heightening recognition that there is a problem associated with drivers under the influence of alcohol which increases political will to address the issue. As these two variables increase, policy action escalates which creates a greater impetus for road safety interventions that aim to prevent drivers from driving under the influence of alcohol.

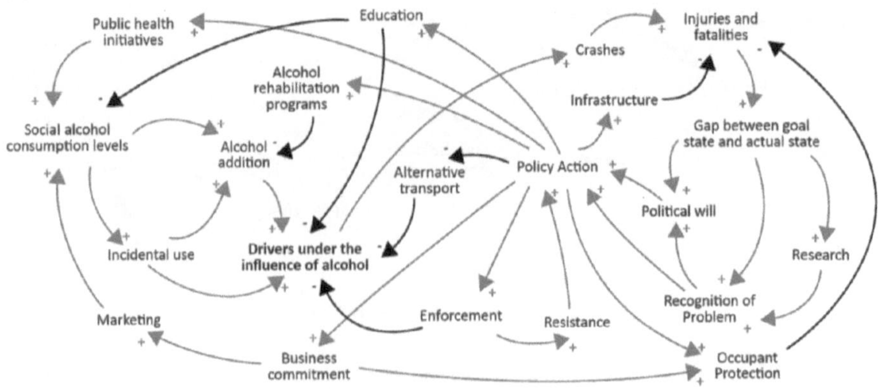

FIGURE 8.8 Causal loop diagram showing the variables which interact to create drink driving-related crashes. (Adapted from Salmon et al., 2020).

The primary prevention measures for driving under the influence of alcohol include enforcement and education. Enforcement includes deterrents such as random breath testing, stricter laws and penalties, and in-vehicle alcohol interlock devices. Education typically includes a range of targeted education programs designed to prevent drivers from driving under the influence in the first place, such as television and social media campaigns and roadside advertising. In addition, remedial education is used to educate drivers who have been previously been caught drink driving. Injury prevention measures designed to limit trauma once crashes have occurred include improved occupant protection in vehicles and safer road infrastructure (e.g. wire rope road safety barriers). Increasing the availability of alternative transport reduces the need for driving by increasing the options available for people under the influence who need to travel.

Outside of standard road safety interventions the driving under the influence of alcohol CLD also includes a wider set of interventions that focus more on the societal issue of alcohol misuse and addiction. These include public health initiatives (e.g. screening for alcohol misuse, health clinics, outreach, education campaigns) and alcohol rehabilitation programs, both of which are designed to reduce alcohol addiction or provide support to those who are addicted. Finally, the nature and level of marketing and advertising used by alcohol companies is included in the CLD as a variable that can influence societal alcohol consumption levels, with aggressive advertising leading to increased sales and consumption (Saffer and Dave, 2002).

According to the CLD, as the number of people addicted to alcohol or using alcohol goes up or down, there is a concomitant increase or decrease in the number of drivers driving under the influence, which in turn reduces the number of crashes and so on. When there is a reduction in addiction levels and road crashes and trauma and the gap between the goal and actual state decreases, resistance then acts to reduce policy action, which leads to a reduction in both road safety and public health initiatives.

The remaining CLDs are presented in Figures 8.9 (illegal drugs), 8.10 (legal drugs), 8.11 (driving while distracted), 8.12 (driving while fatigued), and 8.13 (speeding). All include the generic causal loop structure as described in Figure 8.7, along with specific variables covering factors that either influence drivers'

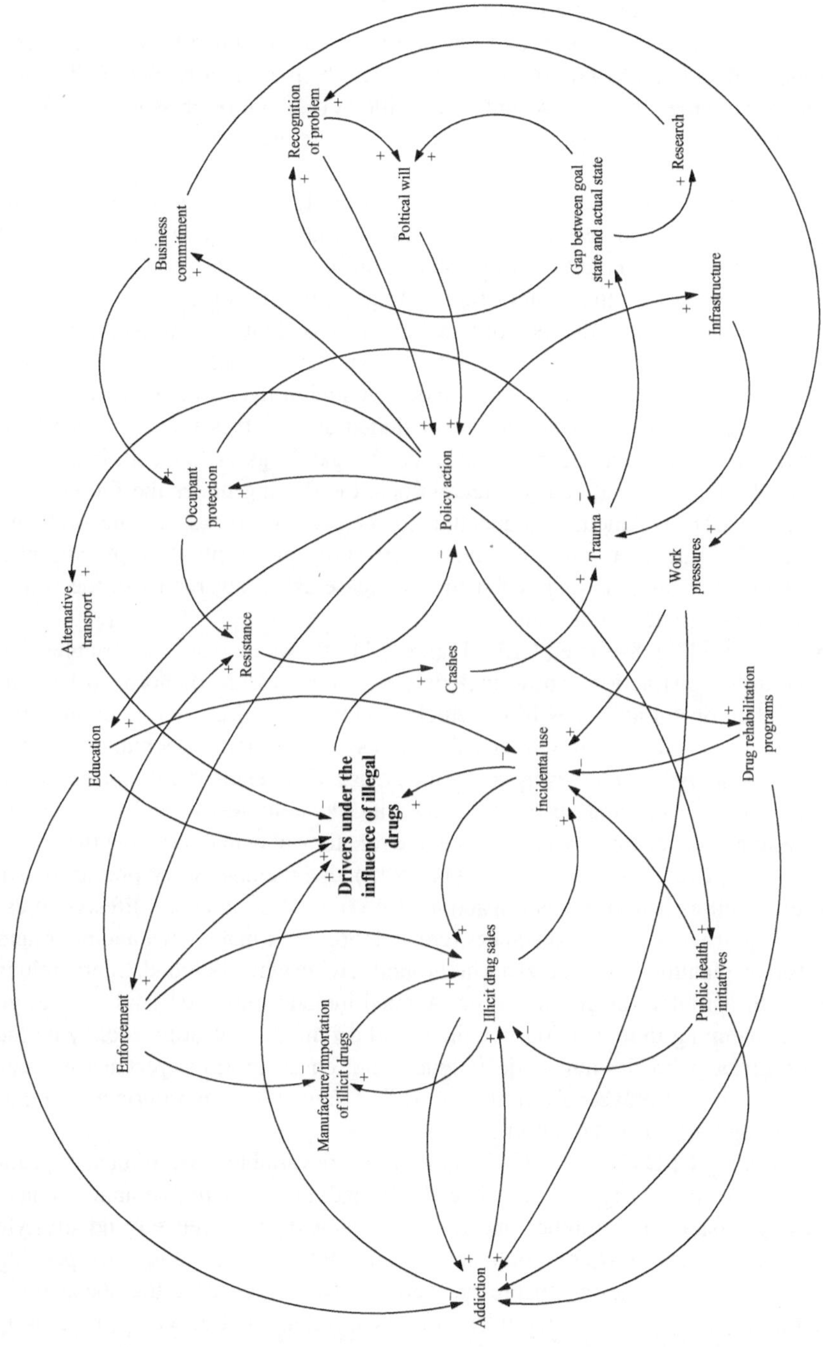

FIGURE 8.9 Driving under the influence of illegal drugs causal loop diagram.

engagement in each behaviour or that can be used to prevent drivers engaging in each behaviour or mitigate the consequences of crashes associated with each behaviour. An overview of the unique features of each CLD is given below.

The driving under the influence of illegal drugs CLD (Figure 8.9) also includes a set of addicted and incidental drug users, however, the main facilitators of addiction and drug use are the manufacture and sale of illicit drugs. Work pressure is included as a variable that also influences illicit drug use. This relates both to those who drive for work (e.g. heavy goods vehicle drivers, taxi drivers) as well as the general working population. The illegal drugs CLD includes a similar set of road safety interventions to the driving under the influence of alcohol CLD; however, additional prevention strategies included drug rehabilitation programs and enforcement designed to prevent both the manufacture and sale of illicit drugs.

The driving under the influence of legal drugs CLD (Figure 8.10) again includes a set of addicted and incidental users but incorporates a broader set of influencing variables. Specifically, society's limited understanding of the problem, an increasing prescription rate, and drug labelling are included as variables that influence both legal drug use and driving under the influence of legal drugs. As a result, in addition to the standard set of road safety interventions, the driving under the influence of legal drugs CLD also includes education to increase awareness of the problem, including both drug users and also those prescribing prescription drugs, and also improvements to drug labelling designed to increase users' awareness of the potentially impairing impacts on driving.

The driving while distracted CLD (Figure 8.11) shows the various variables that influence driver distraction. These include a societal pressure or norm to be connected, technological advances in in-vehicle devices as well as mobile phones and tablets, work-related pressures, increasing access to distracting technologies, distracting passengers, and increasing social media use. Unique interventions relating to distraction include limiting access to distracting technologies through vehicle and device design, and educating passengers in the role they play in distracting the driver.

The driving while fatigued CLD (Figure 8.12) also includes work pressures as a variable as well as health issues impacting sleep (e.g. sleep apnoea), lifestyle pressures, long journey times and distances (created through urban sprawl and poor land use and urban planning), and social norms around driving while fatigued. Interventions unique to the problem of driving while fatigued include improved land use design and urban planning to reduce journey times and distances, educating society on the risks associated with driving while fatigued, and ensuring employers use various means to prevent employees from driving while fatigued (e.g. appropriate shift patterns, limiting long working hours).

The speeding CLD (Figure 8.13) includes five key variables that influence speeding. These are the strong societal view that speeding is not problematic, vehicle design (e.g. faster but smoother and quieter vehicles), work-related and lifestyle pressures, and a lack of awareness of speed limits. Interventions unique to speeding include using vehicle design to prevent vehicles from exceeding the speed limit, attempting to change societal attitudes towards speeding, and educating road users regarding speed limits.

The CLDs demonstrate that crash-causing behaviours have a complex set of interacting variables underpinning them and are a consequence of issues across both road

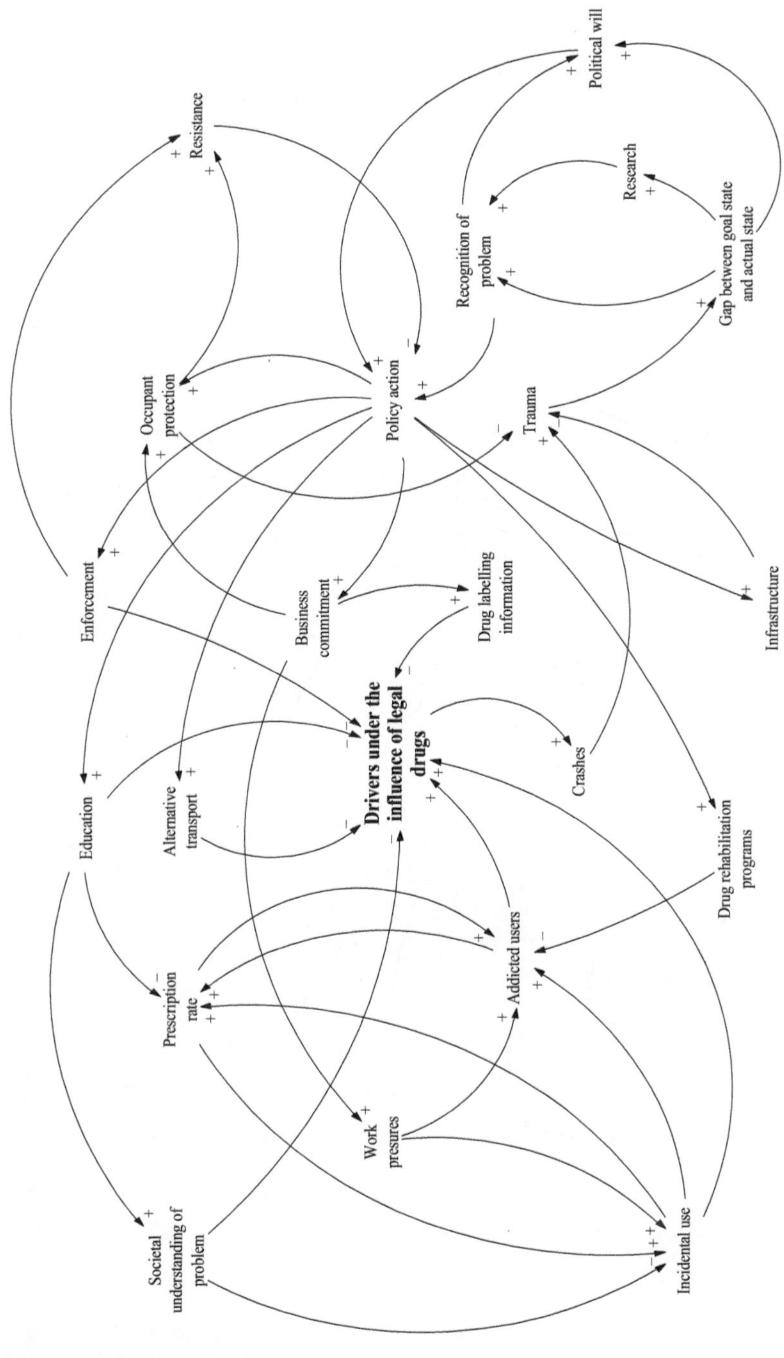

FIGURE 8.10 Driving under the influence of legal drugs causal loop diagram.

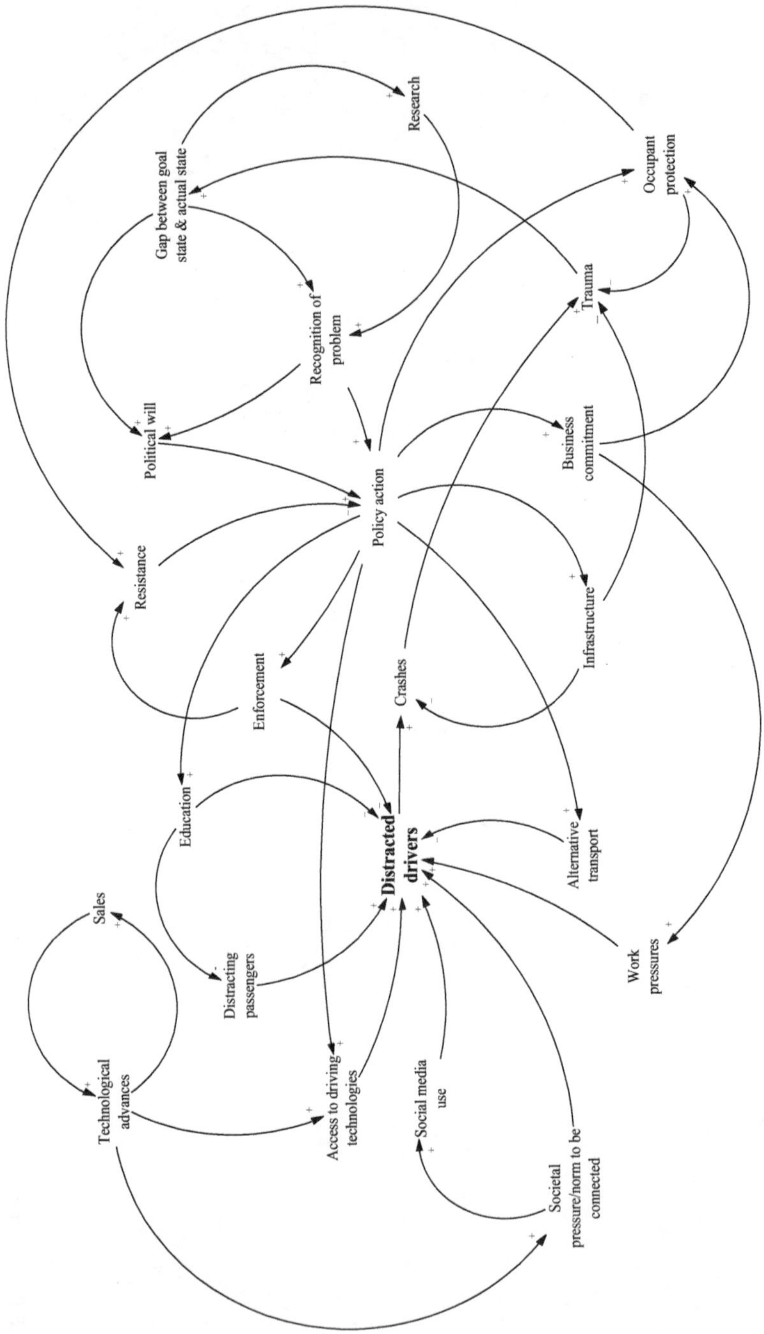

FIGURE 8.11 Driving while distracted causal loop diagram.

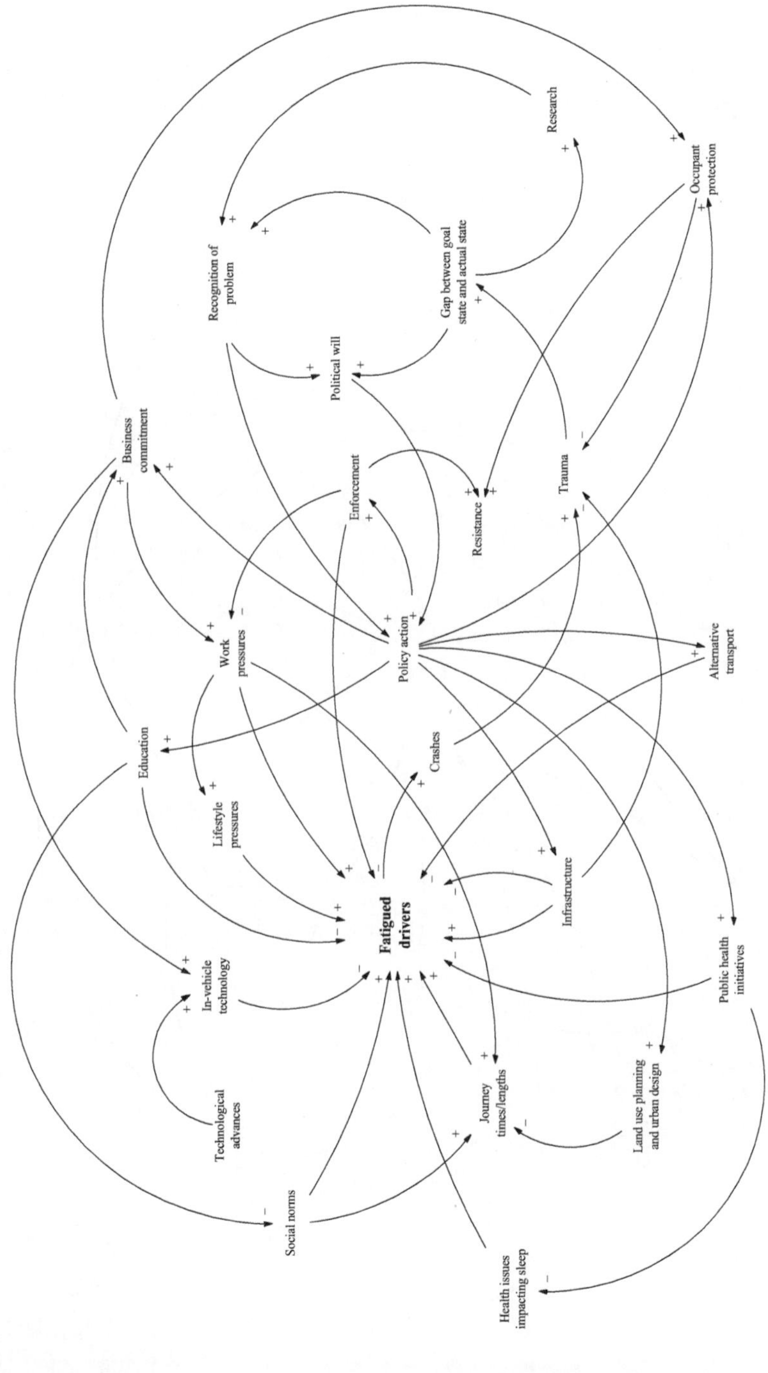

FIGURE 8.12 Driving while fatigued causal loop diagram.

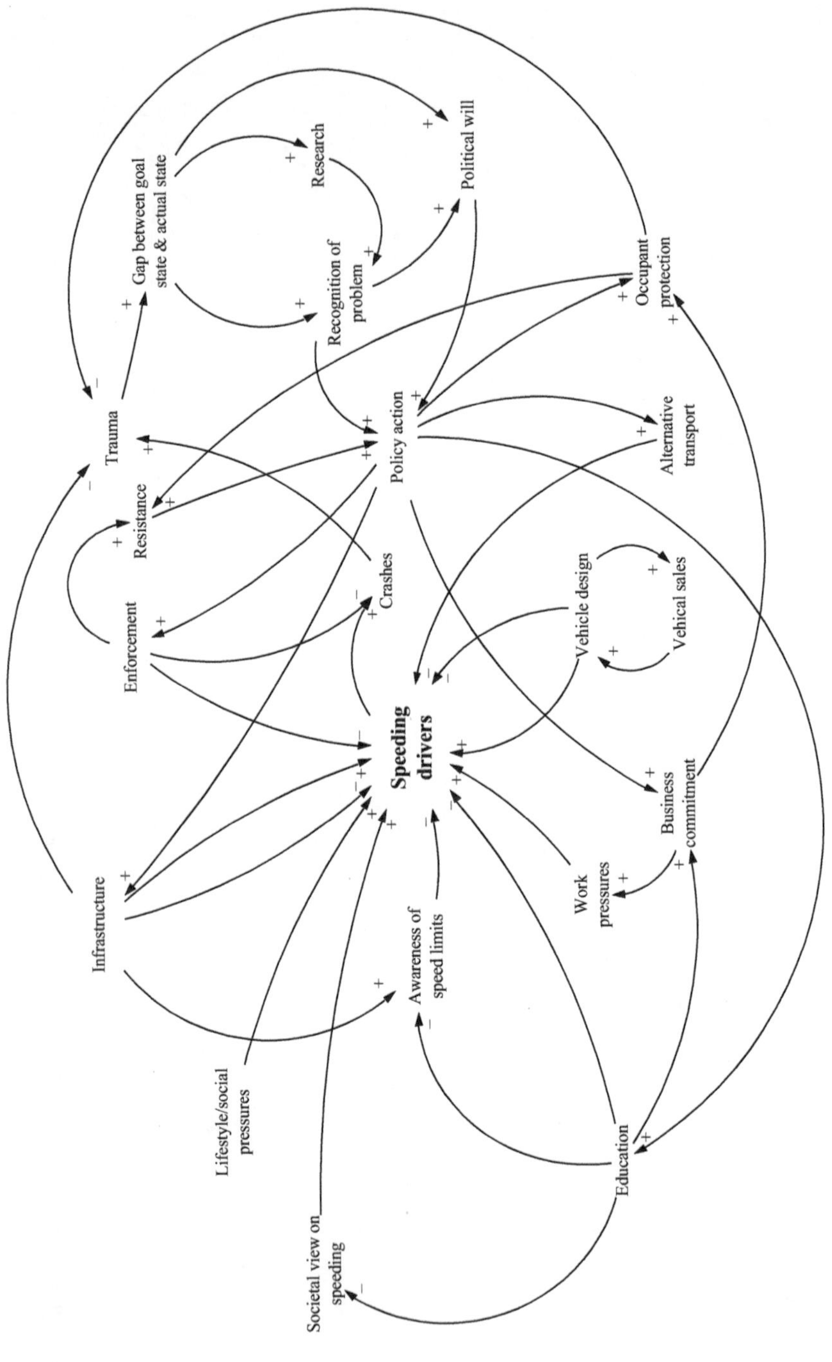

FIGURE 8.13 Speeding causal loop diagram.

transport systems and society generally. Whilst each CLD offers insight into specific driver behaviours, three core leverage points were identified across many of the CLDs. These include the broader societal issues that influence driver behaviour and road trauma, the influence of work pressures on driver behaviour, and the key role of the motor vehicle in road trauma.

THE ROLE OF SOCIETAL ISSUES IN ROAD TRAUMA

The CLDs demonstrate the role that wider societal problems play in creating or facilitating the behaviours known to lead to crashes. In each CLD, drivers' engagement in the behaviours that lead to road crashes stems from broader societal issues that are largely beyond the control of road safety stakeholders. For example, drug driving rates are influenced by increasing rates of illegal and legal drug misuse and addiction (McIntosh et al., 2008; Wickens et al., 2017). Engagement in distracting activities while driving is influenced by a high adoption of technology (Buckley et al., 2014) and an increasing societal pressure to remain connected through technology (Lee et al., 2013). Speeding and driving while fatigued are issues that are exacerbated by increasingly time-poor lifestyles and poor work schedules (Van Den Berg and Landström, 2006). Indeed, the prevalence of driving over other forms of transport is driven by low-density urban form and lack of viable public and alternative transport options (Stevenson et al., 2016).

An important question that arises from this case study, and others, is the extent to which the fatal five behaviours can be effectively managed through road safety interventions alone. Many have previously questioned the utility of a zero-tolerance approach to behaviours underpinned by addiction, such as drink and drug driving (Holmgren et al., 2008; Tassoni et al., 2016). Moreover, the capacity of road safety interventions to deal with complex issues such as addiction has been questioned (Salmon et al., 2019). It seems likely that improved road safety outcomes will be only achieved via the use of appropriate interventions that target the societal issues that lie at the root of the behaviours themselves. This finding is in line with many recent studies that have argued for interventions to incorporate elements such as rehabilitation (McIntosh et al., 2008; Salmon et al., 2019; Stephens et al., 2017; Wickens et al., 2017). Moreover, there is a growing body of work, which argues that an integrated approach to public health, urban planning and transport safety is likely to achieve greater public health and road safety gains than a 'siloed' approach (McClure et al., 2015; Salmon et al., 2020; Stevenson et al., 2016).

THE ROLE OF WORK PRESSURES IN ROAD TRAUMA

Many of the CLDs included variables relating to the role of work pressures in influencing driver behaviour. Previous research has reported a link between work pressures and work stresses in risky driving behaviours (e.g. Rowden et al., 2011; Thompson and Stevenson, 2014). In addition, speeding and fatigue have been identified as prominent factors in work-related driving crashes (Nævestad et al., 2015). The findings indicate the role of work pressures and stresses in influencing drivers to drive under the influence of illegal and legal drugs, speed, engage in distracting activities whilst driving, and drive whilst fatigued. For example, productivity and poor scheduling can lead to drivers speeding, driving whilst fatigued, or using drugs to enable longer driving hours. Similarly, the requirement to be contactable whilst driving for work was identified as a factor that influences drivers in engaging in distracting activities such as phone calls and sending and receiving text messages. These findings add to a growing body of research which highlights the key role that work and employers play in influencing driver behaviour and road trauma (e.g. Caird and Kline, 2004; Havârneanu et al., 2019; Nævestad et al., 2015; Newnam et al., 2017; Thompson et al., 2015; Thompson and Stevenson, 2014)). It should be noted that work-related pressures could be relevant during work-related driving, or where driving was not work-related, such as fatigue when driving home after working excessive hours, or distraction due to expectations to return calls outside of business hours. Further, the fatigue CLD shows how work pressures can increase lifestyle pressures, compounding the effect on opportunity for rest and sleep. That is, longer working hours reduces the time available for household, family, and recreation activities, which may then be undertaken during time previously used for rest. This suggests that employers have a key role to play in preventing road trauma and that interventions need to be developed in collaboration with employees, and indeed are to be implemented by employers.

REDUCING THE REQUIREMENT TO DRIVE

Across the CLDs, alternative forms of transport were included as a variable that could be used to reduce the requirement to drive and in turn reduce the number of drivers engaging in the fatal five behaviours. In a recent survey of drivers, Salmon et al. (2019) reported that participants cited a lack of alternative transport options as one of the reasons for people to drive whilst fatigued and/or under the influence of alcohol or illegal and legal drugs. Ostensibly, better access to safe, efficient, and affordable forms of public transport (including taxis or ride or peer-to-peer ride services) would remove the need for people under the influence of alcohol and drugs to drive. In addition, it would provide transport where engagement with devices such as mobile phones is permitted and safe. Whilst this would likely lead to a reduction in road crashes and trauma, high levels of modal shift to public transport including trains, trams, and buses could also have a series of benefits to health and wellbeing via reduced pollution levels and increased physical activity through walking (Heinrich et al., 2009; Woodcock et al., 2007). The findings therefore suggest that significant reductions in the road toll as well as concomitant benefits to health and wellbeing could be achieved through reducing the requirement to drive through the provision of accessible and efficient alternative public transport options.

RECOMMENDED READING

Sterman, J.D. (2000). *Business Dynamics: Systems Thinking and Modeling for a Complex World*. Boston: Irwin McGraw-Hill.

REFERENCES

Allender, S., Owen, B., Kuhlberg, J., Lowe, J., Nagorcka-Smith, P., Whelan, J., & Bell, C. (2015). A community based systems diagram of obesity causes. *PloS One*, 10(7), e0129683.

Berard, C. (2010). Group model building using system dynamics: An analysis of methodological frameworks. *Journal of Business Research*, 8(1), 13–24

Buckley, L. Chapman, R. L., Sheehan, M. 2014. Young driver distraction: State of the evidence and directions for behavior change programs. *Journal of Adolescent Health*, 54(5), S16–S21

Burrell, M., White, A. M., Frerichs, L., Funchess, M., Cerulli, C., DiGiovanni, L., & Lich, K. H. (2021). Depicting "the system": How structural racism and disenfranchisement in the United States can cause dynamics in community violence among males in urban Black communities. *Social Science & Medicine*, 272, 113469.

Caird, J. K. & Kline, T. J. (2004). The relationships between organizational and individual variables to on-the-job driver accidents and accident-free kilometres, *Ergonomics*, 47, 1598–1613

De Pinho, H., & Larsen, A. M. (2015). *Systems Tools for Complex Health Systems: A Guide to Creating Causal Loop Diagrams*. New York: Mailman School of Public Health, Columbia University, 1.

Havârneanu, C.-E., Măirean, C., & Popuşoi, S.-A. 2019. Workplace stress as predictor of risky driving behavior among taxi drivers: The role of job-related affective state and taxi driving experience. *Safety Science*, 111, 264–270

Heinrich, J., Schwarze, P. E., Stilianakis, N., Momas, I., Medina, S., Totlandsdal, A. I., von Bree, L., Kuna-Dibbert, B., & Krzyzanowski, M. (2009). Studies on health effects of transport-related air pollution, In M. Krzyzanowski, B. Kuna-Dibbert & J. Schneider (Eds.), *Health Effects of Transport-Related Air Pollution*, (pp. 125–165), World Health Organisation Europe Report, ISBN 92 890 1373 7.

Holmgren, A., Holmgren, P., Kugelberg, F. C., Jones, A. W., & Ahlner, J. 2008. High re-arrest rates among drug-impaired drivers despite zero-tolerance legislation. *Accident Analysis and Prevention*, 40(2), 534–540.

Kim, D. H. (1995). Guidelines for drawing causal loop diagrams. *Health*, 1, R2.

Kumar, A., Priya, B, & Srivastava, S. K. (2021). Response to the COVID-19: Understanding implications of government lockdown policies. *Journal of Policy Modeling*, 43(1), 76–94.

Lannon, C. (2012). Causal loop construction: The basics. *The Systems Thinker*, 23(8).

Lee, V. K., Champagne, C. R., & Francescutti, L. H. (2013). Fatal distraction: Cell phone use while driving, *Can Fam Physician*. Jul; 59(7), 723–725.

Lin, C. L., & Chien, C. F. (2019). Systems thinking in a gas explosion accident–lessons learned from Taiwan. *Journal of Loss Prevention in the Process Industries*, 62, 103987.

Marais, K., Saleh, J. H., & Leveson, N. G. 2006. Archetypes for organisational safety. *Safety Science*, 44(7), 565–582.

McClure, R. J., Adriazola-Steil, C., Mulvihill, C., Fitzharris, M., Bonnington, P., Salmon, P. M., & Stevenson, M., 2015. Simulating the dynamic effect of land use and transport policies on the development and health of populations. *American Journal of Public Health*, 105(S2), 223–229.

McIntosh, J., O'Brien, T., & McKeganey, N. (2008). Drug driving and the management of risk: The perspectives and practices of a sample of problem drug users. *International Journal of Drug Policy*, 19(3), 248–254.

McLean, S., Kerhervé, H. A., Stevens, N., & Salmon, P. M. (2021). A systems analysis critique of sport-science research. *International Journal of Sports Physiology and Performance*, 16(10), 1385–1392.

Meadows, D. H. (2008). *Thinking in Systems: A Primer*. Chelsea Green Publishing.

Mirchi, A., Madani, K., Watkins, D., & Ahmad, S. (2012). Synthesis of system dynamics tools for holistic conceptualization of water resources problems. *Water Resources Management*, 26(9), 2421–2442.

Nævestad, P., Phillips, R. O., & Elvebakk, B. (2015). Traffic accidents triggered by drivers at work–a survey and analysis of contributing factors, *Transport Research Part F: Traffic Psychology Behaviour*, 34, 94–107.

National Highway Traffic Safety Administration. (2013). Alcohol-impaired Driving Traffic Safety Facts. *DOT HS*, 811, 870.

Newnam, S., Goode, N., Salmon, P.M., & Stevenson, M., (2017). Reforming the road freight transportation system? An investigation of Coronial inquests in Australia. *Accident Analysis and Prevention*, 101, 28–36.

Richards, C. E., Lupton, R. C., & Allwood, J. M. (2021). Re-framing the threat of global warming: an empirical causal loop diagram of climate change, food insecurity and societal collapse. *Climatic Change*, 164(3), 1–19.

Rowden, P., Matthews, G., Watson, B., & Biggs, H. (2011). The relative impact of work-related stress, life stress and driving environment stress on driving outcomes. *Accident Analysis and Prevention*, 43, 1332–1340.

Saffer, H. & Dave, D. 2002. Alcohol consumption and alcohol advertising bans. *Journal of Applied Economics*, 34(11), 1325–1334.

Sahin, O., Salim, H., Suprun, E., Richards, R., MacAskill, S., Heilgeist, S.,... & Beal, C. D. (2020). Developing a preliminary causal loop diagram for understanding the wicked complexity of the COVID-19 pandemic. *Systems*, 8(2), 20.

Salmon, P. M., Read, G. J. M., Beanland, V., Thompson, J., Filtness, A., Hulme, A., McClure, R., & Johnston, I. (2019). Bad behaviour or societal failure? Perceptions of the factors contributing to drivers' engagement in the fatal five driving behaviours. *Applied Ergonomics*, 74, 162–171.

Salmon, P.M., Read, G.J.M., & Stevens, N., 2016. Who is in control of road safety? A STAMP control structure analysis of the road transport system in Queensland, Australia. *Accident Analysis and Prevention* 96, 140–151.

Salmon, P. M., Read, G. J. M., Thompson, J., McLean, S., & McClure, R. (2020). Computational modelling and systems ergonomics: a system dynamics model of drink driving-related trauma prevention. *Ergonomics*

Shepherd, S. P. 2014. A review of system dynamics models applied in transportation. *Transportmetrica B: Transport Dynamics*, 2(2), 83–105.

Stephens, A. N., Bishop, C. A., Liu, S., & Fitzharris, M. (2017). Alcohol consumption patterns and attitudes toward drink-drive behaviours and road safety enforcement strategies. *Accident Analysis & Prevention*, 98, 241–251.

Sterman, J. D. (2000). *Business Dynamics: Systems Thinking and Modeling for a Complex World*. Boston: Irwin McGraw-Hill.

Stevenson, M., Thompson, J., et al., (2016). Land-use, transport and population health: Estimating the health benefits of compact cities. *The Lancet*, 388(10062), 2925–2935.

Tassoni, G., Cippitelli, M., Mirtella, D., Froldi, R., Ottaviani, G., Zampi, M., Cingolani, M. 2016 Driving under the effect of drugs: Hair analysis in order to evaluate recidivism, *Forensic Science International*, 267, 125–128.

Thompson, J., Newnam, S., & Stevenson, M. (2015). A model for exploring the relationship between payment structures, fatigue, crash risk, and regulatory response in a heavy-vehicle transport system. *Transportation Research Part A: Policy and Practice*, 82, 204–215. doi:10.1016/j.tra.2015.09.016

Thompson, J. & Stevenson, M. (2014). Associations between heavy-vehicle driver compensation methods, fatigue-related driving behavior, and sleepiness. *Traffic Injury Prevention*, 15(sup1), S10–S14. doi:10.1080/15389588.2014.928702

van den Berg, J. & Landström, U. 2006. Symptoms of sleepiness while driving and their relationship to prior sleep, work and individual characteristics. *Transportation Research Part F: Traffic Psychology and Behaviour*, 9(3), 207–226.

Wickens, C. M. et al. 2017. The impact of medical and non-medical prescription opioid use on motor vehicle collision risk, *Transportation Research Part F: Traffic Psychology and Behaviour*, 47, 155–162.

Woodcock, J., Banister, D., Edwards, P., Prentice, A.M., & Roberts I. (2007). Energy and transport. *Lancet*, 370(9592), 1078–1088.

Section IV

Accident Analysis Methods

9 The Accident Mapping (AcciMap) Method

Sharon Newnam
Monash University Accident Research Centre

BACKGROUND

The Accident Mapping (AcciMap) method provides a means with which to describe accidents through the lens of Jens Rasmussen's risk management framework (Rasmussen, 1997; See Figure 9.1). The method was developed in response to the belief that task analysis methods are not capable of modelling the systemic network of factors underpinning accidents (Svedung and Rasmussen, 2002). In line with Rasmussen's risk management framework, AcciMap is based on the principle that that behaviour, safety, and accidents are emergent properties of complex sociotechnical systems. These emergent properties are created by the decisions and actions of all stakeholders within a system – politicians, chief executives, managers, safety officers, and work planners – not just by front line workers alone (Cassano-Piche et al., 2009). Safety is therefore the shared responsibility of all actors within a particular system, and accidents regardless of size or consequence are created by the interacting decisions and actions of multiple actors across the system.

Rasmussen's framework makes the following assertions regarding safety and accident causation:

1. Accidents are emergent properties that are created by the decisions and actions of all actors, not just frontline workers alone;
2. Accidents are caused by multiple contributing factors from across sociotechnical work systems, not just a single poor decision or action at the sharp end;
3. Accidents can result from a lack of poor communication and feedback (or 'vertical integration') across levels of sociotechnical work systems, not just from deficiencies at one level alone;
4. Lack of vertical integration is caused, in part, by lack of feedback within and across levels of sociotechnical work systems;
5. Behaviours within sociotechnical work systems are not static, they migrate over time and under the influence of various pressures such as production, financial, and psychological pressures;
6. Migration occurs at multiple levels of sociotechnical work systems; and
7. Migration of practices cause sociotechnical work system defences to degrade and erode gradually over time, not all at once. Accidents are caused by a combination of this migration and a triggering event(s).

DOI: 10.1201/9780429281624-9

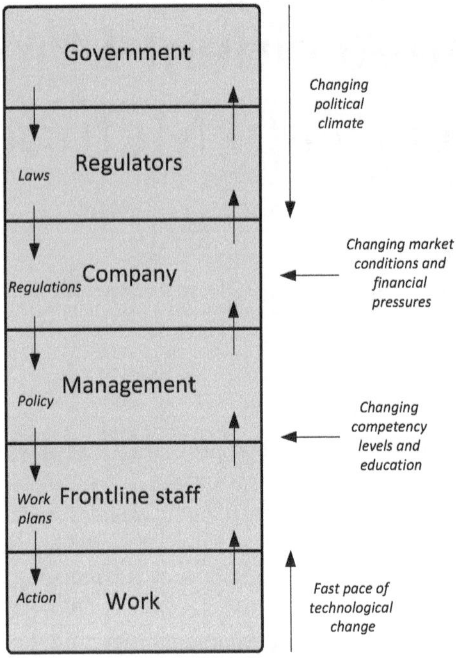

FIGURE 9.1 Rasmussen's risk management framework. (Adapted from Rasmussen, 1997).

Using AcciMap typically involves applying two analysis methods: the ActorMap and AcciMap. ActorMap forms the first analysis component and is used to provide a representation of the stakeholders ('actors') who operate within the system and thus share the responsibility for safety and accidents. The method typically involves identifying relevant actors at the following six hierarchical levels (though these are flexible and can be modified if required): government policy and budgeting; regulatory bodies and associations; local area government planning and budgeting (including company management); technical and operational management; physical processes and actor activities; and equipment and surroundings). The resulting ActorMap shows which actors operate with the system and at which level they reside. For example purposes, a generic ActorMap is presented in Figure 9.2.

Once the ActorMap is developed the AcciMap method is used to identify and represent the network of contributory factors involved in the incident in question across the same hierarchical levels as the ActorMap. Contributory factors are identified, mapped to one of the six levels, and then linked between and across levels based on cause-effect relations. A generic template for an AcciMap diagram is presented in Figure 9.3.

While AcciMap was developed to better understand the mechanisms underpinning large-scale incidents in high-risk industrial settings (Svedung and Rasmussen, 2002), it is highly flexible, and applicable to all types of incidents. To date, AcciMap has been applied to describe and represent large-scale organisational accidents in multiple domains (e.g. Hulme et al., 2021a), as well as multiple incident datasets

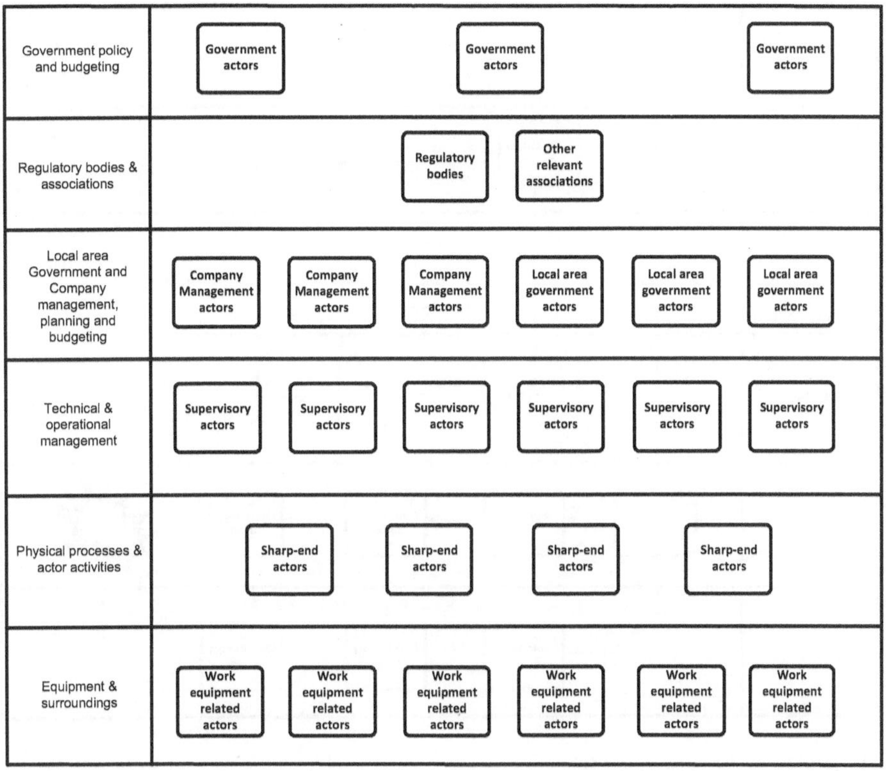

FIGURE 9.2 Generic ActorMap.

(Vicente and Christoffersen, 2006; Salmon et al., 2014a, 2017). In addition, AcciMap has been used to support the analysis of multiple workplace injury incidents (Goode et al., 2014) and can also be used to represent the factors underpinning a general safety issue or problem; for example, Lane et al.'s (2020) AcciMap of the barriers preventing access to eating disorder treatment. This involves placing groups of contributory factors across the hierarchical levels without specifying the relationships between them. Most importantly, it can be used to represent a single incident, or to create an aggregate AcciMap representing recurring contributory factors and relationships across multiple incidents (Svedung & Rasmussen 2002; Salmon et al., 2020).

DOMAIN OF APPLICATION

AcciMap is a generic approach that has been applied in many domains. A recent systematic review undertaken by Hulme et al. (2019) identified applications spanning a range of domains including public health (Woo and Vicente, 2003; Vicente and Christoffersen, 2006; Cassano-Piche et al., 2009; Waterson, 2009; Nayak and Waterson, 2016), counter-terrorism (Jenkins et al., 2010), disaster response (Salmon et al., 2014a), rail transport (Chen et al., 2015; Salmon et al., 2013; Underwood and

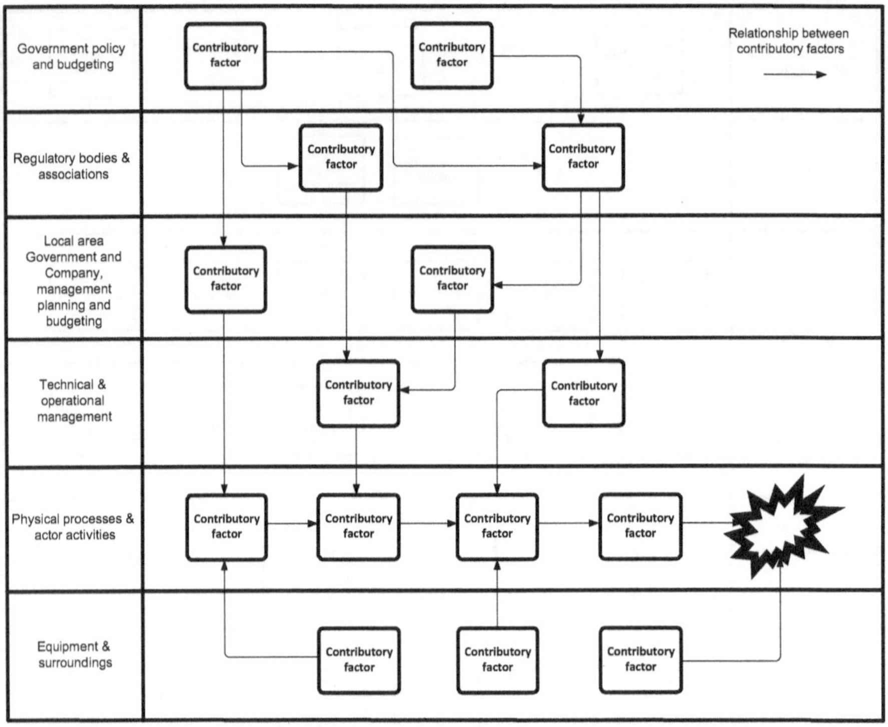

FIGURE 9.3 Generic AcciMap.

Waterson, 2014), road transport (Newnam and Goode, 2015), off-road beach driving (Stevens and Salmon, 2016), maritime (Akyuz, 2015; Kee et al., 2017; Lee et al., 2017) led outdoor recreation (Salmon et al., 2010; Salmon et al., 2012, 2014a; Salmon et al., 2017), aerospace (Johnson and de Almeida, 2008), and civil engineering.

PROCEDURE AND ADVICE

A flowchart depicting the AcciMap procedure is presented in Figure 9.4. Step-by-step guidance is presented below.

STEP 1: DETERMINE ANALYSIS AIMS AND SCOPE

The first step involves clearly defining the aims and scope of the analysis. Typically, the aim is to identify the system wide factors that interacted to create a particular incident. Alternatively, the aim may be to identify recurring contributory factors and relationships across multiple incidents.

Clearly defining the scope or analysis boundaries is important because adverse outcomes (e.g. workplace injuries or fatalities) are often the result of multiple events

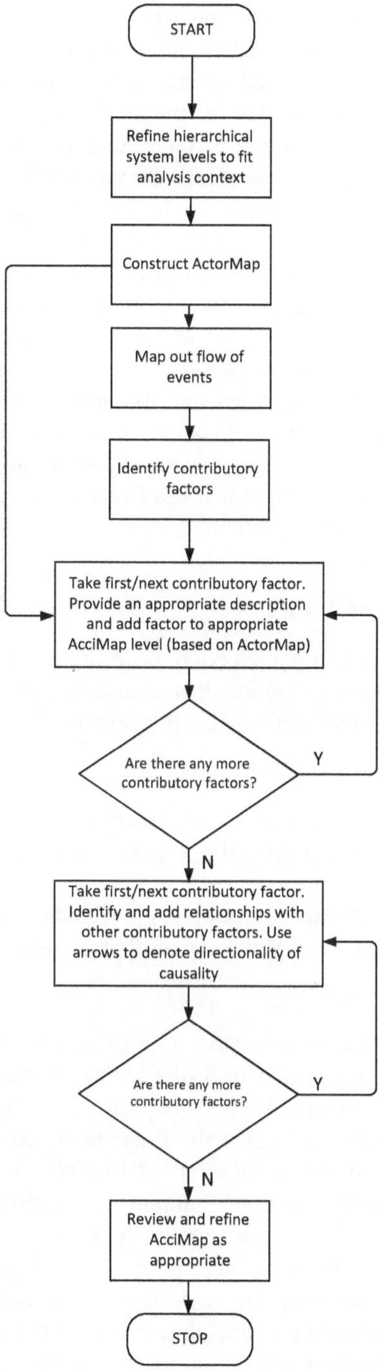

FIGURE 9.4 AcciMap procedure.

or phases. For example, an analysis of the death of a patient in a hospital setting may focus on the decisions and actions leading up to the deterioration of the patient, or it might focus on the decision and actions during the emergency response. The scope will have a significant impact on the recommendations for interventions identified from the analysis. It is not unusual for AcciMap analyses to consider contributory factors months and even many years before the incident in question occurred.

Defining the analysis boundaries is also important as project constraints will dictate how deep the analysis can go in terms of the parts of the system considered and how far back in time the analysis will go. It may be, for example, that an analysis boundary may be set at the organisational level only, or may not extend to international factors outside of the country in which the incident occurred. Further, post incident response may or not be of interest.

At a minimum, it is recommended that the analysis should include influences external to the organisation within state boundaries (e.g. state regulators and governments). For some incidents it is useful to extend these boundaries to consider national (e.g. SafeWork Australia) and sometimes even global (e.g. World Health Organisation) influences that are directly relevant to the events.

STEP 2: DATA COLLECTION

AcciMap is dependent upon the collection of accurate data regarding the decisions and actions that contributed to the incident(s) under analysis. This involves collecting data regarding the system and incident in question. Three broad forms of data are required:

1. Data on the system in terms of who resides in the system and shares the responsibility for safety during the work activities or processes in which the accident occurred;
2. Data on the work activities or processes in which the accident occurred; and
3. Data on the accident itself and any contributory factors that played a role in its occurrence.

Data collection for AcciMap can involve a range of activities, including interviews with those involved in the incident or SMEs for the domain, work, or type of incident under analysis, reviewing reports or inquiries into the incident, and observing recordings of the incident. AcciMap analyses can be based on pre-existing data such as investigation reports (e.g. Salmon et al., 2014a; Newnam et al., 2017), incident reporting and learning system data (e.g. Salmon et al., 2017), or data collected for the purpose of the analysis. This process is typically iterative. Ideally, multiple sources, such as interviews, focus groups, surveys, observations of work activities and relevant documentation, are sourced to develop a draft analysis. The draft AcciMap can then be used to guide a deep understanding of the network of contributory factors involved.

One of the critical elements of developing an AcciMap is to understand the decisions and actions of multiple actors across the sociotechnical work system. This process is not only important for identifying contributory factors at each level of the system, but the relationships between factors within and across system levels. Ideally, interviews should include decision-makers from each system level; however, information about external influences (e.g. the regulatory and legislative context) can usually be gleaned from interviews with senior management and documentation.

The interviews with actors directly involved in the incident should focus on: 1) Developing an incident timeline representing the key decisions and actions; and 2) Understanding why decisions and actions made sense at the time (rather than focusing on errors and failures). The Critical Decision Method (Klein et al., 1989) provides a structured approach for collecting both types of information. Once contributory factors at the lower levels have been identified and represented on the framework, the analysis can be used to structure discussions with actors at the higher levels of the system.

STEP 3: REFINE HIERARCHICAL SYSTEM LEVELS

Before constructing the ActorMap and AcciMap it is first useful to refine the hierarchical system levels to fit the system in which the analysis is taking place. This might include adding additional levels, such as an international level (e.g. Salmon et al., 2021) or modifying the level descriptions to better fit the system under analysis. Table 9.1 presents examples of how the AcciMap framework has previously been adapted for different domains.

STEP 4. CONSTRUCT ACTORMAP

Once data collection is complete and the hierarchical system levels are agreed upon, the next step involves identifying all of the actors and organisations who operate within the system under analysis. These should be annotated onto an ActorMap showing where across the six hierarchical levels the different actors reside. It is also often useful to link actors to one another based on the communications structure of the system; however, this is not a requirement. An example ActorMap for work-related violence in hospital settings is presented in Figure 9.5.

STEP 5: MAP OUT THE FLOW OF EVENTS

When analysing individual events it is useful to map out the flow of proximal events from left to right on the physical processes and activities level of the framework. This is essentially an incident timeline representing key proximal decisions and actions. The descriptions should be short and factual, and based on what actually happened as opposed to what should have happened according to the procedures. If omissions or violations of procedures are important, they should be represented as contributory factors, and include an explanation of why they were omitted.

STEP 6: IDENTIFY CONTRIBUTORY FACTORS

Next the data should be analysed to identify contributory factors which played a causal role in the incident. This involves reviewing the data and recording any factors that the analyst feels played a contributory role in the incident in question. It is important

TABLE 9.1

Examples of Adapting the Levels of the AcciMap Framework to Different Domains

Svedung and Rasmussen (2002)	Waterson (2009)	Jenkins et al. (2010)	Le Coze (2010)	Newnam and Goode (2015)	Goode et al. (2017)	Salmon et al. (2021)
Generic	Public Health (Hospitals)	Policing	Manufacturing	Road Freight Transport	Outdoor Education	Work-related violence in hospital settings
–	–	–	Society, Market	–	–	International
Government Policy and Budgeting	Government	Government Policy and Budgeting	Government, Regulatory System	Government bodies	Government departments	Government
Regulatory Bodies and Associations	Regulatory	Regulatory Bodies and Associations	–	Regulatory bodies	Regulatory Bodies and Associations	External regulation, standardisation and coordination, industry and worker representation
Local Area Government, Company Management, Planning and Budgeting	Trust Governance	Strategic Command	Company	Other Organisations and Clients	Local Area Government, Parents and Schools, Activity centre management planning and budgeting	Healthcare management, Centres of Expertise, Incident response management, Media
Technical and Operational Management	Hospital Management	Tactical Command	Site Management	Heavy Vehicle Companies	Supervisory and management decisions and actions	Healthcare delivery specialists, Coordinators, Work Health and Safety (WHS) specialists
Physical Processes and Actor Activities	Clinical Management	Physical Processes and Actor Activities	Operational Management	Drivers and Other Actors at the Scene of the Incident	Decisions and actions of leaders, participants, and other actors at the scene of the incident	Actors directly involved in violence in hospital settings
Equipment and Surroundings	Equipment and Surroundings	Equipment and Surroundings	Shop Floor and Installations	Equipment, Surroundings, and Metrological Conditions	Equipment, Environment and Metrological Conditions	Hospital environment and related equipment

(Adapted from Goode et al., 2018).

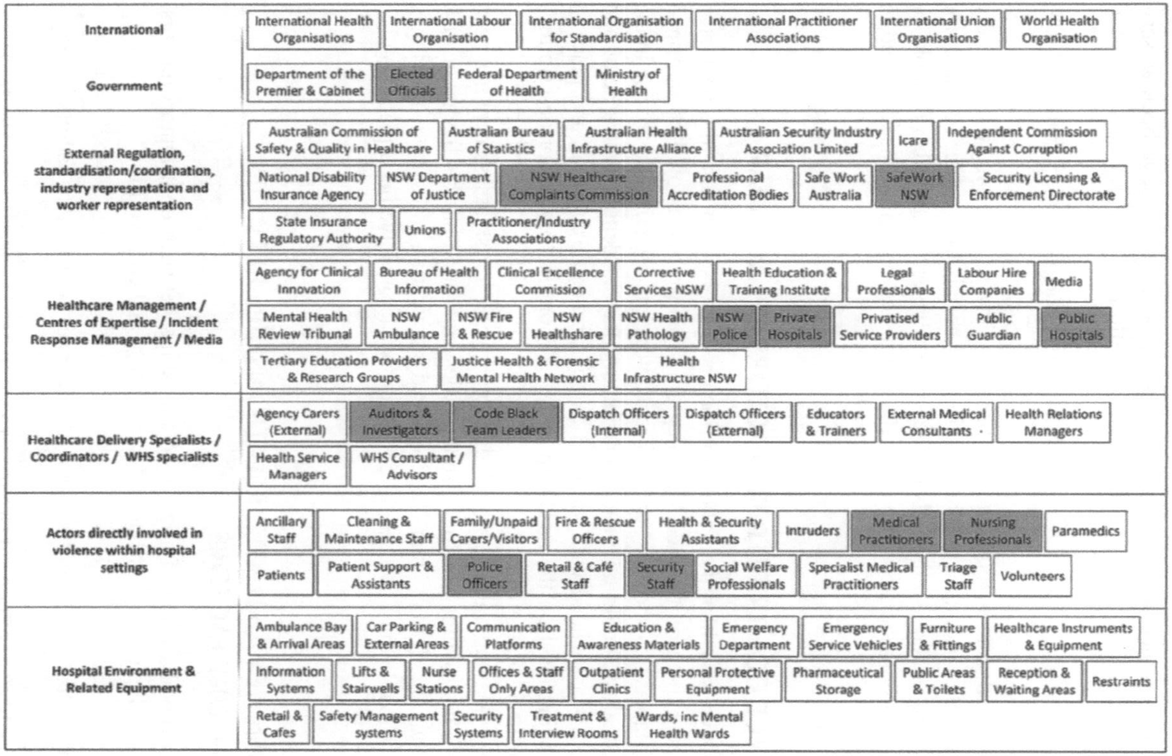

FIGURE 9.5 ActorMap for work-related violence in hospital settings in New South Wales. Shaded boxes indicate actors with formal decision-making authority to prevent or respond to a work-related violence incident.

Source: Salmon et al., 2021.

to clearly define what they consider to be a contributory factor. In our own analyses we have used the following definition which defines contributory factors as factors

> that, if it had not occurred or existed at the relevant time, then either the occurrence would probably not have occurred, adverse consequences associated with the occurrence would probably not have occurred or have been as serious, or another contributing safety factor would probably not have occurred or existed.
>
> (Australian Transport Safety Bureau, 2008)

It is recommended that analysts take as broad a view as possible when initially identifying contributory factors. A key requirement here is to consider or search for contributory factors across all levels of the system hierarchy and also to look for contributory factors associated with all of the actors and organisations identified in the ActorMap.

STEP 7: PLACE CONTRIBUTORY FACTORS ON ACCIMAP

Once the analyst has identified the contributory factors involved, the next step involves placing them on the AcciMap diagram. This process should be informed by the ActorMap as this shows where different actors and organisations reside in the system. The analyst should take each contributory factor, identify which actor and organisation it is associated with, develop an appropriate description, and place at the corresponding level on the AcciMap diagram. Contributory factor descriptions should use neutral language and try to avoid failure-based language where possible. For example, where the training provided around dynamic risk assessment has been identified as a contributory factor, a suitable description would be 'Dynamic risk assessment training' (rather than 'Inadequate dynamic risk assessment training'). This process should continue until all contributory factors have been placed on the AcciMap.

STEP 8: IDENTIFY AND ADD RELATIONSHIPS BETWEEN CONTRIBUTORY FACTORS

The next step involves identifying the relationships between contributory factors. This involves taking each contributory factor in turn and considering:

1. Whether it had an influence on any of the other contributory factors in the AcciMap; and
2. Whether it was influenced by any of the other contributory factors in the AcciMap.

A relationship should be specified if one contributory factor creates, influences, or occurs in sequence with another contributory factor. When a relationship is found the analyst draws a line to depict the relationship on the AcciMap, with an arrow showing the direction of influence. For example:

- A truck driver pulled out of a side road in front of a group of cyclists because he was not aware of how quickly the cyclists were travelling and thought it was safe to do so;

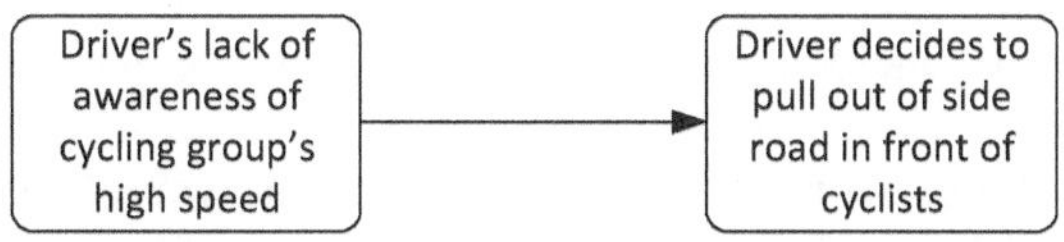

Drowning-related hazards are not identified and managed during a school camp water activity because the outdoor education provider's risk assessment process did not identify any hazards relating to the water activity;

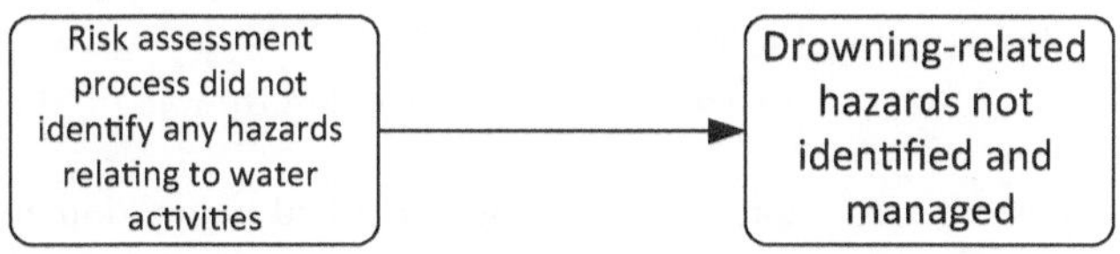

- The ships bow doors were left open because the assistant bosun (whose job it was to close them) was asleep in his cabin due to being fatigued from working a double shift.

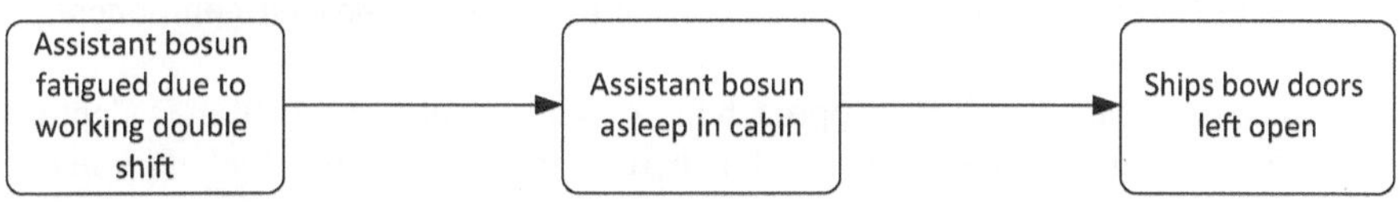

Identifying relationships between contributory factors also acts as a check on the contributory factors identified. If it is not possible to identify a relationship with other contributory factors, the analyst should consider whether the contributory factor should be included within the AcciMap or whether further contributory factors need to be identified.

STEP 9: FINALISE AND REVIEW ACCIMAP DIAGRAM

At this point, a draft AcciMap diagram has been developed. The analyst should next review the AcciMap and ensure that all contributory factors and relationships are included and that the descriptions are appropriate. It is useful during this step to return to the data and review it to verify the contributory factors and relationships identified. It is recommended that multiple reviews are undertaken by multiple analysts during this step.

STEP 10: SME REVIEW

The final stage of the process involves asking appropriate SMEs to review the final ActorMap and AcciMap diagrams. It is best practice to use SMEs who were either involved in the incident or who have extensive knowledge of the system and activities under analysis. The AcciMap should be updated and finalised based on the feedback provided by the SMEs.

ADVANTAGES

- AcciMap enables the identification of the network of contributory factors involved in the incident in question. The complete accident aetiology is exposed;
- AcciMap aligns with state-of-the-art accident causation models (e.g. Rasmussen, 1997) and at the time of writing is arguably the most commonly applied accident analysis method (Salmon et al., 2020);
- The method is simple to learn and use;
- It is possible to consider contributory factors across the overall system of work, up to an including decisions and actions at the regulatory, government, and international levels;
- The output offers an exhaustive and visual description of the accident aetiology;
- AcciMap is a generic approach and can be applied in any domain and any context;
- Using AcciMap prevents a blame approach and instead engenders a focus on systems failure;
- The output encourages analysts to identify safety interventions which tackle broader system issues as opposed to a fixing broken components approach;
- The ActorMap is a useful approach in its own right and provides an in-depth overview of the stakeholders who shared the responsibility for safety and accident causation in a particular system; and
- The approach is extremely powerful when used to look at multiple incidents (e.g. as part of an incident reporting and learning system) as it can be used to identify trends and recurring systemic issues (McLean et al., 2021; Salmon et al., 2017).

DISADVANTAGES

- AcciMap can be time consuming to apply particularly for large and complex incidents;
- The identification of relationships between contributory factors is difficult and analysts have been found to perform poorly in reliability and validity studies (e.g. Hulme et al., 2021a);
- The quality of the analysis produced is entirely dependent upon the quality of the data collected;
- AcciMap does not provide a structured taxonomy or classification scheme to support the classification of contributory factors. This raises concerns regarding reliability when used by different analysts;

- AcciMap does not provide a method to identify and develop corrective measures; these are based on the judgement of the analyst. Goode et al. (2016) describe the PreventiMap approach which can be used to support the development of safety interventions following the use of AcciMap; and
- Its graphical output can become complex and hard to decipher when used to analyse large and complex incidents.

RELATED METHODS

AcciMap involves the use of various data collection methods such as structured or semi-structured interviews (e.g. the Critical Decision Method), direct observation, walkthrough analysis and documentation review (e.g. incident reports, standard operating procedures). The PreventiMap method was developed to work in conjunction with AcciMap and is used to identify and depict the network of interventions that is required to respond to critical safety issues (Goode et al., 2016). PreventiMaps use the same hierarchical structure and levels as ActorMap and AcciMap and show what interventions are required at each level of the system in question to prevent or manage a particular safety issue. The interventions are linked in a network showing how interventions at one level can support those at other levels. A generic PreventiMap is presented in Figure 9.6.

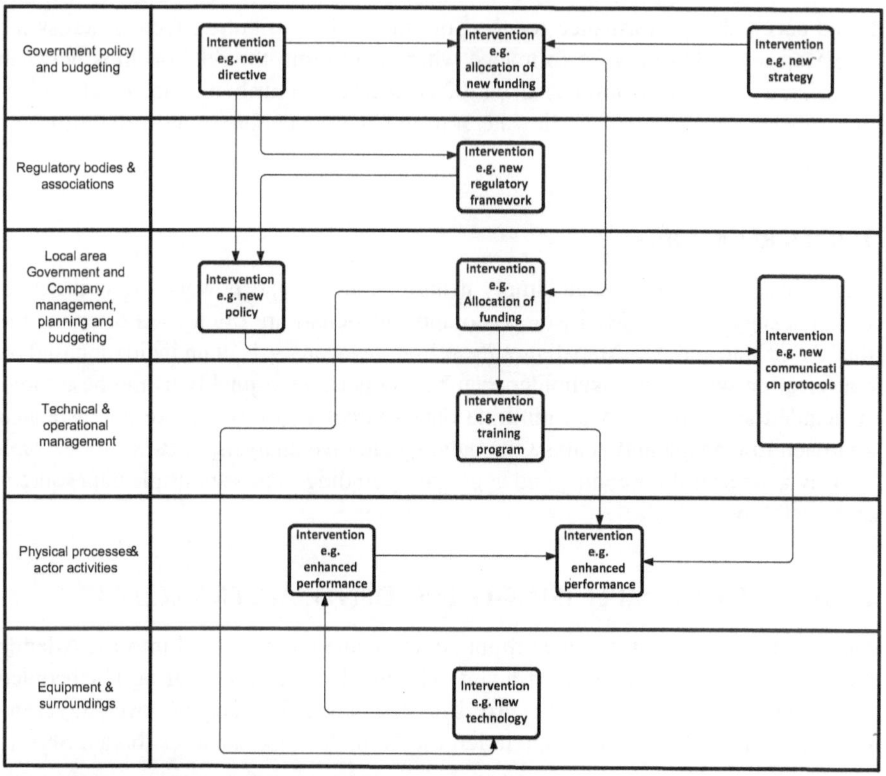

FIGURE 9.6 Example PreventiMap.

APPROXIMATE TRAINING AND APPLICATION TIMES

AcciMap is a simple method to learn and apply, with half or full day training workshop normally sufficient for trainees to pick up the basic concepts and approach. Application time is dependent on the incident(s) under analysis and the method can become time consuming when applied to large-scale complex incidents. For example, an analysis of a workplace injury can be produced during a 1–2-hour focus group with staff, and refined over a couple of hours. For more complex incidents timescales are expected to be around one to two weeks for data collection and a further week for the initial construction of the AcciMap.

RELIABILITY AND VALIDITY

The reliability and validity of AcciMaps has been questioned by a number of authors due to the relatively unconstrained manner in which AcciMaps are constructed as well as the lack of a contributory factor classification scheme (e.g. Johnson and de Almeida, 2008; Branford, 2007; Branford et al., 2009; Salmon et al., 2012; Waterson and Jenkins, 2010, 2011; Waterson et al., 2017). As a result, differences in both the way actual contributory factors are identified and described are likely to emerge across different analyses. Hulme et al. (2021b) compared the criterion-referenced validity of AcciMap, STAMP-CAST (Chapter 11), and AcciNet (Chapter 10) and found acceptable performance for the placement of contributory factors across the AcciMap levels but poor performance when identifying the relationships between contributory factors. Salmon et al.'s (2020) AcciMap contributory factor classification scheme can be used to alleviate some of the concerns around reliability and validity.

TOOLS REQUIRED

AcciMaps can be constructed simply using pen and paper; however, drawing software packages such as Microsoft Visio and Lucidchart are often used to create the final ActorMap and AcciMap diagrams. A high-level analysis of an incident based on conversations with key stakeholders can be produced very quickly using these tools. A detailed analysis based on multiple data sources requires a more sophisticated approach to manage and analyse the data. Qualitative analysis software tools, such as Nvivo, are useful for coding and aggregating findings across multiple data sources and AcciMaps.

CASE STUDY EXAMPLE (SINGLE INCIDENT): AIR FRANCE 447

On the 31 May 2009 Air France flight AF447 stalled and crashed into the Atlantic Ocean, killing all 228 people on board. The incident occurred during a scheduled passenger flight from Rio de Janeiro, Brazil, to Paris, France. Just over three and half hours after take-off, the Captain left the flight deck to take a rest break. Shortly afterwards, the Pilot Flying (PF) noted that the plane had entered the Intertropical Convergence Zone (TCZ), an area close to the equator where Norther and Southern Hemisphere winds come together and generate severe weather. The PF called through to a flight attendant to warn of likely turbulence and the need to take care. The aircraft's anti-icing system was then turned on.

Upon entry into the TCZ, the aircraft's pitot tubes froze due to the air temperature. An alert sounded in the cockpit to notify the pilots that the autopilot was disconnecting and was handing control back to them. At 2.10.06am, the PF remarked, "I have the controls" and subsequently put the aeroplane into a steep climb by pulling back on his sidestick, triggering a stall warning which went on to sound 75 times. The plane gained altitude rapidly but lost speed quickly, with the PF continuing to apply nose up inputs whilst the Pilot Not Flying (PNF) was ostensibly unaware of this. The aeroplane went into a stall and began to lose altitude. Concerned about the unfolding situation, the PNF called the Captain back into the cockpit. At 2.11.32am, the PF announced "I don't have control of the plane". At 2.11.37am the PNF took control of the aeroplane and six seconds later the Captain returned to the cockpit, asked "What the hell are you doing?" and began to attempt to diagnose the situation. The PF and PNF informed the Captain that they had lost control of the aircraft and did not understand why. At 2.13.40am the PF remarked that he had "had the stick back the whole time", at which point the PNF applied nose down inputs in an attempt to prevent the stall and gain speed. Unfortunately, these actions were taken too late, and at 2.14.29am the voice recorder stopped as the plane crashed into the ocean.

The Bureau d'Enquetes et d'Analyses (BEA) pour la sécurité de l'aviation civile (2012) investigation report placed significant emphasis on the aircrew's inability to understand and respond to the situation that arose following the autopilot disconnection. Both the PF and PNF were seemingly unaware that the plane had entered a stall and was descending rapidly towards the ocean, with the BEA report stating that the crew failed to make a link between the loss of indicated airspeeds and the appropriate procedure, made a late identification of the deviation from flight path, and failed to identify the approach to stall and subsequent stall situation itself.

As part of a study examining the generic features of accident causation across domains, Hulme et al. (2021a) developed an AcciMap for the incident based on the BEA's investigation report. The process involved reviewing the investigation report and extracting contributory factors and the relationships between them. The resulting AcciMap is presented in Figure 9.7.

CASE STUDY EXAMPLE (MULTIPLE INCIDENTS): INJURY CAUSATION IN LED OUTDOOR ACTIVITIES

The Understanding and Preventing Led Outdoor Accidents Data System (UPLOADS) incident reporting system (Goode et al., 2018; McLean et al., 2021; Salmon et al., 2017) was developed based on a major program of research which aimed to develop and implement an incident reporting and learning system and

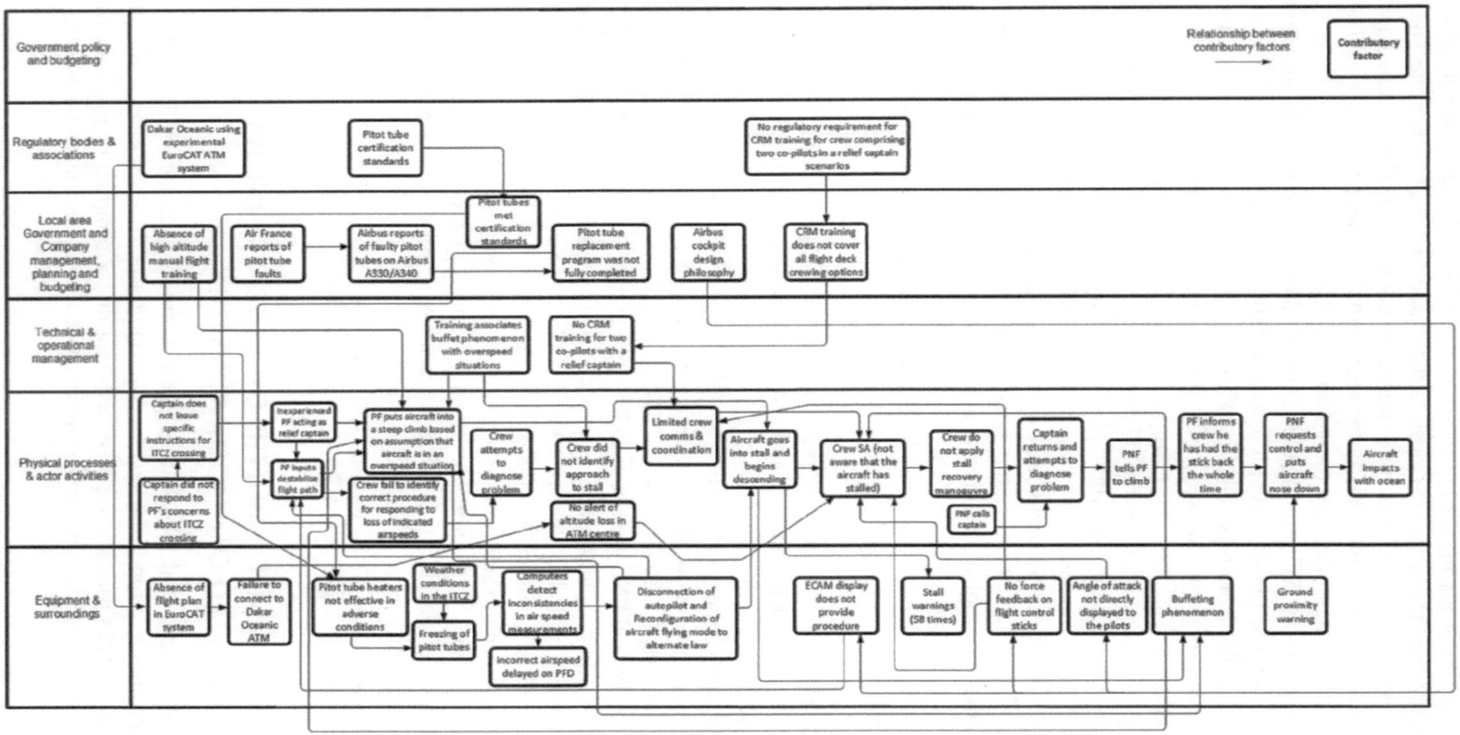

FIGURE 9.7 Air France AcciMap.

national incident dataset underpinned by contemporary models of accident causation (Salmon et al., 2017). UPLOADS provides led outdoor activity providers with a means of reporting and analysing injury and near miss incidents that occur during led outdoor activity programs (e.g. injuries, illnesses, equipment, environmental damage, and psychological impacts).

UPLOADS is based on Rasmussen's risk management framework and uses the AcciMap method (Svedung and Rasmussen, 2002) to analyse reported incidents. Specifically, the reported data is used to produce AcciMaps for single or multiple incidents or for different activities (e.g. Kayaking incidents versus Hiking incidents) and timescales (e.g. all incidents in 2019 versus all incidents in 2018). This allows users to investigate incidents from a systems perspective and identify emergent trends and recurrent issues. UPLOADS also enables users to contribute de-identified data to a national dataset. This goal of this is assist professional associations and government agencies to make evidence-based decisions about issues that affect those involved in the provision of led outdoor activities.

UPLOADS takes the form of a web-based app that enables users to enter detailed data regarding participation and incidents and near miss incidents, including data related to the activity, leaders and participants, injury outcome and severity, the nature of the incident, and the contributory factors involved. Once users enter an incident report, they then identify contributory factors and any relationships between them. UPLOADS supports this process by providing users with a classification scheme comprising contributory factor categories from across the led outdoor activity system.

The classification scheme (Figure 9.8) was developed based on an ActorMap depicting the key led outdoor activity actors, organisations and artefacts at each of Rasmussen's six system levels (modified for the led outdoor activity context), an analysis of led outdoor accident data (e.g. Salmon et al., 2010, 2014a), and SME review. Developing the contributory factor classification scheme involved identifying contributory factors associated with each of the actors, organisations, and artefacts listed within the ActorMap based on analyses of incident data and SME input. The prototype classification scheme was subsequently tested for reliability, validity and usability with end-users (Goode et al., 2014).

Example UPLOADS AcciMaps for 2125 led outdoor activity incidents reported between 14 September 2019 and 13 September 2020 are presented in Figures 9.9 and 9.10. Figure 9.9 includes a summary of the contributory factors identified in the 2125 reported incidents. Figure 9.10 depicts the relationships between the contributory factors identified in the 2125 reported incidents.

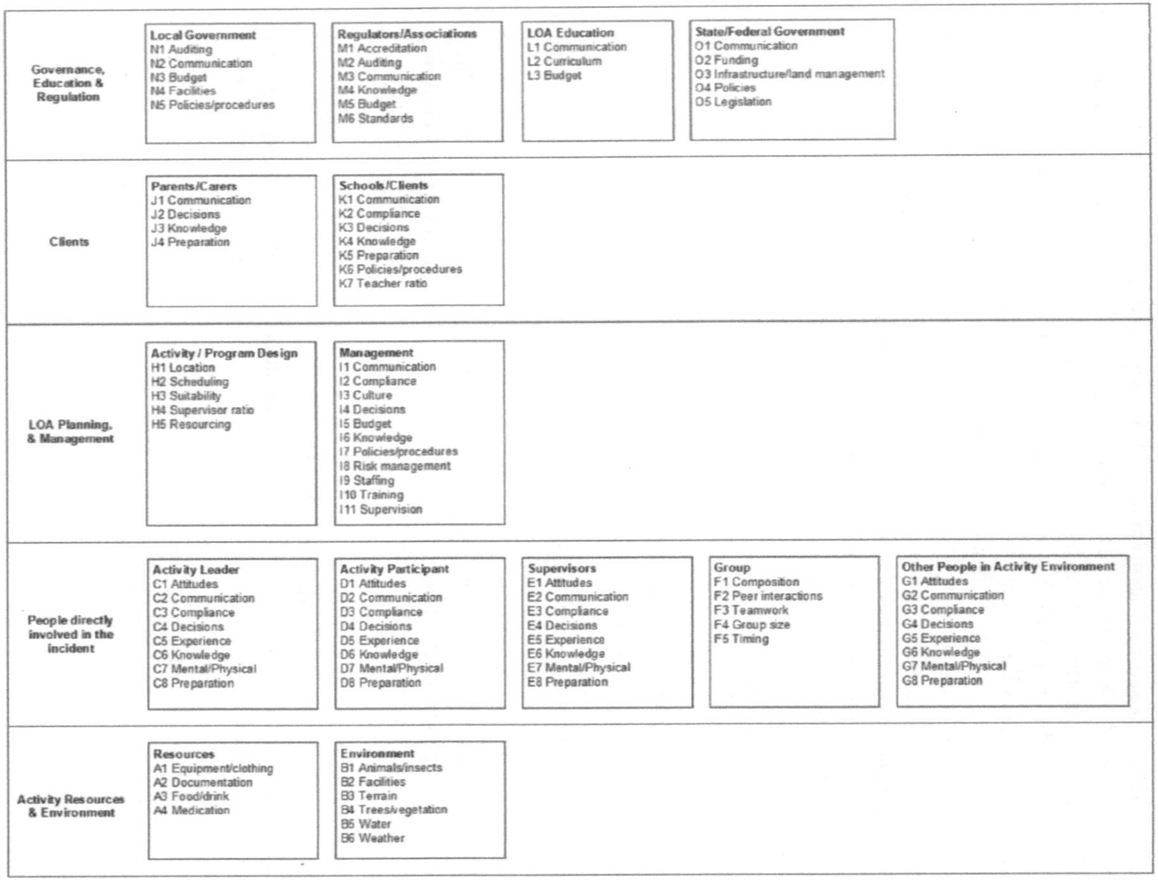

FIGURE 9.8 UPLOADS contributory factor classification scheme (LOA = Led Outdoor Activity).

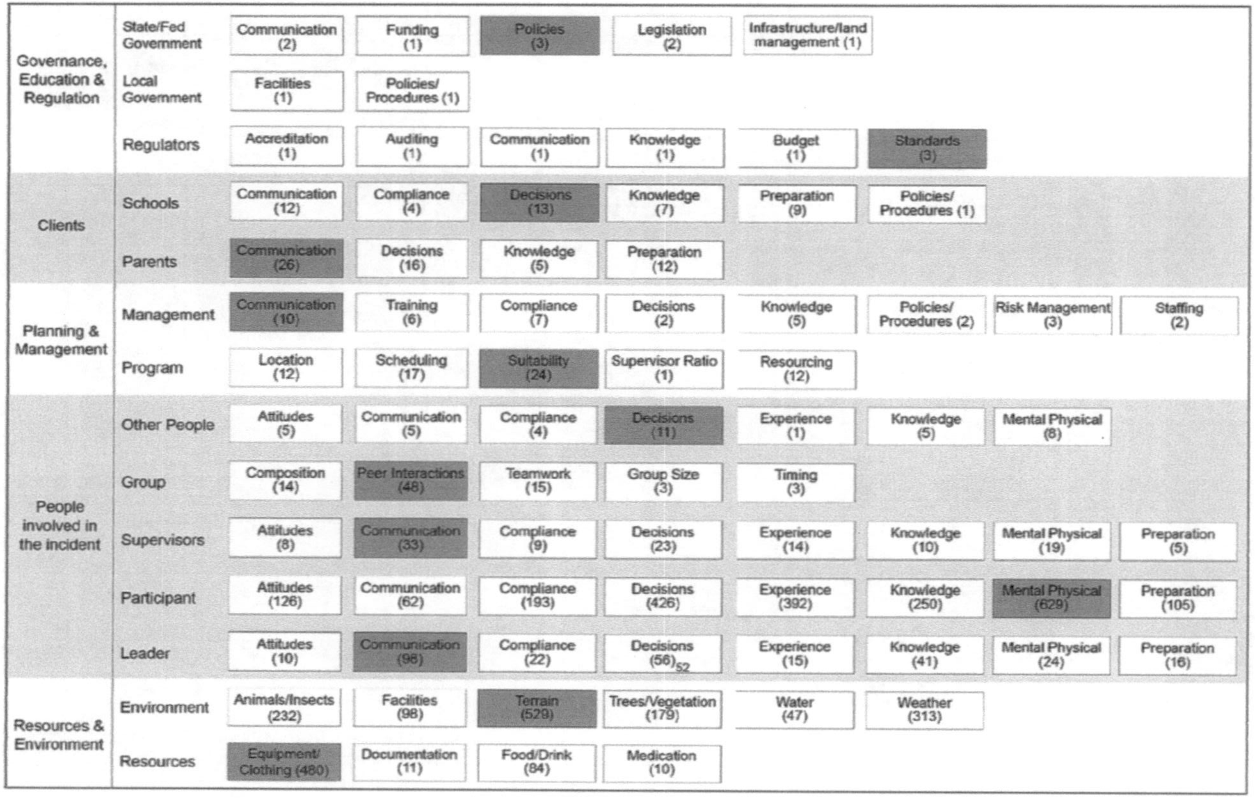

FIGURE 9.9 UPLOADS AcciMap showing the contributory factors involved 2125 reported incidents.

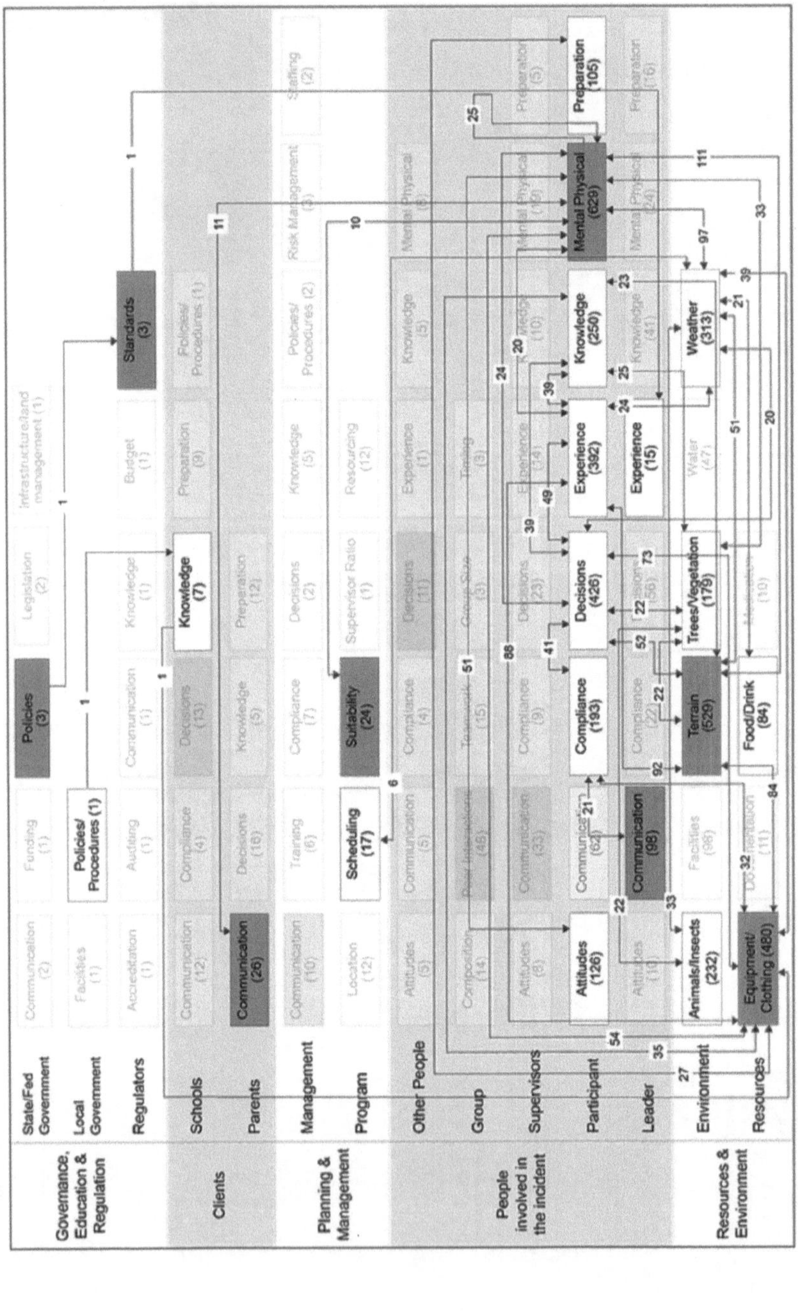

FIGURE 9.10 UPLOADS AcciMap showing the relationships between contributory factors in 2125 reported incidents.

RECOMMENDED READING

Hulme, A., Stanton, N. A., Waterson, P., Walker, G. H., & Salmon, P. M. (2019). What do applications of systems thinking accident analysis methods tell us about accident causation? A systematic review of applications between 1990 and 2018. *Safety Science* 117, 164–183.

Rasmussen, J. (1997). Risk management in a dynamic society: A modelling problem. *Safety Science* 27(2/3), 183–213.

Salmon, P. M., Hulme, A., Walker, G.H., Berber, E., Waterson, P., & Stanton, N. A. (2020). The big picture on accident causation: A review, synthesis and meta-analysis of AcciMap studies. *Safety Science* 126, 1–15.

Svedung, I., & Rasmussen, J. (2002). Graphic representation of accident scenarios: Mapping system structure and the causation of accidents. *Safety Science*.

Waterson, P.E., Jenkins, D.P., Salmon, P.M., & Underwood, P. (2017). 'Remixing Rasmussen': The evolution of AcciMaps within systemic accident analysis. *Applied Ergonomics* 59:B, 489–503.

REFERENCES

Akyuz, E. (2015). A hybrid accident analysis method to assess potential navigational contingencies: The case of ship grounding. *Safety Science* 79, 268–276. doi:10.1016/j.ssci.2015.06.019

Australian Transport Safety Bureau. (2008). *Analysis, Causality and Proof in Safety Investigations. Aviation Research and Analysis Report – AR-2007-053*. Canberra: Australian Transport Safety Bureau.

Branford, K. (2007), An investigation into the validity and reliability of the AcciMap approach. Unpublished doctoral dissertation, Australian National University, Canberra, Australia.

Branford, K., Naikar, N., & Hopkins, A. (2009), Guidelines for AcciMap analysis. In A. Hopkins (Ed.), *Learning from High Reliability Organisations*. Sydney: CCH.

Bureau d'Enquêtes et d'Analyses (BEA) pour la sécurité de l'aviation civile. (2012). Final Report On the accident on 1st June 2009 to the Airbus A330-203 registered F-GZCP operated by Air France flight AF 447 Rio de Janeiro – Paris. http://www.bea.aero/docspa/2009/f-cp090601.en/pdf/f-cp090601.en.pdf

Cassano-Piche, A. L., Vicente, K. J., & Jamieson, G. A. (2009). A test of Rasmussen's risk management framework in the food safety domain: BSE in the UK. *Theoretical Issues in Ergonomics Science* 10(4), 283–304.

Chen, L, Zhao, Y., & Zhao, Z. (2015). An AcciMap analysis on the China-Yongwen railway accident. In P.W. Tse et al. (eds.), *Engineering Asset Management - Systems, Professional Practices and Certification*, Springer: Lecture Notes in Mechanical Engineering.

Goode, N., Read, G. J., van Mulken, M. R., Clacy, A., & Salmon, P. M. (2016). Designing system reforms: Using a systems approach to translate incident analyses into prevention strategies. *Frontiers in Psychology* 7, 1974.

Goode, N., Salmon, P. M., & Lenné, M. G., Finch, C. (2018). *Bridging the Gap Between Accident Prevention Theory and Practice*. Boca Raton, FL: CRC Press.

Goode, N., Salmon, P. M., Lenné, M. G., & Hillard, P. (2014). Systems thinking applied to safety during manual handling tasks in the transport and storage industry. *Accident Analysis & Prevention* 68, 181–191.

Goode, N., Salmon, P. M., Taylor, N. Z., Lenné, M. G., & Finch, C. F. (2017). Developing a contributing factor classification scheme for Rasmussen's AcciMap: Reliability and validity evaluation. *Applied Ergonomics* 64, 14–26.

Hulme, A., Stanton, N. A., Walker, G. H., Waterson, P., & Salmon, P. M. (2021a). Complexity theory in accident causation: Using AcciMap to identify the systems thinking tenets in 11 catastrophes. *Ergonomics*, 64(7), 821–838.

Hulme, A., Stanton, N. A., Walker, G. H., Waterson, P., & Salmon, P. M. (2021b). Are accident analysis methods fit for purpose? Testing the criterion-referenced validity of AcciMap, STAMP-CAST and AcciNet. *Safety Science*. Accepted for publication 16th August 2021.

Hulme, A., Stanton, N. A., Waterson, P., Walker, G. H., & Salmon, P. M. (2019). What do applications of systems thinking accident analysis methods tell us about accident causation? A systematic review of applications between 1990 and 2018. *Safety Science* 117, 164–183.

Jenkins, D. P., Salmon, P. M., Stanton, N. A., & Walker, G. H. (2010). A systemic approach to accident analysis: A case study of the Stockwell shooting. *Ergonomics* 53, 1–17.

Johnson, C. W. & de Almeida, I. M. (2008). An investigation into the loss of the Brazilian space program's launch vehicle VLS-1 V03. *Safety Science* 46(1), 38–53.

Kee, D., Jun, G.T., Waterson, P., & Haslam, R. (2017). A systemic analysis of South Korea Sewol ferry accident – Striking a balance between learning and accountability. *Applied Ergonomics* 59, 504–516. doi:10.1016/j.apergo.2016.07.014.

Klein, G. A., Calderwood, R., & Macgregor, D. (1989). Critical decision method for eliciting knowledge. *IEEE Transactions on Systems, Man, and Cybernetics*, 19(3), 462–472.

Lane, B., Read, G. J. M., Salmon, P. M. (2020). A systems-thinking perspective on the barriers to treatment access for people with eating disorders. *International Journal of Eating Disorders* 53(2), 174–179.

Le Coze, J. C. (2010). Accident in a French dynamite factory: An example of an organisational investigation. *Safety Science* 48(1), 80–90.

Le Coze, J. C. (2015). Reflecting on Jens Rasmussen's legacy: A strong program for a hard problem. *Safety Science* 71, 123–141.

Lee, S., Moh, Y.B., Tabibzadeh, M., & Meshkati, N. (2017). Applying the AcciMap methodology to investigate the tragic Sewol Ferry accident in South Korea. *Applied Ergonomics*, 59, B517–525. doi:10.1016/j.apergo.2016.07.013.

Lintern, G. (2019). Jens Rasmussen's risk management framework. *Theoretical Issues in Ergonomics Science* 1–29.

McLean, S., Finch, C. F., Goode, N., Clacy, A., Coventon, L. J., & Salmon, P. M. (2021). Applying a systems thinking lens to injury causation in the outdoors: Evidence collected during 3 years of the Understanding and Preventing Led Outdoor Accidents Data System. *Injury Prevention*, 27(1), 48–54.

Nayak, R., & Waterson, P. (2016). 'When Food Kills': A socio-technical systems analysis of the UK Pennington 1996 and 2005 E. coli O157 Outbreak reports. *Safety Science*, 86, 36–47.

Newnam, S. & Goode, N. (2015). Do not blame the driver: A systems analysis of the causes of road freight crashes. *Accident Analysis & Prevention* 76, 141–151.

Newnam, S., Goode, N., Salmon, P., & Stevenson, M. (2017). Reforming the road freight transportation system using systems thinking: An investigation of Coronial inquests in Australia. *Accident Analysis & Prevention*, 101, 28–36.

Rasmussen, J. (1997). Risk management in a dynamic society: A modelling problem. *Safety Science* 27, 2/3, 183–213.

Salmon, P. M., Cornelissen, M., & Trotter, M. (2012). Systems-based accident analysis methods: A comparison of AcciMap, HFACS, and STAMP. *Safety Science* 50(4), 1158–1170.

Salmon, P. M., Coventon, L., & Read, G. J. M. (2021). Understanding and preventing work-related violence in hospital settings: A systems thinking approach. *SafeWork New South Wales Final Report*, https://www.safework.nsw.gov.au/__data/assets/pdf_file/0006/964716/understanding-and-preventing-work-related-violence-in-hospital-settings-a-systems-thinking-approach.pdf

Salmon, P. M., Goode, N., Archer, F., Spencer, C., McArdle, D., & McClure, R. J. (2014a). A systems approach to examining disaster response: Using AcciMap to describe the factors influencing bushfire response. *Safety Science*, 70, 114–122.

Salmon, P. M., Goode, N., Lenné, M. G., Finch, C. F., & Cassell, E. (2014b). Injury causation in the great outdoors: A systems analysis of led outdoor activity injury incidents. *Accident Analysis & Prevention*, 63, 111–120.

Salmon, P. M., Goode, N., Taylor, N., Dallat, C., Finch, C., & Lenné, M. G. (2017). Rasmussen's legacy in the great outdoors: A new incident reporting and learning system for led outdoor activities. *Applied Ergonomics*, 59(Part B), 637–648.

Salmon, P. M., Hulme, A., Walker, G. H., Berber, E., Waterson, P., & Stanton, N. A. (2020). The big picture on accident causation: A review, synthesis and meta-analysis of AcciMap studies. *Safety Science*, 126, 1–15.

Salmon, P.M., Read, G.M., Stanton, N.A., & Lenné, M.G. (2013). The crash at Kerang: Investigating systemic and psychological factors leading to unintentional non-compliance at rail level crossings. *Accident Analysis and Prevention* 50, 1278–1288. doi:10.1016/j.aap.2012.09.029.

Salmon, P.M., Williamson, A., Lenné, M., Mitsopoulos-Rubens, E., & Rudin-Brown, C.M. (2010). Systems-based accident analysis in the led outdoor activity domain: Application and evaluation of a risk management framework. *Ergonomics* 53(8), 927–939. doi:10.1080/00140139.2010.489966.

Stevens, N.J., & Salmon, P.M. (2016). Sand, surf and sideways: A systems analysis of beaches as complex roadway environments. *Safety Science* 85, 152–162. doi:10.1016/j.ssci.2016.01.009.

Svedung, I., & Rasmussen, J. (2002). Graphic representation of accident scenarios: Mapping system structure and the causation of accidents. *Safety Science*.

Thoroman, B., Goode, N., Salmon, P., & Wooley, M. (2019). What went right? An analysis of the protective factors in aviation near misses. *Ergonomics*, 62(2), 192–203.

Trotter, M. J., Salmon, P. M., & Lenné, M. G. (2014). Impromaps: Applying Rasmussen's risk management framework to improvisation incidents. *Safety Science*, 64, 60–70.

Underwood, P., & Waterson, P. (2014). Systems thinking, the Swiss Cheese Model and accident analysis: A comparative systemic analysis of the Grayrigg train derailment using the ATSB, AcciMap and STAMP models. *Accident Analysis and Prevention* 68, 75–94. doi:10.1016/j.aap.2013.07.027.

Vicente, K. J., & Christoffersen, K. (2006). The Walkerton E. coli outbreak: A test of Rasmussen's framework for risk management in a dynamic society. *Theoretical Issues in Ergonomics Science*, 7(02), 93–112.

Waterson, P. (2009). A systems ergonomics analysis of the Maidstone and Tunbridge Wells infection outbreaks. *Ergonomics*, 52(10), 1196–1205.

Waterson, P., Jenkins, D. P., Salmon, P. M., & Underwood, P. (2017). 'Remixing Rasmussen': The evolution of AcciMaps within systemic accident analysis. *Applied Ergonomics*, 59, 483–503.

Waterson, P. E., & Jenkins, D. P. (2010). Methodological considerations in using AcciMaps and the Risk Management Framework to analyse large-scale systemic failures.

Waterson, P.E. & Jenkins, D.T. (2011). Lessons learnt from using AcciMaps and the Risk Management Framework to analyse large-scale systemic failures. In M. Anderson (Ed.) *Contemporary Ergonomics 2011*, pp. 6–13, London: Taylor & Francis Group.

Woo, D. M., & Vicente, K. J. (2003). Sociotechnical systems, risk management, and public health: Comparing the North Battleford and Walkerton outbreaks. *Reliability Engineering & System Safety*, 80(3), 253–269.

10 The Accident Network (AcciNet)

BACKGROUND AND APPLICATIONS

Of the existing systems-based accident analysis methods currently used by HFE and safety science practitioners, the AcciMap method (Rasmussen 1997; Svedung and Rasmussen 2002, Chapter 9), is one of the most popular in the peer reviewed literature (Hulme et al. 2019; Hulme et al. 2020; Salmon et al. 2020a). There are, however, known limitations with the AcciMap method. These include the difficulty associated with identifying relationships between contributory factors; an inability to show the interaction between 'normal performance' and failure; an inability to formally represent the ownership of contributory factors; and, the requirement to force fit contributory factors into Rasmussen's (1997) six hierarchical system levels (i.e. the Risk Management Framework; RMF) (Hulme et al. 2020). Previous studies have also identified the lack of a contributory factor classification scheme as a key barrier preventing its use as part of an incident reporting and learning system (Salmon et al. 2012; Goode et al. 2018; Stanton et al. 2019; Salmon et al. 2020a). Finally, there is no proactive risk assessment method associated with AcciMap, and so it is difficult to achieve consistency between risk assessment and accident analysis outputs. In other words, it is challenging to feed accident analysis findings from AcciMap back into risk assessment processes to support organisational learning and the enhancement of risk controls.

Removing the above limitations would enhance the capability of AcciMap to support learning and prevent or manage future incidents. A failure to consider the interaction between work as imagined, normal performance (or work as done) and failures has been identified as a significant issue impacting our capacity to learn from incidents. Dekker (2011), for example, argues that the seeds for failure are often found in the "normal, day-to-day processes" (p. 99) influenced by everyday goal conflicts and production pressures. Normal performance thus includes workarounds, improvisations, and adaptations, but also normal work behaviours routinely undertaken to get the job done (Salmon et al., 2017). Classifying such behaviours as failures is misleading and inhibits learning, yet most accident analysis methods do exactly that. As a result, the role of normal performance in accident causation remains unclear.

AcciNet (Salmon et al., 2020c) is a new systems theory-based accident analysis method that was developed to address the known methodological limitations associated with existing incident analyses methods. AcciNet uses a task network and associated classification scheme to identify the network of contributory factors, including normal performance, involved in accident events. Specifically, AcciNet uses the

DOI: 10.1201/9780429281624-10

same task network as Net-HARMS (Chapter 2) to identify contributory factors and describe incident aetiology within complex systems. From a theoretical standpoint, AcciNet is based on the following three fundamental tenets of accident causation (Dekker, 2011; Rasmussen, 1997; Leveson, 2004):

1. That all accidents are created by an interacting network of behaviours associated with multiple actors, both human and non-human, who reside across the sociotechnical system of interest (Rasmussen, 1997);
2. That the interacting network of contributory factors involved in accidents includes 'work as imagined' (i.e. undertaken in line with procedures), 'normal performance' (i.e. 'work as done' where performance was appropriate and no discernible failure occurred), and decisions and actions whereby performance can reasonably be classified as sub-optimal (Dekker, 2011; Hollnagel, 2012); and
3. That emergent risks play a critical role in accident causation. Emergent risks occur when multiple behaviours interact with one another to create unexpected and difficult to foresee behaviours. These emergent behaviours occur across the sociotechnical system and may be proximal or distal to the accident event.

AcciNet was developed to support analysts in identifying and depicting the three tenets of accident causation described above. The method uses a task network and actor network (ActorNet) for the system under analysis as its primary input, with analysts subsequently using both networks to identify contributory factors, their interrelations, and their associated actors. A classification scheme (Dallat et al., 2018; Salmon et al., 2020a) can then be used to classify contributory factors further if needed; however, this is included only as an optional step if required as part of the investigation or analysis aims.

AcciNet outputs include a task network showing the interaction between tasks and contributory factors and a table describing each contributory factor and potential safety interventions. A key feature of the AcciNet method is the use of the task network to identify contributory factors which removes of the need for analysts to identify the relationships between contributory factors (as the relationships between tasks are already specified within the task network). A final notable feature of AcciNet is it was developed to be applied in conjunction with the Networked Hazard Analysis and Risk Management (NET-HARMS; Dallat et al., 2018) risk assessment method (see Chapter 2) whereby the task network supports both risk assessment and accident analysis for the same work system.

DOMAIN OF APPLICATION

AcciNet is a generic approach that is designed to be applied in all safety-critical domains, including construction, healthcare, manufacturing, mining, transportation, and general workplace safety. Initial test applications of the method occurred in road (Uber Volvo collision), rail (Kerang rail level crossing crash), and aviation (Lion Air Flight 610 crash).

PROCEDURE AND ADVICE

A flowchart depicting the AcciNet procedure is presented in Figure 10.1. Step-by-step guidance is presented below.

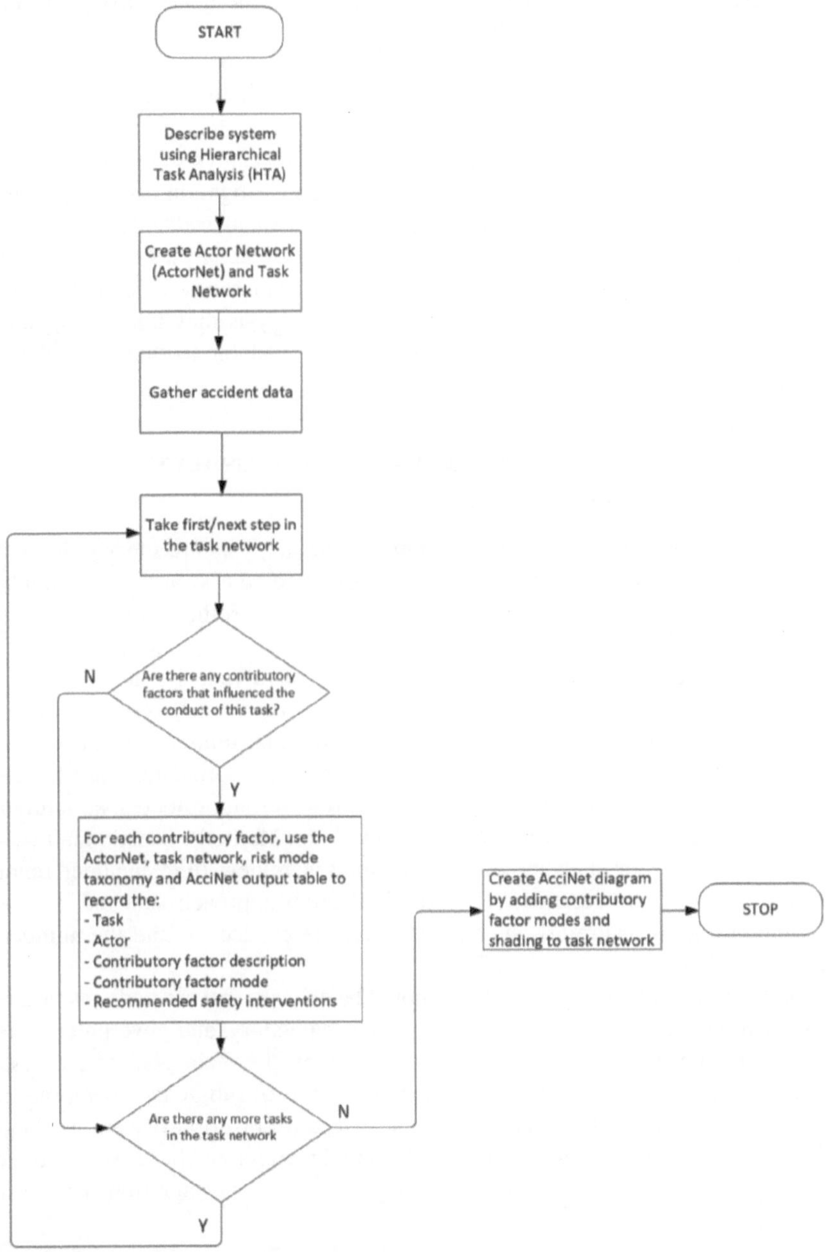

FIGURE 10.1 AcciNet procedure.

STEP 1: DEFINE AIMS OF THE ANALYSIS AND THE INCIDENT AND SYSTEM UNDER ANALYSIS

The first step in an applying AcciNet involves clearly defining the aims of the analysis, the incident to be analysed, and any analysis boundaries. For example, in the case study presented in this chapter the aim was to analyse the Uber Volvo collision to identify the contributory factors that interacted to create the incident. Whilst applications of AcciNET will always involve the analysis of an incident of some sort, the aims may differ in terms of whether the intention is to understand the contributory factors involved or whether it is to develop safety interventions designed to prevent similar incidents from occurring in future. Defining the analysis boundary is important as project constraints will dictate how deep the investigation can go in terms of the parts of the system considered and the incident to be analysed and how far back in time is required to address the study aims. Whilst AcciNet analyses typically include government and international activities, project and resource constraints may result in the boundary being set lower; for example the analysis may not go beyond the organisation involved if it is known that data is not available to support analysis of activities outside of the organisation.

STEP 2: CONSTRUCT A HIERARCHICAL TASK ANALYSIS FOR THE SYSTEM UNDER ANALYSIS

Once the incident under analysis and aims of the analysis are clearly defined, a Hierarchical Task Analysis (HTA; See Chapter 5) for the system under analysis should be created. Initially, this involves collecting specific data regarding the goals, tasks, and operations undertaken within the system under analysis. Data should be collected regarding the goals and tasks involved, the human and non-human agents involved, the interactions between humans and non-human agents, the ordering of tasks and information on the factors that influence behaviour, task performance, and task outcome. A number of different approaches can be used to collect this data, including observations, concurrent verbal protocols, structured or semi-structured interviews (e.g. the Critical Decision Method), questionnaires, and surveys, walkthrough analysis and documentation review (e.g. incident reports, standard operating procedures). The data collection approach selected is dependent upon project constraints, such as time, resources, access, and the number of analysts available.

A key focus of any AcciNet analysis should be on tasks that are undertaken across the system, including supervisory, management, regulatory, and government activities. Developing the HTA itself involves identifying the main goal of the system under analysis and then decomposing this into a series of sub-goals, operations, and plans. Once the initial draft HTA is complete it is useful to have various Subject Matter Experts (SMEs) review it. The HTA should then refined based on SME feedback. It is normal practice for the HTA to go through many iterations before it is finalised.

STEP 3: CREATE TASK NETWORK

Once the HTA is finalised, the next step involves constructing a task network to support the identification of task and emergent risks. Task networks (see Figures 10.2) are used to represent HTA outputs in the form of a network which shows key tasks and the relationships between them (Stanton et al., 2013; 2019). This enables analysts to understand the interactions and coupling that exists between tasks across the work system. Within Figure 10.2 the circular nodes represent tasks and the arrows linking the tasks represent relationships between tasks. Relationships between tasks are included in the task network when:

1. Tasks are undertaken sequentially (e.g. task three is undertaken after completion of task two);
2. Task are undertaken together (e.g. tasks one and two are concurrent);
3. The outcomes of one task influence the conduct of another (e.g. the outcomes of task four influence how task five is undertaken); or
4. The conduct of one task is dependent on completion of the other (e.g. task five cannot be undertaken until task four is completed).

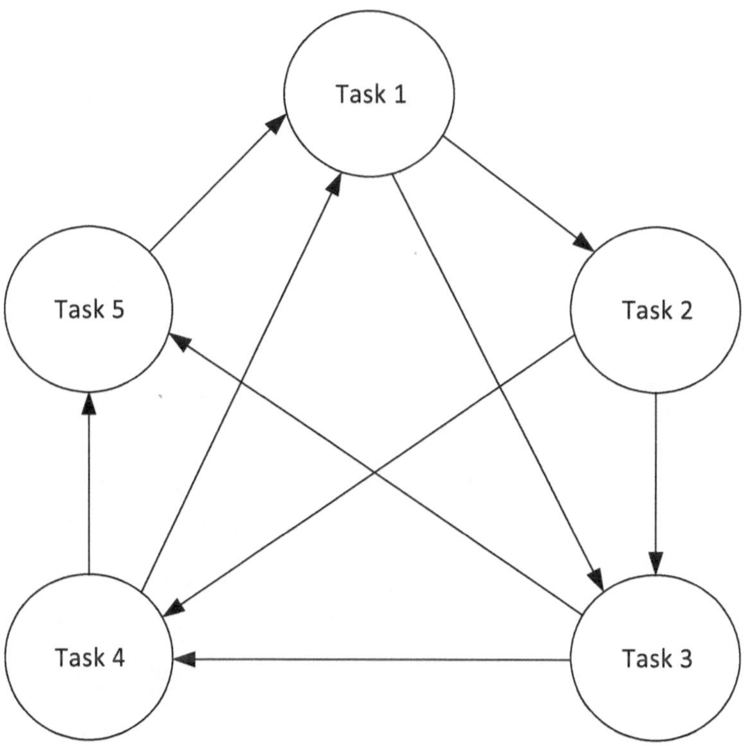

FIGURE 10.2 Task network example showing the relationships between five tasks.

Task networks are constructed by taking the first layer of sub-goals from the HTA and identifying which of them are related with one another based on relationship types described above. For example, in the Uber Volvo case study presented at the end of this document, the tasks 'Train vehicle operators' and 'Drive vehicle' are linked because the Uber test vehicle operators can only operate the test vehicle once they have completed the vehicle operator training program. The 'Conduct safe autonomous vehicle testing' task network is presented in Figure 10.3.

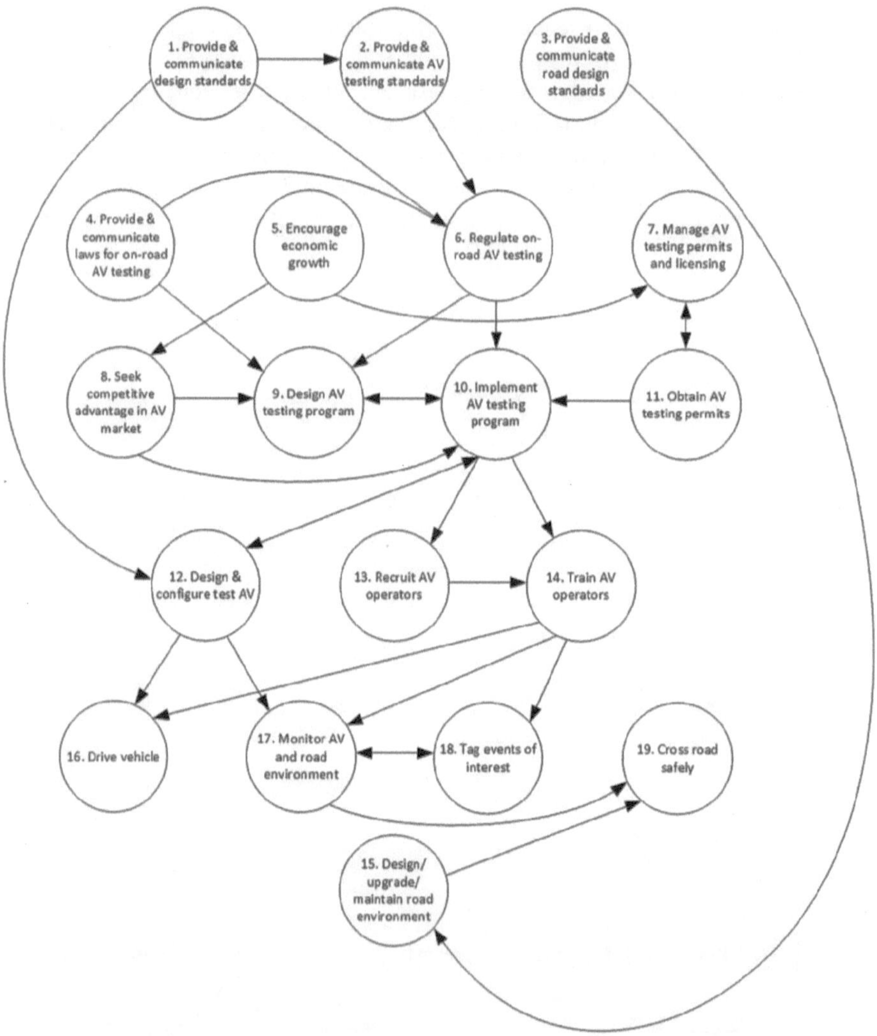

FIGURE 10.3 'Conduct safe autonomous vehicle testing' task network. Relationships in the task network represent instances where the conduct of one task influences, is undertaken in combination with, or is dependent on, another task.

The task network should then be refined based on the SMEs feedback. Changes at this stage typically involve the addition and/or removal of relationships between tasks as the tasks or the modification of task descriptions. Like HTA, it is normal practice for the task network to go through various iterations before it is finalised.

It should also be noted that whilst we recommend that the development of a task network should be based on HTA, it is possible to develop a task network based directly on raw data collection activities, SME input and verification, and document reviews (Stanton, 2014). This is often the case where project constraints do not enable development of an initial HTA. If the HTA step is indeed not utilised, then we recommend that analysts use SME review to ensure that the core tasks of interest are included in the task network.

STEP 4: GATHER DATA REGARDING INCIDENT UNDER ANALYSIS

AcciNet is dependent upon the provision of detailed and accurate data regarding the incident under analysis. The next step therefore involves collecting data regarding the incident in question as well as the system in which the incident occurred. Three broad forms of data are required:

1. Data on the work activities or processes which occur across the system in which the accident occurred;
2. Data on the incident itself and any contributory factors that played a role in its occurrence; and
3. Data on the system in terms of who resides in the system and shares the responsibility for safe completion of the work activities or processes relating to the incident.

Data collection for AcciNet can involve a range of activities, including interviews with those involved in the incident or SMEs for the domain, work, or type of incident under analysis, reviewing reports or inquiries into the incident, and observing recordings of the incident. It is important during data collection to consider all levels of the system in question. For example, if using interviews as the primary data collection approach it is useful to interview relevant actors from all levels of the system. Likewise, when reviewing procedures this should include the procedures for different actors across all levels of the system.

STEP 5: REVIEW AND REFINE TASK NETWORK

Once sufficient data regarding the incident and system in question is collected, the next step involves reviewing and refining the draft task network. This involves ensuring that the existing tasks and relationships are appropriate and also adding any new tasks and relationships identified during data collection activities.

STEP 6: DEVELOP ACTORNET (OPTIONAL)

The ActorNet shows the different people and organisations who share responsibility for the incident in question and specifically which tasks within the task network

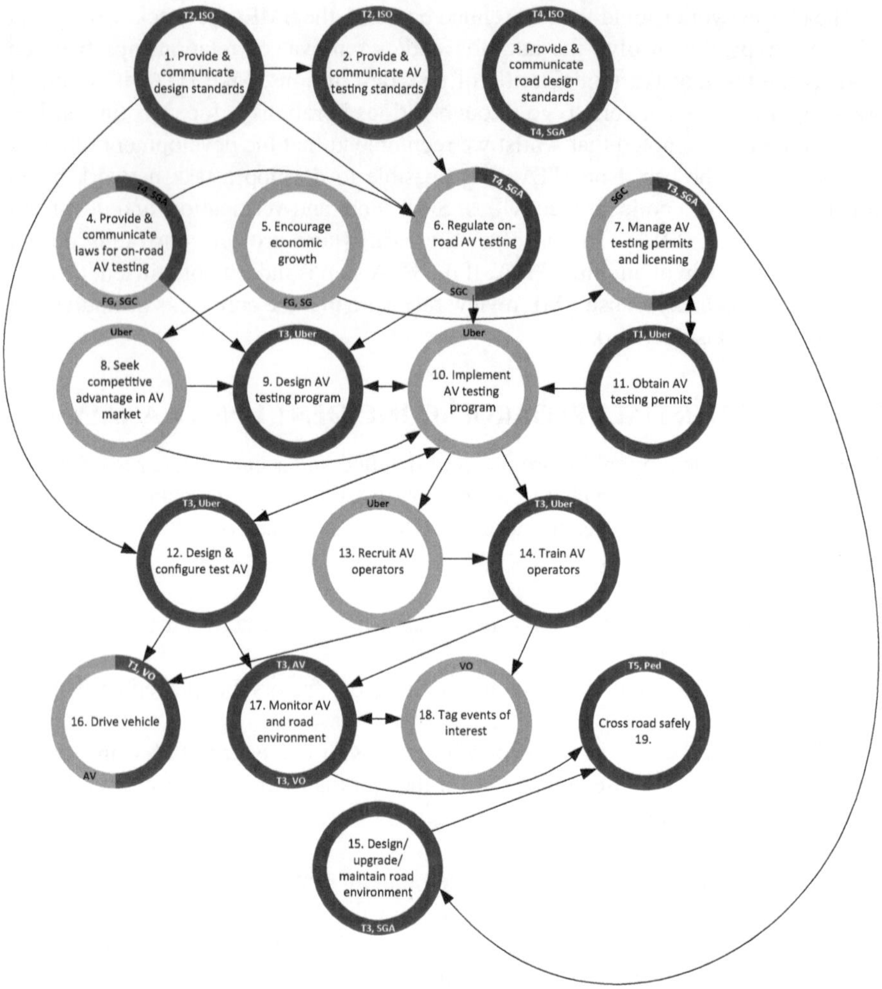

FIGURE 10.4 Uber Volvo AcciNet. Red shading represents failure; Green shading represents normal performance. ISO = International Standards Organisation; SGA = State Government Road Arizona; SGC = State Government California; FG = Federal Government; AV = Autonomous Vehicle; VSO = Vehicle safety operator; Ped = Pedestrian. (Adapted from Salmon et al., 2020b).

they are undertake. Once an in-depth understanding of the system and relevant tasks is achieved via the refined task network, an ActorNet should be constructed to show which actors are involved in completion of the tasks within the task network. This involves identifying all of the relevant actors who contribute to the tasks and then adding them to the task network. For the Uber Volvo case study, the following relevant actors were identified and mapped onto the task network (see Figure 10.4):

- Autonomous vehicle (AV);
- Vehicle Operator (VO);

- Pedestrian (Ped);
- Uber;
- State Government California (SGC);
- State Government Arizona (SGA);
- State GC;
- Federal Government (FD); and
- International Standards Organisation (ISO).

STEP 7: REVIEW TASKS AND IDENTIFY RELEVANT CONTRIBUTORY FACTORS

Next the data gathered during Step 4 should be reviewed in conjunction with the task network to identify relevant contributory factors that played a role in creating the incident. For the purposes of accident analysis, contributory factors can be defined as

> that, if it had not occurred or existed at the relevant time, then either the occurrence would probably not have occurred, adverse consequences associated with the occurrence would probably not have occurred or have been as serious, or another contributing safety factor would probably not have occurred or existed.
>
> (Australian Transport Safety Bureau, 2008)

It is recommended that analysts take as broad a view as possible when identifying contributory factors. A key requirement here is to consider or search for contributory factors across tasks within the task network, and also to look for contributory factors associated with all of the actors and organisations identified in the ActorNet. As such, the recommended process is to take each task in the task network and ask whether the conduct of this task or its outcomes play a role in creating the incident. If the answer is yes, then the analysts should identify the actors associated with the task and document the contributory factor in a table which includes a description of each task and its associated contributory factors. Here is also a good time to start to populate the AcciNet output table (Table 10.1). The AcciNet output table develops incrementally as

TABLE 10.1
AcciNet Output Table

Task	Actor(s)	Contributory Factor Description	Contributory Factor Classification	Recommended Safety Intervention
Insert from task network	Insert from ActorNet	Identify from data	See Step 8: use the risk modes taxonomy or classification scheme	e.g. SME group decision-making; Delphi approach
Insert				
Insert				

the analysis process unfolds (i.e. at this stage we are only up to column three – the classification scheme and safety interventions will be added in the subsequent phases below). A final consideration when identifying contributory factors is, again, to perform this step as a team, using multiple SMEs and practitioners from the organisation/system.

STEP 8: CLASSIFY CONTRIBUTORY FACTORS

Next, all identified contributory factors should be classified using an appropriate contributory factor classification scheme. Whilst early versions of the method used a taxonomy from the Net-HARMS method (Dallat et al., 2018, see Table 10.2), Salmon et al.'s (2020a) classification scheme (Table 10.3) can also be used. The process involves taking each contributory factor identified during Step 7 and classifying into a relevant contributory factor type from the classification scheme. It is normally useful to use multiple analysts for this process and discuss each contributory factor until consensus is achieved. In cases where only one analyst is used, it is useful to have a second analyst conduct a portion of the classification and compare levels of agreement between the two analysts.

STEP 9: OVERLAY CONTRIBUTORY FACTORS ONTO TASK NETWORK

The next step involves overlaying all identified contributory factors onto the task network. This involves adding contributory factors and actors to the relevant tasks in the task network. For example, in the Uber Volvo analysis presented in the case

TABLE 10.2
Net-HARMS Risk Mode Taxonomy

Classification	Risk Mode	Description
Task	T1: Task mistimed	Task is undertaken too early or too late within process
	T2: Task omitted	Task is not undertaken
	T3: Task completed inadequately	Task is undertaken but is completed in an inadequate manner
	T4: Inadequate task object	The object(s) used to complete the task are inadequate
	T5: Inappropriate task	An inappropriate task is performed instead of the required task
Communication	C1: Information not communicated	Information required to complete the task is not communicated
	C2: Wrong information communicated	The wrong information is communicated
	C3: Inadequate information communicated	Information is communicated but is inadequate (e.g. incomplete communication with missing information)
	C4: Communication mistimed	Communication is undertaken too early or too late within process
Environment	E1: Adverse environmental conditions	Adverse environmental conditions influence task performance

TABLE 10.3
Generic AcciNet Classification Scheme

Equipment, Environment and Surroundings	**Local Area Government, Planning and Budgeting & Company Management**
1. Animal, plant & biological hazards	44. Communication & coordination
2. Built environment & infrastructure	45. Compliance with procedures, violations & unsafe acts
3. Equipment, technology & resources	46. Culture
4. Information & data	47. Financial pressures
5. Noise & visibility	48. Judgement & decision-making
6. Other	49. Other
7. Physical & natural environment	50. Personnel management & recruitment
8. Time-related	51. Planning & preparation
9. Weather & climate	52. Policy & procedures
10. Work environment	53. Qualification, training, experience & competence
Physical processes and actor activities	54. Risk assessment & management
11. Accident event	55. Supervision
12. Activity, work & operations	56. Time-related
13. Adverse events	**Regulatory bodies and associations**
14. Communication & coordination	57. Audits & inspections
15. Compliance with procedures, violations & unsafe acts	58. Communication & coordination
16. Delayed discovery & response	59. Compliance with procedures, violations & unsafe acts
17. Equipment, technology & environment	60. Culture
18. Group & teamwork	61. Financial pressures
19. Judgement & decision-making	62. Judgement & decision-making
20. Other	63. Planning & preparation
21. Personnel management & workloads	64. Qualification, training, experience & competence
22. Physical & mental condition	65. Regulatory structures & services
23. Planning & preparation	66. Risk assessment & management
24. Qualification, training, experience & competence	67. Standards, policy & regulations
25. Risk assessment & management	68. Time-related
26. Situation awareness	69. Unclear roles & responsibilities
27. Supervision & leadership	**Government policy and budgeting**
28. Time-related	70. Action omitted & failure to act
29. Weather, climate & natural processes	71. Budget & finance
Technical and operational management	72. Communication & coordination
30. Communication & coordination	73. Culture
31. Compliance with procedures, violations & unsafe acts	74. Judgement & decision-making
32. Culture	75. Policy, legislation & regulation
33. Equipment & environmental design	76. Political structures & services
34. Financial pressures	77. Priorities
35. Judgement & decision-making	78. Qualification, training, experience & competence
36. Other	79. Supervision & enforcement
37. Personnel management & recruitment	
38. Planning & preparation	
39. Policy & procedures	
40. Qualification, training, experience & competence	
41. Risk assessment & management	
42. Supervision	
43. Time-related	

(Salmon et al., 2020a)

study example the contributory factor of the pedestrian failing to check the oncoming traffic before crossing the road was added to the task 'Cross road safely'. If a classification scheme has been used the relevant code should be added to the task network. For example, in Figure 10.2 all contributory factors include the relevant Net-HARMS risk mode (e.g. T1 = Task mistimed, T2 = Task omitted, T3 = Task completed inadequately).

Contributory factors and their associated actor should be shaded red. Tasks within the AcciNet which represent 'work as imagined' or 'normal performance' are shaded green to show that no failure sub-optimal performance, or inappropriate behaviours occurred. Relevant actors associated with each instance of work as imagined or normal performance are also added to the task network.

STEP 10: COMPLETE ACCINET CONTRIBUTORY FACTORS TABLE

The next step involves completing the AcciNet output table, ensuring all columns are complete (Table 10.1). The AcciNet output table is important for two main reasons. First, it supplements and qualitatively explains graphical outputs and figures, including the ActorNet and task network. Second, an important aspect to accident analysis is not the analysis of individual adverse incidents and loss events *per se*, but rather it is the capacity to generate patterns and identify themes and trends across multiple incidents. Aggregating codes from the taxonomy or classification scheme (Table 10.2 or Table 10.3) can indicate common failure types and tasks which support and inform learning.

STEP 11: SME REVIEW

It is important to have appropriate SMEs review the final AcciNet diagram and output table. It is best practice to use SMEs who were either involved in the incident or who have extensive knowledge of the system and work in question. The AcciNet and contributory factors table should be updated and finalised based on the feedback provided by the SMEs.

STEP 12: IDENTIFY APPROPRIATE SAFETY INTERVENTIONS

The final step in the AcciNet process (if required as part of the analysis) involves developing appropriate safety interventions designed to present similar future occurrences. It is recommended that a group involving the original analysts and appropriate SMEs discuss each contributory factor and identify relevant interventions to prevent future occurrences. The interventions identified should be added to the contributory factors table as shown in Table 10.4.

ADVANTAGES

- AcciNet considers contributory factors across the overall system of work, providing an exhaustive analysis of adverse incidents;
- AcciNet provides a clear visual representation of the accident aetiology;

TABLE 10.4
Uber Volvo AcciNet Contributory Factors Output Table Extract Showing Example Contributory Factors Across the Task Network

Task	Actor(s)	Contributory Factor Description	Contributory Factor Classification	Recommended Safety Intervention
1. Provide and communicate AV design standards	International Standards Organisation	Appropriate AV design standards not provided	T2: Task Omitted	Develop and communicate AV design standards
2. Provide and communicate AV testing guidelines and standards	International Standards Organisation	Appropriate AV design testing guidelines and standards not provided	T2: Task Omitted	Develop and communicate AV testing guidelines and standards
12. Design and configure test autonomous vehicle	Uber	Decision to disable the Volvo CitySafe automatic emergency braking system to enable smoother ride	T3: Task completed inadequately	Design decisions involving vehicle systems to be discussed and approved with vehicle manufacturer
17. Monitor AV systems and road environment	AV Vehicle operator	AVs failure to identify and classify pedestrian in a timely manner Vehicle operator's failure to monitor the road environment and late identification of pedestrian Vehicle operator's use of mobile phone while operating test vehicle	T3: Task completed inadequately T3: Task completed inadequately T3: Task completed inadequately	Improved hazard identification and classification system Vehicle operator training on monitoring roadway environment System which disables vehicle operator mobile phones when vehicle is driving
19. Cross road safely	Pedestrian	Pedestrian's decision to cross road via crosswalk and not pedestrian crossing	T5: Inappropriate task	Improved signage prohibiting pedestrian crossings via crosswalk Roadside barriers to prevent pedestrians from crossing outside of pedestrian crossing

- AcciNet is a generic approach which can be applied in any domain;
- The systems perspective offered ensures that there is a focus on systematic improvements rather than on blaming individuals;
- Analysts are not required to describe relationships among individual contributory factors as the task network supports this feature by default. The identification of relationships between contributory factors is difficult and analysts have been found to perform poorly in AcciMap reliability and validity studies (e.g. Hulme et al., 2021a);
- AcciNet provides a taxonomy and classification scheme which supports the aggregation of data across multiple incidents;
- The task network is a useful modelling approach in its own right and provides an in-depth overview of the network of tasks undertaken in a particular system;
- AcciNet does not focus exclusively on failure, considering both normal and abnormal performance and how both interact to create adverse events; and
- AcciNet was designed to be used with Net-HARMS (Chapter 2) as part of an integrated safety management framework.

DISADVANTAGES

- AcciNet can be time-consuming to apply particularly for large and complex incidents;
- The quality of the analysis produced is entirely dependent upon the quality of the data collected;
- AcciNet does not provide a method to identify and develop corrective measures; these are based on the judgement of the analyst. The SafetyNet method was recently developed to assist in the development of safety interventions following AcciNet analyses (see Related methods section);
- The graphical output can become complex and hard to decipher when used to analyse large and complex incidents;
- Developing the initial task network requires significant work and can be difficult for novice analysts; and
- The concept of normal performance requires some introduction and study. It is a newer concept in HFE and safety science.

RELATED METHODS

AcciNet uses a task network which is developed based on a HTA (see Chapter 5) of the system under analysis. AcciNet was designed to be used in conjunction with Net-HARMS (Chapter 2) and SafetyNet as part of the Systems-Thinking Accident and Risk Toolkit (START). The integrated toolkit is driven by two components that are shared across the three methods: the task network and the risk mode classification scheme. The task network provides the description of the work system that is the basis on which to identify risks during risk assessment, to identify contributory factors during accident analysis, and to test and refine new controls or safety interventions developed based on risk assessment or accident analysis outputs. The task network is then used as the basis for the Net-HARMS risk assessment, whereby tasks and their interactions are examined to identify potential safety risks. The outputs are used to inform the development of risk controls designed to manage the risks identified. When an adverse incident occurs within the work activity, scenario, or system, the task network is then used as the basis for the AcciNet incident analysis. Here contributory factors associated with the different tasks are identified and safety interventions are proposed. The risk controls and safety interventions identified during risk assessment and accident analysis methods are tested initially by inserting them into the task network to identify the likelihood of success and any unwanted adverse impacts via SafetyNet. The task network thus ensures that organisational risk assessments, accident analyses, and the evaluation of new risk controls and safety interventions are undertaken based on the same description of the work system.

As with the task network, the risk mode taxonomy is used to support both the proactive identification of risks (Net-HARMS), the retrospective classification of contributory factors (AcciNet), and the evaluation of risk controls and safety interventions (SafetyNet). The task network and risk mode taxonomy ensure that there is consistency between all three analyses and enables integration of the outputs to support safety management. For example, accident analysis data can be fed into risk assessments either to add to the risks identified or to provide an indication of where identified risks have occurred previously. Moreover, risk controls that have failed (as identified during accident analysis) can be noted during risk assessment activities. During accident analysis activities the relevant risk assessment outputs, based on the same task network, can support the identification of contributory factors. Specifically, analysts can first review the risks identified and then seek to determine whether they played a role in the accident under analysis. This also enables the accuracy of risk assessments to be verified as well as the effectiveness of risk controls.

APPROXIMATE TRAINING AND APPLICATION TIMES

AcciNet is a relatively simple method to learn and apply but does involve multiple steps, some of which can take longer to complete than others. For example, the identification of data sources and the recording of data and information can be the most time-consuming phases of analysis. For the analysis of large complex incidents, estimated timescales are expected to be around one to two weeks for data collection and a further week for the initial construction of the AcciNet. However, the final procedural stage of review can take additional time. For smaller incidents it is often possible to construct draft AcciNet outputs in one to two hours. These estimated timeframes are dependent on other factors as well, such as analyst skill and experience.

RELIABILITY AND VALIDITY

AcciNet was only recently introduced (Salmon et al., 2020c) and hence there is little reliability and validity evidence associated with the method. Hulme et al. 2021b tested the criterion-referenced concurrent validity of AcciNet, AcciMap (Chapter 9) and STAMP-CAST (Chapter 11) when used by 34 HFE and safety practitioners to analyse the Uber Volvo collision presented in the case study example in this chapter. The findings revealed that AcciNet achieved a moderate positive correlation coefficient against the expert-derived incident analysis (Hulme et al. 2021b). This is promising and means that AcciNet achieved respectable levels of validity overall when assigning and classifying contributory factors to tasks within the task network.

TOOLS NEEDED

AcciNet can be conducted using pen and paper; however, the AcciNet output table is normally undertaken using a Microsoft Excel spreadsheet or Microsoft Word document with the table function selected. Task networks are typically drawn in Microsoft Visio or a related drawing program.

CASE STUDY APPLICATION: UBER VOLVO INCIDENT ANALYSIS

At around 9:58 pm on Sunday 18 March 2018, a Volvo XC90 Sport Utility Vehicle (SUV) fitted with Uber's self-driving system struck and killed a pedestrian on Mill Avenue, in Tempe, Maricopa County, Arizona (NTSB, 2018). The vehicle was being tested as part of Uber's Arizona testing program and was occupied by a vehicle operator at the time of the collision. The following description of the incident is adapted from the NTSB's preliminary investigation report (NTSB, 2018).

The test vehicle was travelling northbound along Mill Avenue, a two-lane road with a posted speed limit of 45 mph. The test vehicle collided with a 49-year-old female pedestrian who attempted to cross Mill Avenue from a centre median strip rather than via a pedestrian crosswalk around 360 feet north of the collision site. Although the test vehicle detected an obstacle in the road ahead, it failed to initiate an emerging braking manoeuvre, and the test vehicle safety operator failed to intervene until it was too late, only seeing the pedestrian immediately prior to impact.

The test vehicle was equipped with Uber's developmental self-driving system which comprised forward- and side-facing cameras, radars, LIDAR, navigation sensors, and a computing and data storage unit (NTSB, 2018). Within the vehicle a monitor located on the centre console presented diagnostic information to the vehicle operator. The test vehicle had two control modes: computer and manual. It was also equipped with various Volvo driver assistance functions, including the City Safety system which provides collision avoidance and automatic emergency braking. These functions, however, were disabled due to the vehicle being driven in the computer control mode and a desire to avoid an erratic ride in the vehicle, such as the vehicle braking in the event that objects in the vehicles path were falsely detected.

At the time of the collision (approximately 9:58 pm), the vehicle had been in self-driving mode for around 19 minutes and was negotiating the second loop of the test-route. According to the NTSB report, the Uber system first registered radar and LIDAR observations of the pedestrian around 6 seconds prior to impact. The self-driving system initially classified the pedestrian as an unknown object and then as a bicycle, but could not identify its intended path. Around 1.3 seconds before impact, the self-driving system determined that an emergency braking manoeuvre was required. Such a manoeuvre could not be initiated by the vehicle due to the City Safety system being disabled.

The vehicle safety operator only noticed the pedestrian and intervened less than a second before impact. Initially she engaged the steering wheel but did not brake until after the vehicle hit the pedestrian. More recent accounts of the incident have suggested that the vehicle safety operator was watching the Hulu streaming service on her mobile phone. The role of the vehicle safety operators was to observe the vehicle and to note events of interest on a central tablet. Vehicle operators were also supposed to monitor the environment for hazards and to regain control of the vehicle in the event of an emergency.

Initially four analysts developed an ActorNet for the task ‘Conduct on-road autonomous vehicle testing program’. This involved reviewing the NTSB report and associated documentation to identify all relevant stakeholders that shared the responsibility for the safe completion of Uber’s vehicle test program. Following this, the same four analysts developed a task network for the ‘Conduct on-road autonomous vehicle testing program’ task. This involved identifying the core tasks that each actor in the ActorNet undertook in relation to the safe completion of Uber’s vehicle test program. Again, the primary source of information for the task network was the NTSB report; however additional documentation was also used including academic articles reporting on the incident (e.g. Stanton et al., 2019).

Once the ActorNet and task network were complete, a workshop was held to identify relevant contributory factors. During the workshop the four authors reviewed the NTSB reports and existing analyses of the incident and identified contributory factors and instances of normal performance that played a role in the incident. For each contributory factor, the relevant actor from the ActorNet was identified based on who or what (i.e. human or non-human agent) was responsible for creating contributory factor. Each contributory factor was then coded into one of the following contributory factor types:

1. Task
    i. T1: Task mistimed
    ii. T2: Task omitted
    iii. T3: Task completed inadequately
    iv. T4: Inadequate task object
    v. T5: Inappropriate task

2. **Communication**
    i. C1: Information not communicated
    ii. C2: Wrong information communicated
    iii. C3: Inadequate information communicated
    iv. C4: Communication mistimed
3. **Environmental**
    i. E1: Adverse environmental conditions

All actors and associated contributory factors were subsequently mapped onto the task network and shaded red. Relevant actors associated with each contributory factor were also added to the task network diagram. Tasks within the AcciNet which represent 'normal performance' were shaded green to show that no failure or inappropriate behaviour occurred. Nodes that are shaded both red and green represent tasks where selected agents performed the task successfully as required (green) but other agents performed them sub-optimally or not at all (red). Relevant actors associated with each instance of normal performance were added to the task network diagram. The resulting annotated task network, presented in Figure 10.4, represents an AcciNet analysis of the Uber Volvo incident. The AcciNet output table is presented in Table 10.4.

As shown in the task network multiple interacting tasks were required to achieve safe on-road testing of AVs. These include tasks relating to the design of the test vehicles and their safe operation when being tested on road, but also tasks relating to the design of the testing program itself and the recruitment and training of test vehicle operators. Beyond the vehicle and testing program, other important tasks included the regulation of on-road AV testing, the management of test permits and licensing, and the provision and communication of AV design and testing standards. Finally, designing and upgrading the road environment in line with road design standards is also included in the task network as this is important to ensure safe interactions between autonomous test vehicles and other road users including drivers, pedestrians, and cyclists.

As shown in Figure 10.4, the Uber Volvo incident was created by a network of contributory factors that included both failures, work as imagined, and multiple instances of normal performance. For example, the AcciNet shows multiple instances where tasks were completed either inadequately or inappropriately, including for example the tasks:

1. **Monitor autonomous vehicle and road environment**. This task was performed sub-optimally by both the autonomous vehicle and the vehicle operator. The autonomous vehicle failed initially to classify pedestrian correctly, and could not identify her intended path. It is alleged that the vehicle operator was interacting with her phone immediately prior to the collision.
2. **Cross road safely**. The pedestrian was not crossing the road at the crossing. Although the area at which she crossed had the appearance of a pedestrian

crossing, there were small, unlit, signs stating that the real crossing was further up the road (Stanton et al., 2019).

3. **Design and configure test autonomous vehicle**. Uber engineers disabled the Volvo City Safety system, including the automatic emergency braking system, to prevent an erratic ride experience.
4. **Design AV testing program**. The design of the testing program was problematic. Vehicle operators worked eight-hour shifts and were required to monitor the vehicle as it drove around a preset route and take over control if necessary (Stanton et al., 2019).

At the higher levels of the Uber Volvo system, the lack of international standards (International Organisation for Standardization (ISO)) and national committee standards (Society of Automotive Engineering (SAE)) for automation design and testing meant that Uber had no technical guidance for appropriate interfaces, safety standards, or testing regimes. A legal framework permitted the testing of automated vehicles on public roads at a federal level and so state governments allowed such testing to go ahead.

In addition, the AcciNet also shows the role of tasks that were completed appropriately. For example, the tasks ‘Regulate on-road autonomous vehicle testing’ and ‘Manage autonomous vehicles testing permits and licensing’ were undertaken as required by the Californian government. Initially Uber was planning to undertake its testing in California but the dispute over the need for permits led to Uber’s vehicle registrations being revoked. As a result, Uber moved its testing program to the more lenient state of Arizona. The move was encouraged by the Arizona Governor, likely due to the perceived economic growth expected to follow investment in the development of autonomous vehicles (Stanton et al., 2019). As such, this represents a task that was undertaken as required, but played a key contributory role in the incident, as it ultimately led to Uber’s testing program being moved to Arizona (Table 10.5)

TABLE 10.5
AcciNet Output Table

Task	Actor(s)	Contributory Factor Description	Contributory Factor Classification	Recommended Safety Intervention
1. Provide and communicate autonomous vehicle design standards	International Standards Organisation	Appropriate AV design standards not provided	T2: Task Omitted	Develop and communicate AV design standards
2. Provide and communicate AV testing guidelines and standards	International Standards Organisation	Appropriate AV design testing guidelines and standards not provided	T2: Task Omitted	Develop and communicate AV testing guidelines and standards
4. Provide & communicate laws for on-road AV testing	State Government Arizona	On-road AV testing laws were inadequate	T4: Inadequate task object	Develop and risk assess comprehensive AV testing laws
6. Regulate on-road AV testing	State Government Arizona	On-road AV testing regulations inadequate	T4: Inadequate task object	Improve regulations and regulatory oversight for AV testing
7. Manage AV permits and licensing	State Government Arizona	Management of licensing and permits was inadequate	T3: Task completed inadequately	Improve licensing and permit requirements, process, and enforcement
9. Design AV testing program	Uber	Testing program included long test drives on a loop with a requirement for passive vehicle monitoring	T3: Task completed inadequately	Redesign testing program with input from human factors specialist with knowledge of issues such as underload, distraction, and fatigue
11. Obtain AV testing permits	Uber	Testing permits were not obtained	T1: Task omitted	Agreed requirement for permits across road system stakeholders
12. Design and configure test autonomous vehicle	Uber	Decision to disable the Volvo CitySafe automatic emergency braking system to enable smoother ride	T3: Task completed inadequately	Design decisions involving vehicle systems to be discussed and approved with vehicle manufacturer
14. Train AV operators	Uber	Extent to which training program incorporates advice and guidance around underload, distraction, etc.	T3: Task completed inadequately	Redesign training program with input from human factors specialist with knowledge of issues such as underload, distraction, and fatigue

15. Design/upgrade maintain road environment	State Government Arizona	Design of road environment including ineffective signage and pathway which did not permit pedestrians	T3: Task completed inadequately	Improved design of signage and centre median strip
17. Monitor AV systems and road environment	AV	Autonomous vehicles failure to identify and classify pedestrian in a timely manner	T3: Task completed inadequately	Improved hazard identification and classification system
	Vehicle safety operator	Vehicle safety operator's failure to monitor the road environment and late identification of pedestrian	T3: Task completed inadequately	Vehicle safety operator training on monitoring roadway environment
		Vehicle safety operator's use of mobile phone while operating test vehicle	T3: Task completed inadequately	System which disables vehicle safety operator mobile phones when vehicle is driving
19. Cross road safely	Pedestrian	Pedestrian's decision to cross road via crosswalk and not pedestrian crossing	T5: Inappropriate task	Improved signage prohibiting pedestrian crossings via crosswalk Roadside barriers to prevent pedestrians from crossing outside of pedestrian crossing

REFERENCES

Australian Transport Safety Bureau. (2008). Analysis, Causality and Proof in Safety Investigations. Aviation Research and Analysis Report – AR-2007-053. Canberra: Australian Transport Safety Bureau.

Dallat, C., Salmon, P. M., & Goode, N. (2018). Identifying risks and emergent risks across sociotechnical systems: The NETworked hazard analysis and risk management system (NET-HARMS). *Theoretical Issues in Ergonomics Science*, 19(4), 456–482.

Dekker, S. (2011). *Drift Into Failure: From Hunting Broken Parts to Understanding Complex Systems*. Surrey, UK: Ashgate.

Goode, N., Salmon, P.M., Lenne, M., & Finch, C. (2018). *Translating Systems Thinking into Practice: A Guide to Developing Incident Reporting Systems*. Boca Raton, FL: CRC Press.

Hollnagel, E. (2012). *FRAM, the Functional Resonance Analysis Method: Modelling Complex Socio-Technical Systems*. Surrey, UK: Ashgate Publishing, Ltd.

Hulme, A., Stanton, N.A., Walker, G.H., Waterson, P., & Salmon, P.M. (2019). What do applications of systems thinking accident analysis methods tell us about accident causation? A systematic review of applications between 1990 and 2018. *Safety Science*, 117, 164–183.

Hulme, A., Stanton, N.A., Walker, G.H., Waterson, P., & Salmon, P.M. (2021a). Complexity theory in accident causation: Using AcciMap to identify the systems thinking tenets in 11 catastrophes. *Ergonomics* in Press, 1–45.

Hulme, A., Stanton, N.A., Walker, G.H., Waterson, P., & Salmon, P.M. (2021b). *Are Accident Analysis Methods Fit for Purpose? Testing the Criterion-Referenced Concurrent Validity of AcciMap*, STAMP-CAST and AcciNet. *Safety Science* Under review.

Leveson, N.J. (2004). A new accident model for engineering safer systems. *Safety Science*, 42(4), 237–270.

NTSB. (2018). National Transportation Safety Board Preliminary Report Highway: HWY18MH010. Retrieved 15 June from https://www.ntsb.gov/investigations/AccidentReports/Pages/HWY18MH010-prelim.aspx

Rasmussen, J. (1997). Risk management in a dynamic society: A modelling problem. *Safety Science*, 27(2–3), 183–213.

Salmon, P., Hulme, A., Walker, G. H., Waterson, P., & Stanton, N. A. (2020b). The Accident Network (AcciNet): A new accident analysis method for describing the interaction between normal performance and failure. In *Proceedings of the Human Factors and Ergonomics Society Annual Meeting* (Vol. 64, No. 1, pp. 1676–1680). Sage CA: Los Angeles, CA: SAGE Publications.

Salmon, P.M., Cornelissen, M., & Trotter, M.J. (2012). Systems-based accident analysis methods: A comparison of AcciMap, HFACS, and STAMP. *Safety Science*, 50(4), 1158–1170.

Salmon, P.M., Hulme, A., Walker, G.H., Waterson, P., Berber, E., & Stanton, N.A. (2020a). The big picture on accident causation: A review, synthesis and meta-analysis of AcciMap studies. *Safety Science*, 126, 104650.

Salmon, P.M., Hulme, A., Walker, G.H., Waterson, P., & Stanton, N.A. (2020c). The Accident Network (AcciNet): A new accident analysis method for describing the interaction between normal performance and failure. *Proceedings of the Human Factors and Ergonomics Society Annual Meeting*. Los Angeles, CA: SAGE Publications Sage CA.

Salmon, P.M., Walker, G.H., Read, G.J.M., Goode, N., & Stanton, N.A. (2017). Fitting methods to paradigms: Are ergonomics methods fit for systems thinking? *Ergonomics*, 60(2), 194–205.

Stanton, N.A. (2014). Representing distributed cognition in complex systems: How a submarine returns to periscope depth. *Applied Ergonomics*, 57(3), 403–418.

Stanton, N. A., Salmon, P. M., Rafferty, L. A., Walker, G. H., Baber, C., & Jenkins, D. P. (2013). *Human Factors Methods: A Practical Guide for Engineering and Design*: CRC Press.

Stanton, N.A., Salmon, P.M., Walker, G.H., & Stanton, M.J. (2019). Models and methods for collision analysis: A comparison study based on the Uber collision with a pedestrian. *Safety Science*, 120, 117–128.

Svedung, I. & Rasmussen, J. (2002). Graphic representation of accident scenarios: Mapping system structure and the causation of accidents. *Safety Science*, 40(5), 397–417.

11 The Systems Theoretic Accident Model and Process-Causal Analysis based on Systems Theory (STAMP-CAST) Method

BACKGROUND

The Systems Theoretic Accident Model and Process (STAMP) (Leveson, 2004) is a model of accident causation which has an associated proactive risk assessment method, the Systems Theoretic Process Analysis (STPA; Chapter 4) (Leveson, 2011; Leveson and Thomas, 2018), and a retrospective incident analysis method, the Causal Analysis based on Systems Theory (CAST) (Leveson, 2004). STAMP is based on systems and control theory and suggests that adverse events occur when interactions between system components are not controlled through managerial, organisational, physical, operational, and manufacturing-based controls. According to Leveson (2004), safety risks are managed through a hierarchy of controls and feedback mechanisms, and adverse events emerge when behaviour and emergent properties are not adequately controlled.

STAMP-CAST is an accident analysis method that is used to identify the control and feedback failures which enabled the incident under analysis to occur. When using STAMP-CAST, the first phase involves using the control structure modelling technique to describe the system under analysis and the control and feedback relationships that exist across its different levels (see Salmon et al., 2016). The control structure model is used to describe the control and feedback mechanisms that exist within the system during both system development and operation. A generic control structure is presented in Figure 11.1. The control structure takes inspiration from Rasmussen's (1997) risk management framework and incorporates a series of hierarchical system levels to represent the structure of the overall work system. Levels of the work system include congress and legislatures at the top (i.e. government and policy) all the way down to the operations management and physical and technical processes that occur at the bottom. Accordingly, systems are viewed as comprising interrelated components that maintain a state of dynamic equilibrium through feedback loops of control and information across all levels (Leveson, 2004). Each level therefore includes a description of the relevant agents and organisations that play a

DOI: 10.1201/9780429281624-11

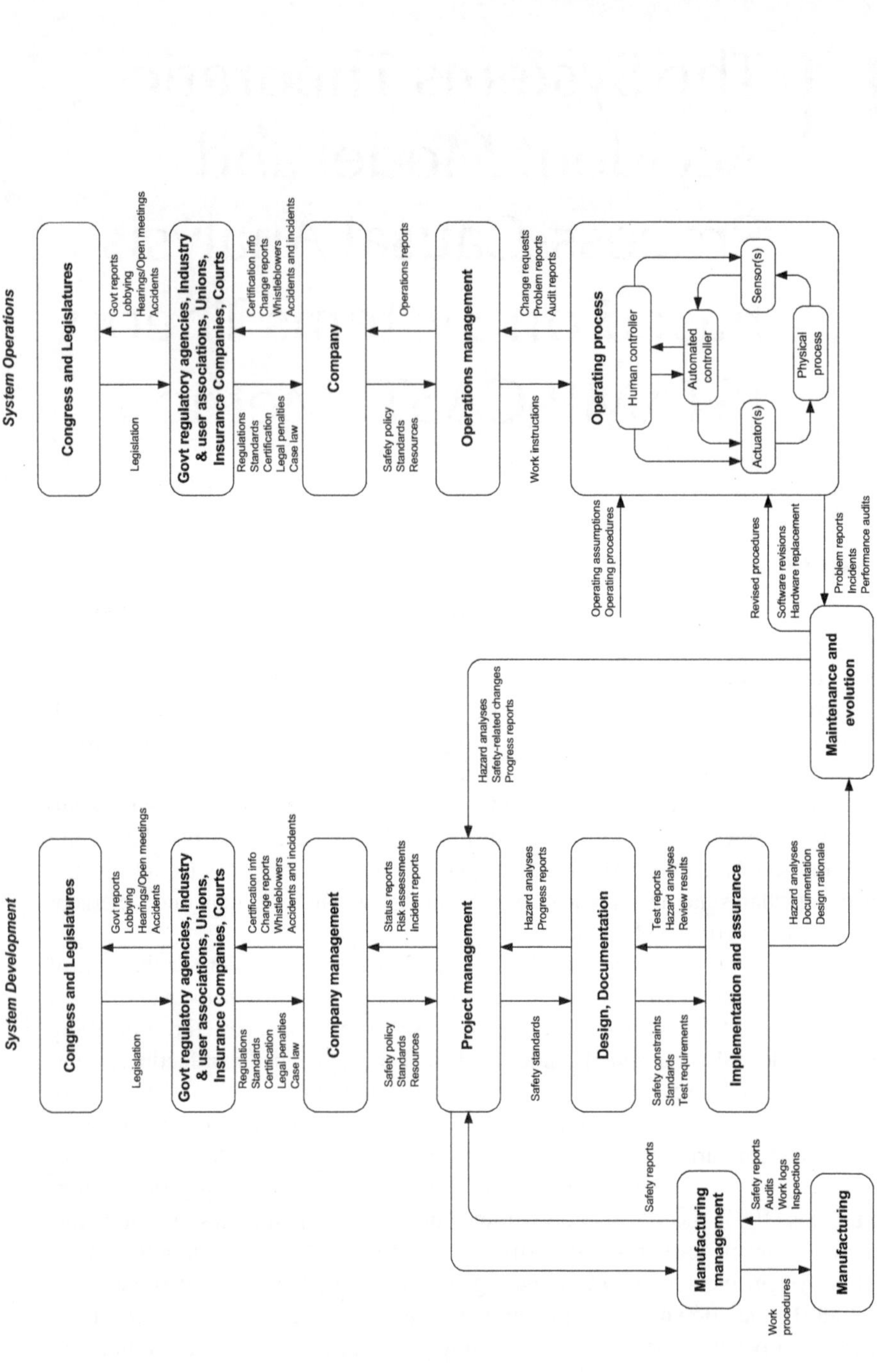

FIGURE 11.1 Generic control structure (adapted from Leveson (2004)). On the left of the control structure hierarchy is system development, and on the right system operation.

role in system design or operation. Control and feedback mechanisms are included to show what controls are enacted down the hierarchy and what information about the status of the system is sent back up the hierarchy. Within Figure 11.1, the arrows flowing down the hierarchy represent controls (or reference channels) and the arrows flowing up the hierarchy represent feedback loops (or measuring channels). In other words, each entity has control and authority over the entities immediately below it, and each entity is likewise subject to control and authority from the entities immediately above.

STAMP-CAST works by identifying where control and feedback mechanisms failed and then classifying these failures using an associated control and feedback failure taxonomy. The control and feedback failures can then either be placed directly onto the control structure or into an output table describing each contributory factor. One of the main goals of STAMP-CAST is to move focus away from human error, instead shifting emphasis to the entire system to better understand why the incident occurred.

DOMAIN OF APPLICATION

STAMP-CAST is a generic approach and can be applied in any domain for accident analysis purposes. Based on a systematic review, Hulme et al. (2019) identified STAMP-CAST applications in a range of domains including rail, emergency response, oil and gas, missile defence systems, maritime, and healthcare.

PROCEDURE AND ADVICE (STAMP-CAST)

A flowchart depicting the STAMP-CAST procedure is presented in Figure 11.2. Step-by-step guidance is presented below.

STEP 1: DETERMINE ANALYSIS AIMS AND SCOPE

The first step involves clearly defining the aims and scope of the analysis. Typically, the aim is to identify the system wide control and feedback failures that interacted to create a particular incident. Clearly defining the scope or analysis boundaries is important because adverse outcomes (e.g. workplace injuries or fatalities) are often the result of multiple events or phases. For example, an analysis of the death of a patient in a hospital setting may focus on the decisions and actions leading up to the deterioration of the patient, or it might focus on the decision and actions during the emergency response. The scope will have a significant impact on the recommendations for interventions identified from the analysis. It is not unusual for STAMP-CAST analyses to consider contributory factors months and even many years before the incident in question occurred.

Defining the analysis boundaries is also important as project constraints will dictate how deep the analysis can go in terms of the parts of the system considered and how far back in time the analysis will go. With STAMP-CAST specifically, project constraints may influence how detailed the control structure is in terms of the levels,

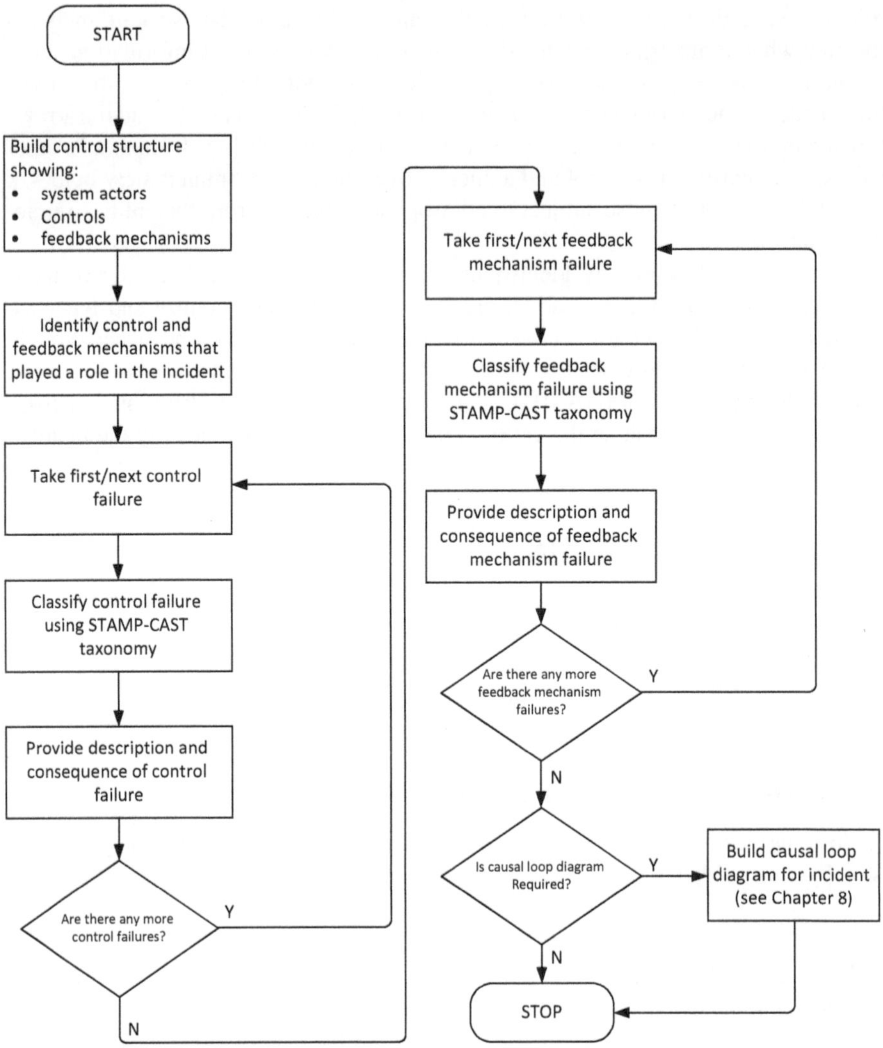

FIGURE 11.2 STAMP-CAST procedure.

actors, and organisations included. It may be, for example, that an analysis boundary may be set at the organisational level only or may not extend to international factors outside of the country in which the incident occurred. Further, post incident response may or not be of interest.

At a minimum, it is recommended that STAMP-CAST analyses should include the levels specified in the system operations side of the generic control structure presented in Figure 11.1. For some incidents it can be useful to extend these boundaries to consider international influences (e.g. International Standards Organisation, World Health Organisation) that are directly relevant to the incident in question.

STEP 2: DATA COLLECTION

STAMP-CAST is dependent upon the collection of accurate data regarding the decisions and actions that contributed to the incident(s) under analysis. This involves collecting data regarding the system and incident in question. Four broad forms of data are required, and these cover:

1. The system, in terms of who resides in the system and shares the responsibility for safety during the work activities or processes in which the accident occurred;
2. The control and feedback mechanisms that are used to manage safety during system design and operation;
3. The work activities or processes in which the incident occurred; and
4. The incident and any contributory factors that played a role in its occurrence.

Data collection for STAMP-CAST analyses can involve a range of activities, including interviews with those involved in the incident or Subject Matter Experts (SMEs) for the domain, work, or type of incident under analysis; reviewing official reports, coronial investigations or inquiries into the incident; and observing recordings of the incident. It is important during data collection to consider all levels of the system, up to and including regulatory, government, and even international levels and influences. For example, if using interviews as the primary data collection approach it is useful to interview relevant actors from all levels of the system. Likewise, when reviewing procedures this should include the procedures for different actors across all levels of the system.

STEP 3: CONSTRUCT AND VERIFY CONTROL STRUCTURE

Once data collection is complete, the next step involves developing a control structure showing the hierarchical system levels to be considered, relevant actors and organisations involved in the work system, and the controls and feedback mechanisms used to manage safety. As shown in Figure 11.1, the generic control structure specifies a set of hierarchical levels for both system design and operation; however, these are flexible and can be modified depending on the aims of the analysis and the resources available.

Building the control structure first involves specifying the levels to be used, identifying relevant actors and organisations at each level, and then specifying the controls and feedback mechanisms that are currently in place to manage safety. The data collected during Step 2 should be used for this purpose; however, the process is iterative and often involves the identification of new data requirements. Development of the control structure usually begins at a very abstract level and is gradually refined to capture more detail about the system (Leveson and Thomas 2018). For example, in cases where more than one agent or organisation is responsible for safety and performance, it can be useful to collapse these into a single entity where necessary. Likewise, instead of listing each and every control mechanism, it can be useful to start at a more abstract level and refine from there. This may mean breaking down an initial control or feedback description into multiple further controls or feedback mechanisms. There is also the matter of determining whether the control structure in question requires a strict linear hierarchy, or in cases where control structures and sub-systems are detailed, a hierarchy that combines both linear and horizontal control and feedback mechanisms in more complex configurations (Figure 11.3).

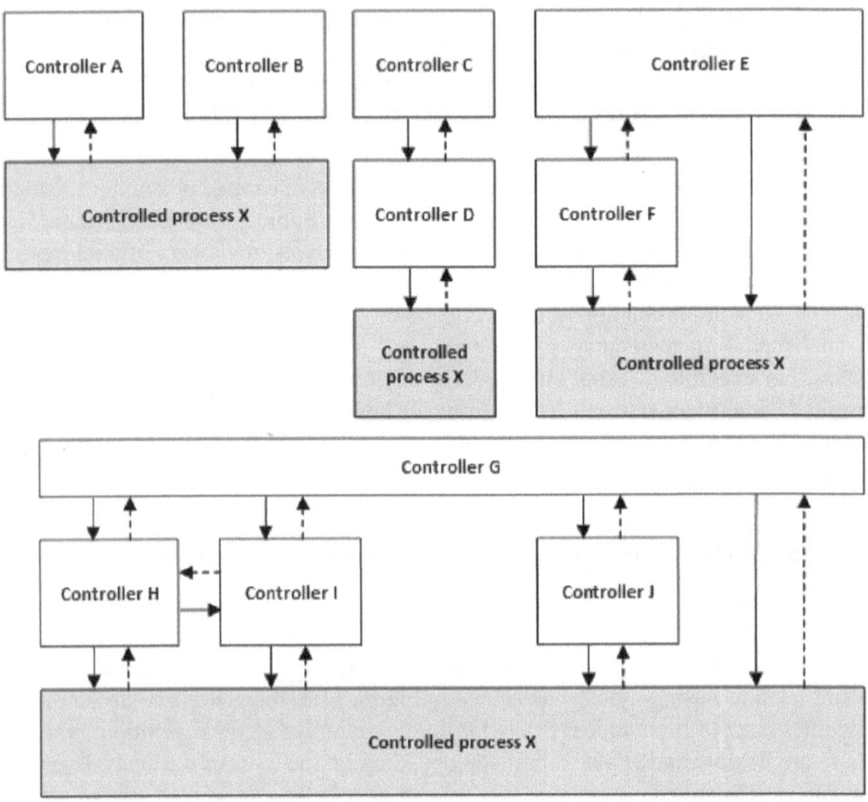

FIGURE 11.3 A comparison of different control structure types adapted from Leveson and Thomas (2018). Controller A and B both control and monitor shared control process X and do not interact with one another. Controller C and D are integrated linearly, the effects of controller C of which are exerted through controller D on process X. Control and feedback mechanisms may also bypass levels as shown in the control structure including controller E, controller F, and process X. In this case, emergency situations or unusual circumstances may mean that controller E can directly control and demonstrate authority over process X, thus circumventing controller F. Finally, more complex control structures as is the case between controller G, controller H, controller I, and controller J are possible. Controller J may operate independent of controller H and I, which, both exert their influence over process X whilst being controlled by controller G.

It is most useful to include both system design and system operation within the control structure; however, most analyses typically focus on the operations control structure. First, relevant actors and organisations should be placed at the appropriate level in the system hierarchy. The name and number of levels used can be flexible. The following levels represent those that we have used previously in safety analyses:

- Level 0: International influences;
- Level 1: Parliament and legislatures;
- Level 2: Government agencies, user groups, industry associations, courts, universities;
- Level 3: Operational delivery and management;
- Level 4: Local management and supervision; and
- Level 5: Operating process and environment.

Once the relevant actors and organisations have been placed at the appropriate level, control mechanisms should be added. It is worth noting that a broad view of controls should be adopted. Leveson (2004) describes various forms of control, including managerial, organisational, physical, operational, and manufacturing-based controls. In addition, external influences such as social norms and culture are often included. That is, system behaviour is controlled not only by engineered systems and direct intervention, but also by policies, procedures, shared values, and other aspects of the surrounding organisational and social culture.

Once the control mechanisms have been added, feedback mechanisms flowing back up the hierarchy should be added. Again, a broad view of feedback mechanisms is adopted. Examples of feedback mechanisms include incident reports, accident investigations, key performance indicators, leading and lagging safety indicators, audit and inspection reports, information regarding certification, and lobbying.

Once developed, it is useful to subject the control structure to a validation exercise whereby it is reviewed and refined by appropriate SMEs. This can take on various forms, including an SME workshop or Delphi study (e.g. Salmon et al., 2016; Hulme et al., 2017). It can be useful to supplement the control structure model with further information about the included controllers such as their description, purpose, responsibilities, actions, and process models.

STEP 4: MAP OUT THE FLOW OF EVENTS

When analysing individual incidents it is useful to map out the flow of proximal events at the operating process and environment level of the control structure. This provides an incident timeline representing key proximal decisions and actions which can be used to help identify broader systemic contributory factors. It is useful to include both physical and technological as well as operator-based factors at this stage, all of which are typically a manifestation of higher-level system decisions and actions (Sebnem Düzgüna and Leveson, 2018).

STEP 5: IDENTIFY CONTRIBUTORY FACTORS

Next the data and flow of events description should be analysed to identify contributory factors which played a causal role in the incident. This involves reviewing the data and recording any factors that the analyst feels played a contributory role in the incident in question. It is important to clearly define what they consider to be a contributory factor. In our own analyses we have used the following definition of contributory factors as factors:

> that, if it had not occurred or existed at the relevant time, then either the occurrence would probably not have occurred, adverse consequences associated with the occurrence would probably not have occurred or have been as serious, or another contributing safety factor would probably not have occurred or existed.
>
> (Australian Transport Safety Bureau, 2008)

It is recommended that analysts take as broad a view as possible when initially identifying contributory factors. A key requirement here is to consider or search for contributory factors across all levels of the control structure and to look for contributory factors associated with all of the actors, organisations, and control and feedback mechanisms identified within the control structure.

STEP 6: CLASSIFY CONTRIBUTORY FACTORS USING THE STAMP-CAST TAXONOMY

Once all relevant contributory factors are identified, analysts should use the STAMP-CAST taxonomy (see Table 11.1) to classify them into specific control and feedback failures. This involves using the STAMP-CAST taxonomy to determine the type of control or feedback failure for each identified contributory factor. The control and feedback failures can then either be placed directly onto the control structure or into an output table describing each contributory factor.

Whilst not included in Table 11.1, the STAMP-CAST taxonomy can include context in which decision-making occurred and the mental model flaws on behalf of human and organisational agents (e.g. Li et al., 2020).

STEP 7: CONSTRUCT CAUSAL LOOP DIAGRAM FOR INCIDENT

Step 7 is optional, with a less used element of STAMP-CAST involving the development to of a Causal Loop Diagram (CLD) of the incident in question. Guidance on how to construct CLDs is presented in Chapter 8.

STEP 8: REVIEW AND REFINE ANALYSIS

Once the draft STAMP-CAST analysis is complete it is useful to have various SMEs review it. It is best practice to use SMEs who were either involved in the incident or who have extensive knowledge of the system and work in question. The STAMP-CAST analysis should be updated and finalised based on the feedback provided by the SMEs.

TABLE 11.1
STAMP-CAST Taxonomy

1. **Inadequate enforcement of constraints**
   1.1. Unidentified hazards
   1.2. Inappropriate, ineffective, or missing control actions for identified hazards
      1.2.1. Design of control process does not enforce constraints
         – Flaws in creation process
         – Process changes without appropriate change in control process
         – Incorrect modification or adaptation
      1.2.2. Process models inconsistent, incomplete, or incorrect
         – Flaws in creation process
         – Flaws in updating process
         – Time lags and measurement inaccuracies not accounted for
      1.2.3. Inadequate coordination among controllers and decision makers
2. **Inadequate execution of control actions**
   2.1. Communication flaw
   2.2. Inadequate actuator operation
   2.3. Time lag
3. **Inadequate or missing feedback**
   3.1. Not provided in system design
   3.2. Communication flaw
   3.3. Time lag
   3.4. Inadequate sensor operation (incorrect or no information provided)

Adapted from (Leveson, 2004).

ADVANTAGES

- STAMP-CAST output offers an exhaustive analysis of the incident in question, identifying control and feedback failures across the overall system of work, up to and including government and international influences;
- STAMP-CAST is based upon a sound theoretical model and is aligned with state-of-the-art accident causation theory;
- The consideration of feedback failures and their associated risks is novel and is a key strength over and above other accident analysis methods.
- STAMP-CAST is relatively simple to learn and use;
- The use of the control failure taxonomy enhances reliability and validity.
- STAMP-CAST has been applied across many domains, suggesting that it has the capacity to be generalisable;
- The output encourages analysts to identify safety interventions which tackle broader system issues as opposed to a fixing broken components approach; and
- The control structure is a useful approach in its own right and provides an in-depth overview of the stakeholders who shared the responsibility for safety and accident causation in a particular system and the control and feedback mechanisms used to maintain system safety.

DISADVANTAGES

- STAMP-CAST can be time consuming to apply, involving the development of a system control structure followed by the identification of control and feedback failures;
- STAMP-CAST does not consider the interactions or relationships between contributory factors;
- The STAMP-CAST taxonomy is not straightforward, and it can be difficult to accurately classify contributory factors;
- The quality of the outputs is highly dependent on the quality of the data available and the expertise of the analysts;
- There is little reliability and validity evidence associated with STAMP-CAST;
- STAMP-CAST does not provide a method to support identification and development of corrective measures; these are based on the judgement of the analyst; and
- The outputs can be complex, large, and unwieldy, particularly for large-scale incidents.

RELATED METHODS

Various data collection methods can be used to gather the data required for STAMP-CAST analyses, including interviews, observation, documentation review, and surveys. STAMP-CAST requires a control structure model of the system in question. The control structure method is useful in its own right for developing models of complex sociotechnical systems (Read et al., 2019; Salmon et al., 2016). Previous applications have used the Delphi method or SME workshops to verify control structure diagrams (Hulme et al., 2017; Read et al., 2019; Salmon et al., 2016). CLDs (Chapter 8) can also be used as part of STAMP-CAST analyses to provide an overview of the positive and negative feedback loops which played a role in the incident under analysis (Leveson, 2004).

Leveson's (2004) STAMP is unique as it provides both a proactive risk assessment method (STAMP-STPA, Chapter 4) and a retrospective accident analysis method (STAMP-CAST). Both analyses are undertaken on a control structure model of the system in question, with STAMP-STPA analyses seeking to establish where controls and feedback mechanisms could fail, and STAMP-CAST analyses exploring what control and feedback failures actually did fail during a particular incident.

APPROXIMATE TRAINING AND APPLICATION TIMES

STAMP-CAST is a simple method to learn and apply, with half- or full-day training workshop normally sufficient for trainees to pick up the basic concepts and approach. It can become time-consuming when applied to large and complex incidents. Development of the control structure in particular can be time consuming, with timescales of around one to two weeks for data collection and construction of the control structure, and further time required for model validation. Identification of contributory factors can also be time consuming and is dependent on the complexity of the incident under analysis as well as the amount of investigation resources and data available. Classification of contributory factors using the STAMP-CAST taxonomy is relatively quick, however it is often difficult to fit contributory factors to items in the taxonomy.

RELIABILITY AND VALIDITY

Reliability and validity evidence for systems thinking-based accident analysis methods is scarce. A recent study by Goncalves Filho et al. (2019) compared AcciMap and STAMP-CAST using percentage overlap of the identified casual factors from independently conducted analyses as the main measure of inter-rater stability and validity. The results indicate that the reliability (degree of overlap of causal factors identified from the same method; i.e. inter-analyst overlap) of STAMP-CAST (65%) was higher than AcciMap (38%). Conversely, the validity (degree of overlap of causal factors identified from two different methods) was as low as 8% between STAMP-CAST and AcciMap (Goncalves Filho et al., 2019). The study by Goncalves Filho et al. (2019) did not, however, consider the potential for false positives, false negatives or true negatives when determining relative agreement between analyst results and methods. Further, the validity of STAMP-CAST and AcciMap was extrapolated based on the similarity between results obtained from different methods (Goncalves Filho et al., 2019), rather than against an expert or gold standard criterion which represents the ideal analytical option.

Hulme et al. (2021) recently applied Signal Detection Theory (SDT) (Green and Swets, 1966) to evaluate the criterion-referenced concurrent validity of STAMP-CAST when used by 34 HFE and safety practitioners to analyse a fatal automated vehicle/pedestrian collision. Criterion-referenced concurrent validity compares participant STAMP-CAST analyses against an expert criterion analysis to assess whether new users of the method can achieve accurate results following a period of training. The SDT paradigm was used to identify hits, misses, false alarms, and correct rejections, which in turn were used to compute a correlation coefficient between negative one (−1; perfect negative correlation) and positive one (+1; perfect positive correlation), with zero marking no correlation either way (Chicco and Jurman, 2020). The findings revealed that STAMP-CAST performed achieved a weak to moderate positive correlation coefficient against the expert-derived incident analysis (Hulme et al., 2021). This is promising and means that STAMP-CAST achieved respectable levels of validity overall when assigning and classifying contributory factors.

TOOLS NEEDED

STAMP-CAST can be conducted using pen and paper; however, the control structure is normally created in a software drawing package such as Microsoft Visio and the output table is typically constructed in Microsoft Word or Excel.

CASE STUDY EXAMPLE: MEDICATION ADMINISTRATION ERROR

Medication administration control structure

Medication administration failure remains a key threat to patient safety. In this case study we developed a control structure of the medication administration system at an Acute Services Unit (ASU) at a regional hospital in Queensland, Australia. The case study is intended to demonstrate the utility of applying the control structure analysis method on its own, as well showcase the STAMP-CAST method.

The control structure was developed based on multiple sources of data, including documentation, observations of staff and patients during medication rounds, and interviews. The study was approved by the Royal Brisbane & Women's Hospital Human Research Ethics Committee, and the Sunshine Coast Hospital and Health Service (SCHHS). The data were collected from October to December 2018.

In total, nine medication rounds on ASU were observed over a two-week period, including a mix of 8am, 6pm, and 8pm rounds. The observations involved two researchers following each nurse taking structured notes on: 1) tasks, people and equipment involved in medication rounds; 2) controls and feedback mechanisms used to manage the process; 3) interruptions; 4) problems that prevented medications being administered; and 5) any comments to researchers during the observations. Nurses were instructed to ignore the researchers as far as possible and undertake their work as per normal practice.

Twenty-six participants were interviewed, including Enrolled Nurses (EN), Registered Nurses (RN), Team Leaders (TL), Clinical Coaches (CC), a Medication Safety Portfolio Holder, the Nurse Unit Manager, Medical Team Junior Doctors, Medical Team Consultants, Pharmacists, Safety and Quality staff, Consumer Representatives, and Hospital Senior Managers. The interview questions for nursing staff focussed on understanding the processes involved in the medication rounds, problems encountered that prevented medications being administered, and suggestions for improving medication safety. This supported the information collected during the observations of medication rounds; all nurses who were observed participated in follow-up interviews. The interview questions for other staff focussed on their role, their safety responsibilities, who they communicate with, how they influence the medication administration process (e.g. controls), and what information they use to evaluate medication safety (e.g. feedback mechanisms).

An analysis of actors, controls, and feedback mechanisms was undertaken using a data triangulation process, where each component of the analysis was verified by a second source, whether through observation, a document, or an interview transcript. The analyses were validated by representatives of all actors in subsequent workshops; their feedback was incorporated into the analyses. The final control structure is presented in Figures 11.4 (full control structure) and 11.5 (operating process level).

Figure 11.4 shows the control structure used to manage safety during the prescription, preparation, and administration process. Based on the observations, interviews, and control structure, nine issues were identified:

1. **A large burden of responsibility is placed on Administrators (i.e. Enrolled and Registered Nurses) in avoiding the high-level hazards.** As well as correctly administering medications, nursing staff are required to identify incorrect prescriptions and transcriptions, obtain medication stock (i.e. avoiding omissions) and secure access to medications at the bedside (i.e. avoiding misappropriation). While these responsibilities are clearly stated in the

core procedure for prescribing, administering, and ceasing medications, the interviews and observations of the medication rounds suggests that the workload associated with the responsibilities beyond administering medications to patients is poorly recognised. Almost all nurses observed on morning shift arrived at 6.45am (15 minutes early) to review charts and identify problems prior to handover with the night shift at 7am. In contrast, the oncoming afternoon shift have 1 hour of overlap time with the morning shift. This allows for a reasonable amount of time to review charts and handover patients. Almost all nurses interviewed commented that "protected time" or "more time" was needed for both shift planning and medication rounds, particularly in the morning.

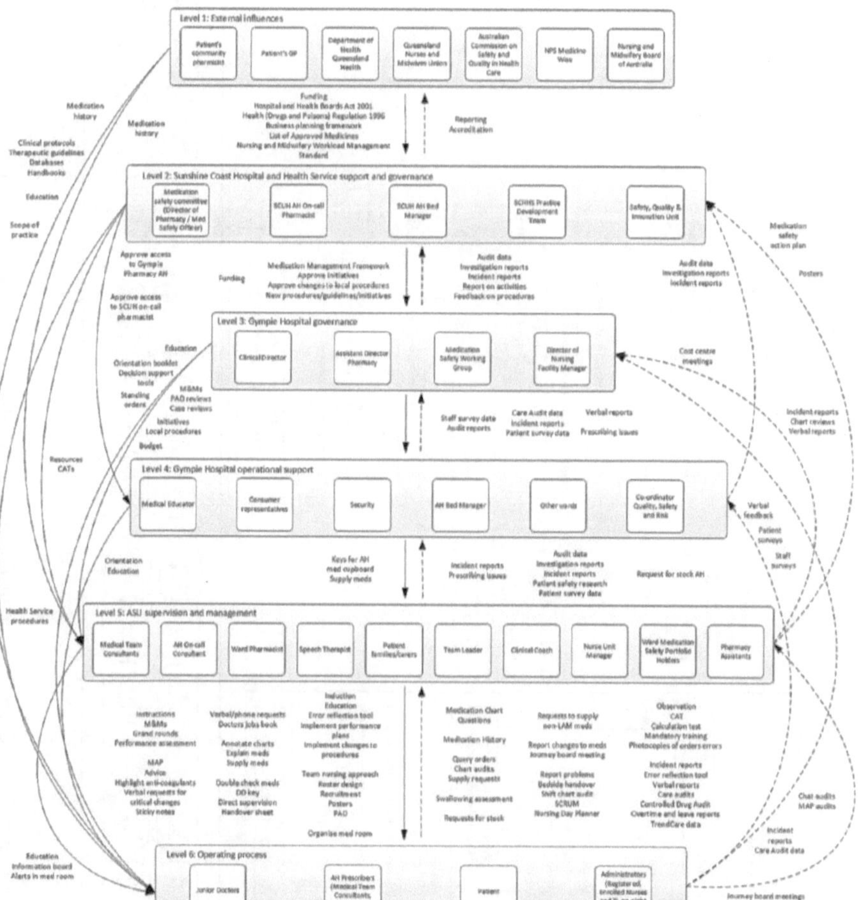

FIGURE 11.4 Control structure for managing the prescribing, preparation, and process. Controls used to manage medication administration are represented in bold, while feedback mechanisms are represented in italics.

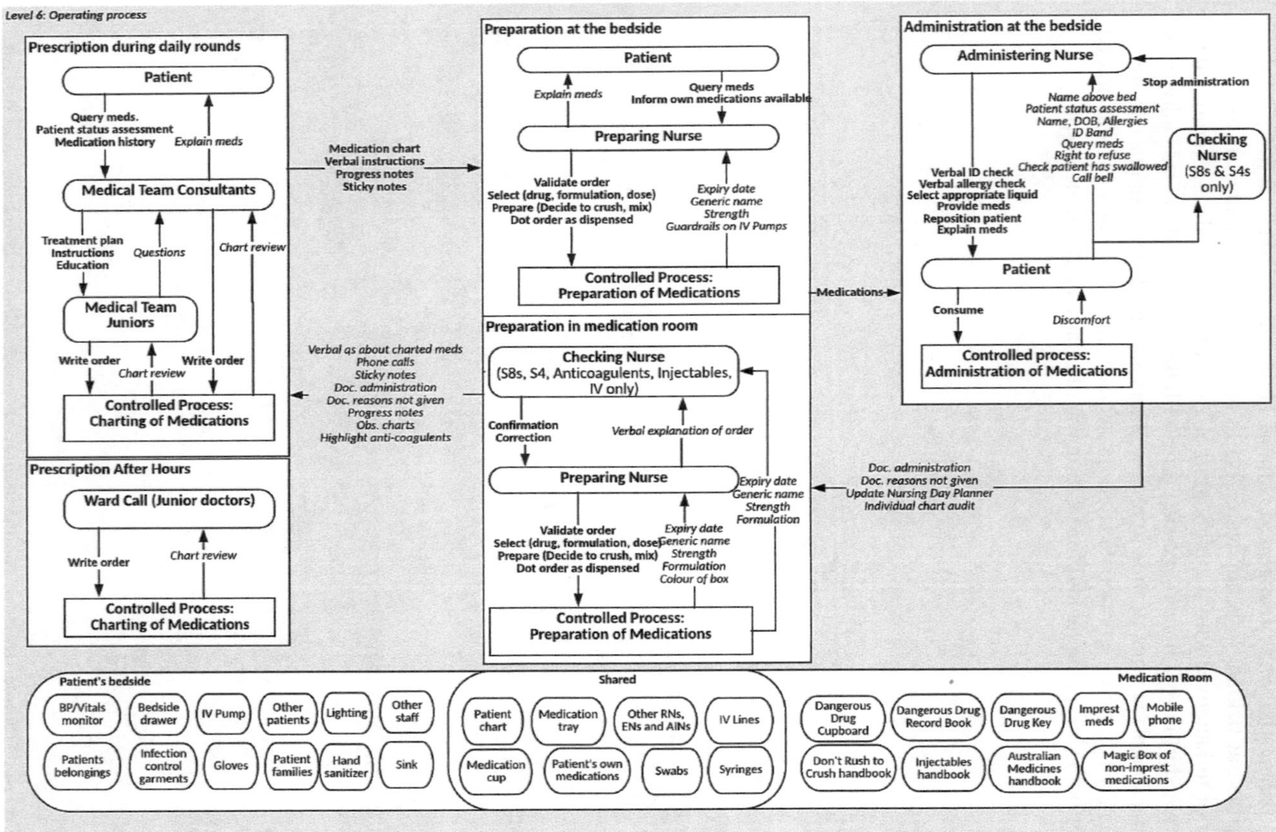

FIGURE 11.5 Operating process level of the medication administration system control structure showing the actors, work processes, equipment and environment involved in medication administration, and the interactions between them (operating processes). Controls used to manage medication administration are represented in bold, while feedback mechanisms are represented in italics.

2. **Consumers (i.e. patients and their families/carers) play a key role in the medication administration system.**
Currently, no formal information is provided to consumers about their responsibilities regarding medication administration. These responsibilities are also not recognised in the core procedure for prescribing, administering, and ceasing medications.
3. **The critical supervisory role of the Ward Pharmacist in the prescribing process is not adequately recognised in workload allocations.**
Following admission, the patient's medication history and charted medications are routinely reviewed and reconciled by the Ward Pharmacist. The Ward Pharmacist immediately notifies the Medical Team of any critical errors and creates a Medication Action Plan outlining suggested changes. Although this process is critical in preventing prescribing and transcribing errors, the Ward Pharmacist does not work after hours or on weekends. This further exacerbates the lack of supervision and support available for Junior Doctors after hours and on weekends.
4. **In comparison to nursing staff, little direct supervision and support is available for Junior Doctors.**
Nursing staff are always directly supervised by a Team Leader (who is only allocated patients during night shift). During business hours, nurses are also supported by a Clinical Coach and the Nurse Unit Manager. In comparison, while Junior Doctors are under the direct supervision and guidance of a Consultant with a heavy clinical load. Unlike the Clinical Coach, the Medical Educator does not directly supervise work practices, and is only on site on Thursdays. Similarly, while the Ward Pharmacist plays a critical role in the prescribing process, they are only available during business hours. There is no direct supervision of Junior Doctors between 11pm and 8am during weekdays. On weekends, the morning rounds are completed with a Consultant, but otherwise they are unsupervised from 11pm Friday night to 8am Monday morning. In addition, only one Junior doctor is rostered on to cover the whole of the Hospital for most of these unsupervised shifts.
5. **The process for obtaining stock after hours prioritises controlling access to medications over timely access to medications.**
During business hours, the process for obtaining stock is simple and efficient. The nursing team informs the Ward Pharmacist, who then obtains supply from the Pharmacy. After hours, Team Leaders are first required to contact other wards in the Hospital to request supply. Team Leaders commented that they often spent many hours calling other wards. If they cannot obtain supply from other wards, they are required to contact the After Hours Bed Manager. The After Hours Bed Manager must then search the afterhours cupboard on the Women's and Families ward. If supply cannot be obtained, they need to contact a local University Hospital (SCUH) After Hours Bed Manager to ask the SCUH On-call Pharmacist for permission to access the pharmacy. Once permission is granted, the After-Hours Bed Manager may then access the pharmacy accompanied by security. This process has primarily been implemented to control after-hours access to the pharmacy.

As a result of this lengthy process, ASU nurses have developed several workarounds to avoid having to access the pharmacy after hours. These include using the "Mystery Box" of medications and other patient's bedside drawers to obtain supply.

6. **Most of the controls are administrative.**
   It is widely recognised that the most effective and robust controls eliminate the risk of harm by removing the hazard, while intermediate controls reduce the risk through changes to the work environment (e.g. forcing functions, physical barriers, changes to the work environment). In the current system, most of the controls are administrative and rely on people to closely follow a procedure or verbally communicate important information.
7. **There is a reliance on education to address the inconsistent implementation of controls.**
   There is also a heavy reliance on education to address reported incidents. For example, participants reported that following a series of prescribing errors, pharmacy would usually organise an education session for Junior Doctors. However, it was acknowledged that only approximately a quarter of the medical workforce attend these sessions and that repeated sessions were required every ten weeks for the new rotation of Junior Doctors. This exacerbates the limitations of the administrative controls discussed above.
8. **Incident reports are the primary source of feedback for the higher levels of the system.**
   There is a heavy reliance on incident reports to monitor the effectiveness of the controls within the operating process. This is problematic because many interview participants reported that medical staff rarely, if ever, report incidents, so issues with prescribing are rarely recorded. Nursing staff reported that they were also unlikely to report a prescribing error if it was resolved without harm to a patient. This is a widely known limitation of healthcare incident reporting systems in Australia (Westbrook et al., 2015), and internationally (Pham et al., 2013).

   The Medication Action Plan (MAP) produced by Ward Pharmacists provides an invaluable source of data for monitoring the implementation of the controls within the prescribing process; however, currently the data is only available in paper-based format and audits are highly resource intensive. As a result, MAPs are only audited on an infrequent, ad hoc basis by the Assistant Director of Pharmacy.
9. **The design of the system does not adequately support feedback and consultation.**
   The actors involved in designing systems, procedures, guidelines, and clinical protocols are far removed from the operating process (e.g. Medication Safety Committee and the Department of Health), with few feedback links to the operating processes level. Several interview participants reported that new systems and procedures were often introduced with little consultation, consideration of the impact on workload, or piloting them in practice.

   Similarly, within the hospital, there are several important actors at the lower levels that are not involved in the design of system improvements. For example,

Clinical Coaches, Junior Doctors, Patients and Patient's families/carers are not represented on the Medication Safety Working Group, which plays a critical role in the design of interventions to improve medication safety.

PRESCRIBING OPERATING PROCESS

The prescribing process is presented on the left-hand side of Figure 11.5. This shows the interactions between the Patient, Medical Team Consultants, and Junior Doctors to chart the medications during daily rounds. After hours, this task is typically only undertaken by the Junior Doctor rostered onto Ward Call. This highlights the lack of direct supervision for Junior Doctors in charting medications after hours. The interviews highlighted several additional factors impacting on the safety of the prescribing process (all these issues were identified by two or more participants):

- No electronic systems are provided to support prescribing or ensure that orders are clear and legally prescribed;
- Medical staff rely on their mobile phone to access medication information databases, as only one computer is available on wheels, and no computers are available at the bedside;
- While prescribing decision support tools have been implemented in the past (e.g. Gentamicin dosing), no formal processes are in place to ensure that these are updated to meet current guidelines;
- There is no formal requirement to double check medication charts during the prescribing process;
- There is intense workload pressure on Consultants as they are typically allocated at least 15 acute patients and supervise a team of Junior Doctors;
- Junior Doctors are not consistently supervised by a specific Consultant and are on 10-week rotations. This makes is difficult for Consultants to assess the skills and knowledge of Junior Doctors and develop an effective team;
- Junior Doctors are often unfamiliar with the patients they are treating because they are not allocated to specific patients;
- Only one Consultant is rostered on to round all patients in the Hospital on weekends;
- No sick leave cover or backfill is provided for Junior Doctors or Consultants;
- Prescribers are limited to prescribing medications on the List of Approved Medications or asking patients to bring in their medications. As a result, patients may not be prescribed the most safe or effective medication; and
- There is no afterhours pathology service to monitor levels for certain medications, so patients may not be prescribed the most safe or effective medication, or they may be transferred to another facility.

PREPARATION AND ADMINISTRATION OPERATING PROCESSES

The preparation and administration processes shown in the middle and right-hand side of Figure 11.5 highlights the distribution of these processes across the ward environment. The findings from the observations suggest that the distribution of medications between the bedside drawers and medication room makes medication rounds highly inefficient, and prone to error. On average, nurses prepared and administered 18 medications to 5 patients, moving 41 times between different areas of the wards. During the interviews, many nurses also commented on the poor ward design and the impact on workload (e.g. "We do a lot of running backwards and forwards between the rooms and the medication room… there's no point saying, oh, slow down, slow down… I've still got all these medications to give and it's nearly half past nine"). The ward design introduces an additional source of workload to a task that already requires intense concentration to detect and resolve prescribing errors, and correctly prepare and administer multiple medications to multiple patients.

Figures 11.4 and 11.5 both depict the controls and feedback mechanisms that are used to ensure safe administration of medication to patients in the ASU. Observations of medication rounds suggest that many of the controls and feedback mechanism are implemented consistently; however, inconsistencies around implementation of the following controls and feedback mechanisms were identified:

- Crushing of medications separately;
- Verbal allergy check;
- Using the nursing day planner;
- Reviewing charts prior to medication round;
- Reviewing charts at end of medication round;
- Restocking medications in the bedside drawer;
- Removing discontinued medications from the bedside drawer;
- The type of information provided to patients about their medications (e.g. "I'm going to give you an injection in your tummy, it's to prevent blood clots" as opposed to "I need to give you a Clexane injection"); and
- Patient's queries about medications were usually brushed over (e.g. Patient: "I don't think that's what I normally take, it's the wrong colour." Nurse: "Don't worry, the doctors must have changed your meds.").

The observations suggest that the nursing day planner is a critical memory aid when nurses are required to administer multiple medications to multiple patients and obtain out of stock medications. It also has the potential to be an effective time management and communication tool if used consistently by all nursing staff; however, whether and how it is used is largely up to the individual nurse. It should also be noted that unless nurses arrive 15 minutes early for their morning shift, no time is allocated to review charts and utilise the nursing day planner.

STAMP-CAST ANALYSIS OF A MEDICATION ADMINISTRATION FAILURE

A simplified version of the medication administration control structure is used below to demonstrate STAMP-CAST. Specifically, we used STAMP-CAST to analyse a hypothetical incident involving an intravenous Heparin overdose whereby a patient was administered Heparin (an anticoagulant) at an inappropriate rate of infusion for a prolonged period of time. The simplified control structure is presented in Figure 11.6 and an extract of the STAMP-CAST table is presented in Table 11.2.

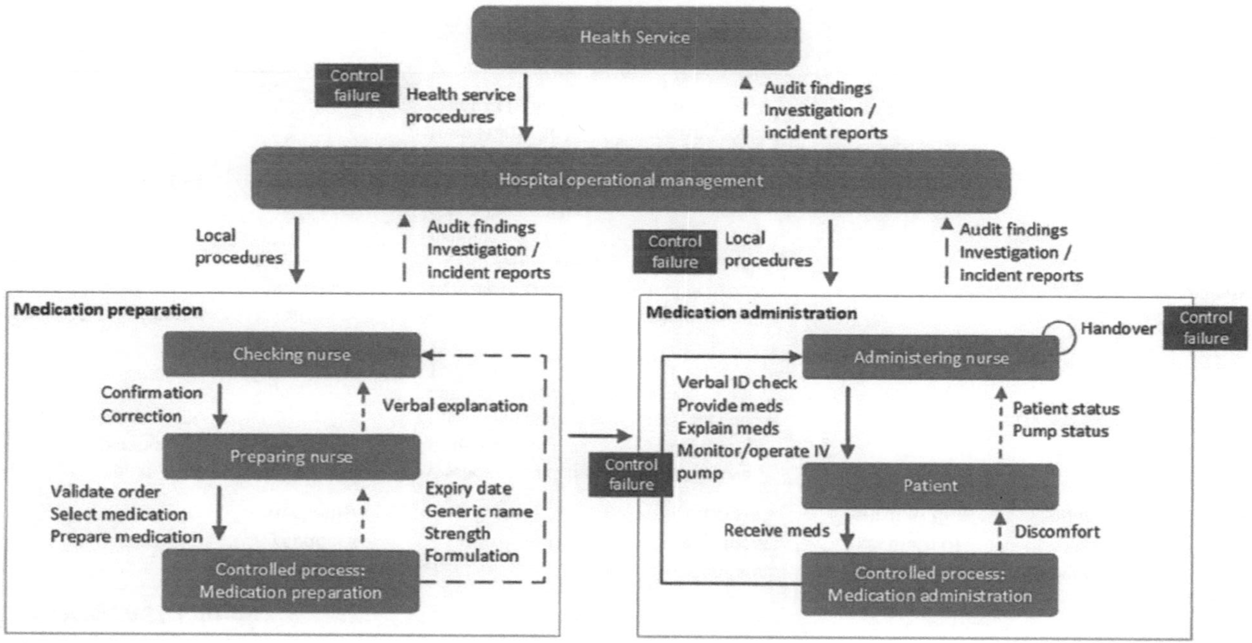

FIGURE 11.6 Simplified control structure for intravenous Heparin overdose incident. Solid arrows represent safety controls and dashed arrows represent feedback mechanisms regarding safety performance.

TABLE 11.2
Extract of STAMP-CAST Output

From	To	Control/ Feedback Mechanism	Contributory Factors	Inadequate Enforcement of Constraints	Inadequate Execution of Control Actions	Inadequate or Missing Feedback	Remedial Measures
Health service	Hospital operational management	Health service procedure	Absence of guideline or procedures for prescribing and administering IV Heparin	Inappropriate, ineffective, or missing control action			Develop and communicate guidelines for prescribing and administering IV Heparin
Hospital operational management	Administering nurse	Local procedures	No formal process in place to ensure face-to-face clinical handover occurs following transfers	Inappropriate, ineffective, or missing control action			Establish process and procedure for face-to-face clinical handover
Administering nurse (Ward 1)	Administering nurse (Ward 2)	Handover	Rate not checked at transfer of patient from Ward 1 to Ward 2 (1430)	Inappropriate, ineffective, or missing control action			Implement rate check during handover process

As shown in Figure 11.6 and Table 11.2, identified control failures included the absence of a guideline or procedure for prescribing and administering Heparin, the lack of a formal process to ensure that a face to face clinical handover occurs following patient transfers, and the failure to check the infusion rate when transferring the patient between wards.

RECOMMENDED READING

Leveson, N. G. (2004). A new accident model for engineering safer systems. *Safety Science*, 42, 237–270. doi:10.1016/S0925-7535(03)00047-X

Leveson, N. G. (2011). Applying systems thinking to analyze and learn from events. *Safety Science*, 49, 55–64. doi:10.1016/j.ssci.2009.12.021

Leveson, N. G. (2017). Rasmussen's legacy: A paradigm change in engineering for safety. *Applied Ergonomics*, 59, 581–591. doi:10.1016/j.apergo.2016.01.015

Rasmussen, J. (1997). Risk management in a dynamic society: A modelling problem. *Safety Science* 27(2–3), 183–213.

REFERENCES

Australian Transport Safety Bureau. (2008). Analysis, Causality and Proof in Safety Investigations. Aviation Research and Analysis Report – AR-2007-053. Canberra: Australian Transport Safety Bureau.

Chicco, D. & Jurman, G. (2020). The advantages of the Matthews correlation coefficient (MCC) over F1 score and accuracy in binary classification evaluation. *BMC Genomics*, 21(1), 1–13.

Goncalves Filho, A. P., Jun, G. T., & Waterson, P. (2019). Four studies, two methods, one accident–An examination of the reliability and validity of AcciMap and STAMP for accident analysis. *Safety Science*, 113, 310–317.

Green, D.M. & Swets, J.A. (1966). *Signal Detection Theory and Psychophysics*. New York: Wiley.

Hulme, A., Salmon, P. M., Nielsen, R. O., Read, G. R., & Finch, C. (2017). From control to causation: Validating a 'complex systems model' of running-related injury development and prevention. *Applied Ergonomics*, 65, 345–354.

Hulme, A., Stanton, N.A., Walker, G.H., Waterson, P., & Salmon, P.M. (2019). What do applications of systems thinking accident analysis methods tell us about accident causation? A systematic review of applications between 1990 and 2018. *Safety Science*, 117, 164–183.

Hulme A., Stanton, N. A., Walker, G. H., Waterson, P., & Salmon, P.M. (2021). Are accident analysis methods fit for purpose? Testing the criterion-referenced concurrent validity of AcciMap STAMP-CAST and AcciNet. *Safety Science*, 144, doi:10.1016/j.ssci.2021.105454

Leveson, N. G. (2004). A new accident model for engineering safer systems. *Safety Science*, 42, 237–270. doi:10.1016/S0925-7535(03)00047-X

Leveson, N. G. (2011). Applying systems thinking to analyze and learn from events. *Safety Science*, 49, 55–64. doi:10.1016/j.ssci.2009.12.021

Leveson, N.G. & Thomas, J.P. (2018). *STPA Handbook*. Cambridge, MA, USA: MIT.

Li, F., Wang, W., Xu, J., Dubljevic, S., & Khan, F., Yi, J. (2020). A CAST-based causal analysis of the catastrophic underground pipeline gas explosion in Taiwan. *Engineering Failure Analysis*. 108; January, doi:10.1016/j.engfailanal.2019.104343

Pham, J. C., Girard, T., & Pronovost, P. J. (2013). What to do with healthcare incident reporting systems. *Journal of Public Health Research*, 2(3).

Rasmussen, J. (1997). Risk management in a dynamic society: A modelling problem. *Safety Science*, 27(2–3), 183–213.

Read, G. J. M., Naweed, A., & Salmon, P. M. (2019). Complexity on the rails: A systems-based approach to understanding safety management in rail transport. *Reliability Engineering and System Safety*, 188, 352–365.

Salmon, P. M., Read, G. J. M., & Stevens, N. (2016). Who is in control of road safety? A STAMP control structure analysis of the road transport system in Queensland, Australia. *Accident Analysis and Prevention*, 96, 140–151.

Sebnem Düzgüna, H. & Leveson, N. (2018). Analysis of soma mine disaster using causal analysis based on systems theory (CAST). *Safety Science*, 110, 37–57.

Westbrook, J. I., Li, L., Lehnbom, E. C., Baysari, M. T., Braithwaite, J., Burke, R., … & Day, R. O. (2015). What are incident reports telling us? A comparative study at two Australian hospitals of medication errors identified at audit, detected by staff and reported to an incident system. *International Journal for Quality in Health Care*, 27(1), 1–9.

Section V

Computational Modelling Methods

12 Agent-Based Modelling (ABM)

BACKGROUND

Agent-Based Modelling (ABM) is perhaps the most 'computational' tool used in systems thinking and research – this can have its advantages and disadvantages. The major advantage of ABM is that it is almost limitless in its application and flexibility. The downside and major disadvantage of ABM is that this flexibility means that ABMs applied to different (or similar) problems rarely look similar or behave the same way. This can leave researchers new to ABM and interested in exploring its use confused about what an ABM actually is, how to conduct one, or whether it is 'right' for the purposes and problems they imagine using it for.

In this chapter we will introduce the systems researcher to the concept of ABM, how ABMs are regularly used, what they are best used for and what they aren't. The structure of this chapter may diverge somewhat from others in that, beyond broad principles, there is no one-way to develop and deploy an ABM because it is essentially a piece of software. ABMs can be designed to represent almost anything; from acting as a simple calculator, system dynamics model, or genetic algorithm, to representing neural networks or reinforcement learning algorithms represented within the 'brain' of thousands of simulated people or animals. They can be bent almost at will by anyone who has the requisite skill and patience to learn their intricacies as well as the imagination to envisage their application.

DOMAIN OF APPLICATION

A common occurrence among researchers who are advocates of ABM, is that once you have seen them, they can't be unseen. Suddenly everything can be modelled as an ABM. You have a powerful, insightful, and exciting research hammer and every unexplained systems problem now looks like a nail; their advocates can become zealots. ABMs are therefore regularly applied across a wide range of disciplines focused on the study of sociotechnical systems. They are common within economics, ecology, biology, geography, epidemiology, mathematics, engineering, anthropology, and sociology. Primarily, however, their technical development occurs in the realms of computer science where advances in the algorithms that drive programs underlying ABM applications and behaviours are made. Figure 12.1 shows the intersection between disciplines that have a stake in the broad area of computational social science. As can be seen, just a single step away from one's comfortable scientific discipline, can be found the area of computational science where you may suddenly be met with researchers from a blend of computer science and mathematics.

DOI: 10.1201/9780429281624-12

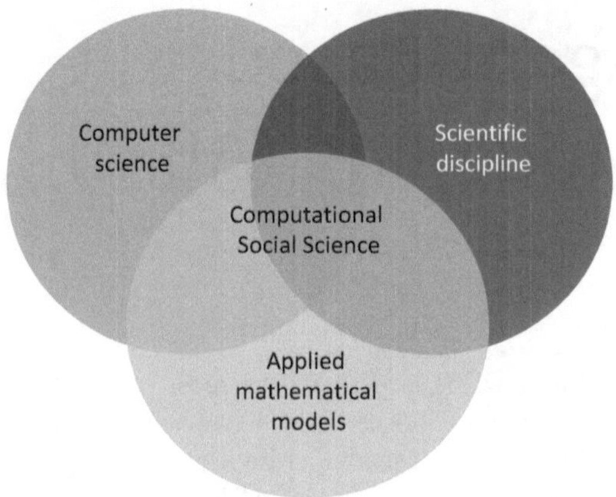

FIGURE 12.1 A conceptual diagram of the relationship between computational social science (including ABM), and other domain-specific areas.

For the social scientist interested in application of ABMs to real-world systems, the computer science and mathematical literature can be a challenging environment replete with seemingly impregnable language and concepts. For example, the first AAMAS (Agent-based, Autonomous and Multi-Agent Systems) conference contained five streams of ABMs devoted solely to game-theory, and its key-note address was given by the world-famous artificial intelligence (AI) researcher, Andrew Ng. As a developer of ABMs focused on transport psychology and safety applications, it was easy to feel out of one's depth. Despite this, Andrew led us through a fascinating discussion of AI development and applications in autonomous vehicles, that was only upstaged by a second keynote showing a video-history of advances in AAMAS' robot soccer competition, RoboCup (https://www.robocup.org/), which has been running since 1997.

"What have autonomous cars and robot soccer got to do with agent-based modelling?" you may ask. Well, quite a lot, as it turns out. In fact, the overlap between AI researchers and ABM or 'multiagent system' researchers is great. Increasingly, AI algorithms such as neural networks or reinforcement learning algorithms are being place inside ABM agents, enabling them to learn, behave, and react in ways that reflect learning and resultant behaviour within the social-technical, socio-environmental, or other systems under study. The term 'agent' is also appearing more frequently in the AI literature, referring often to individual entities housing virtual brains. The convergence of AI and ABM is evident in the literature as both ABM researchers and more 'traditional' (if that can be said) AI researchers begin to understand the benefits of combining both platforms to create artificial societies and tackle some of the world's most pressing (and interesting) social and systems issues. Issues you might also be interested to explore.

However, as intimidating and confusing as a 5-day AAMAS conference might be, I have learned that social scientists can have a genuine role in contributing to the discussion and development of the field. Firstly, if you are female, it is important to be there. Over 90% of the ABM development crowd is male. There is nothing wrong with being male, however (and I don't think this is a controversial position) males tend to have a male perspective when it comes to thinking about and modelling social interactions and systems leading to the establishment of 'truth' in scientific literature. If females are not involved in the development of actual truth, then the truth about how people and systems operate will be constructed from an overwhelmingly male perspective. Involvement with and of female social scientists is therefore paramount. The second (and not unrelated) issue is that the sector is understandably 'nerdy' – and by this, I mean fairly insular, and hesitant to engage with the world outside its domain. Again, this has the potential to push the research agenda (and therefore establishment of research priorities and truth) down a potentially unrepresentative path. Here is but one example of how this plays out.

At the 2019 AAMAS conference in Montreal, I spent most of my time in sessions devoted to human–computer interaction. This is because I am particularly interested in the issue of how policymakers and system managers can learn to trust the outputs of ABMs as decision-support. Attending the sessions, I therefore expected to be presented with world-class research related to the difficult task of tackling humans 'in the loop'. However, what I got was quite different. I am not exaggerating when I tell you that one of the presentations explained the methods of their human-computer interaction in the following way; "We tried working with humans, but they were too difficult, so instead, we simulated them using computers." Without batting an eyelid, the researchers then went on to explain the results of their model and its implications for the real world. They were not alone in taking this approach.[1]

Social and systems scientists including those in HFE therefore have a considerable role to play in the development and application of ABMs. Above, I made the statement that '*ABMs can be bent at will by anyone who has the requisite skills and patience to learn their intricacies*'. While this is true, I also left out a crucial element that renders any ABM practically useless unless it is also embedded in the developed model – domain knowledge gathered from subject matter experts. Here is where domain experts (e.g. clinicians or business operators), social scientists and systems researchers can have their biggest impact; by assisting to translate their tacit domain knowledge into computational representations of real systems. Ultimately, ABMs need to provide transparent and understandable representations of real social phenomena under study. They need to explicitly describe out how they have been constructed, what their assumptions are, which entities and relationships are included, and which are left out. Without this, any ABM is simply an academic exercise with limited validity and real-world implication. Worse, it could be entirely misleading.

So, I say all this because I want to convey that ABM is not a single approach used for obtaining desired outcomes or answers using well-defined steps that progress in clearly defined stages (see below). Talking about 'how to build' an ABM is a little like talking about how to build a phone app – they may all have a common set of capabilities based around their architecture, software capabilities and interface, but within these restrictions, freedom reigns. What's most important for an app developer to understand is the problem you (and your audience) need solving. ABM is similar. It is not a model per se, it is a platform upon which you can build a model.

CAN YOU PLEASE JUST TELL ME WHAT ABMS ARE?

ABMs are computer models composed of individual computational entities called 'agents' who behave based on a set of pre-defined rules of potentially infinite complexity – hence ‘agent-based’ model. These agents can represent any individual entity that you can imagine, and they interact with each other ‘inside’ the model over a time period set by the model designer. Further, each agent can be infused with their own goals, beliefs, desires, intentions or preferences that influence how they behave, and how they behave *in response* to the behaviour of other agents or changes to the environment within which they live. Just like a real person, company, institution, society, or any other individual entity that can be defined within a real system, each agent in an ABM has its own unique set of characteristics that determine the rules by which it chooses to live its virtual life. So, unlike many other systems modelling frameworks that are a) static, b) do not distinguish between individuals, and c) tend to pre-determine the proportion of individuals that act in certain ways based on top-down policy adjustments (see system dynamics modelling in Chapter 13), in ABM, the individual agent is in charge of how it lives its life. You, the modeller, merely design the range of agents you are interested in, set them free in your artificial world, and watch what happens.

In this sense, ABM is a true ‘bottom-up’ approach. You do not first construct the behaviour of the society that agents live within and ask agents to mimic society’s apparent principles on how to behave. Instead, the structure of the society emerges or ‘grows’ from the behaviour that agents decide to take and their interaction with one another in the world – or at least it should (more on the difference between ‘interesting’ and ‘boring’ ABMs later). To explain how a macro societal pattern of behaviour comes about, your ABM should be able to ‘grow’ that behaviour simply from the interaction of individuals within your artificial world (Joshua M. Epstein & Axtell, 1996).[2] In this sense, ABMs differ from microsimulations, which are data driven. While ABMs can use exiting data to inform their structures and behaviour, they are not reliant on data to operate. Instead, ABMs *create* data. Artificial societies generated through ABMs *become* the society of interest. They are what is being studied.

THINKING ABM-STYLE

To help bring the approach to life, let’s practice ‘thinking in ABM’.

First, take a moment to imagine you hold in your hand a set of agents that represent a population of things that interests you. If you are having trouble imagining this, think about collecting a set of 100 random bugs from your garden and then putting them in a jar which represents your artificial world. Pick up the jar and carefully study the bugs – what characteristics set them apart from one another? For example, each bug might have its own colour, size, level of aggression, apparent health, speed, ability to fly, energy, desire to escape, or any number of other characteristics. Similarly, other bugs might have qualities that determine how they behave and react to those around them – perhaps when approached by other bugs, some want to go to the corner of the jar to be alone. But what if 50 bugs all get the same idea to go to the corner – what then? How do the first bugs respond?

The analogy we are using is bugs, and any entomologist will tell you that bugs are perfectly reasonable agents to model, and demonstrate some fascinating collective behaviour that is emblematic of research in complex systems (Foster et al., 2014). However, you can also broaden your imagination. Imagine you are now holding a jar of miniature people, vehicles, cyclists, cats, birds, soldiers, workers, businesses, or classes of any of the above. For example, you might have a jar of people who are police (people class 1), suspected criminals (people class 2), and potential victims (people class 3). You might even imagine a jar of workers (agents class 1) together with their managers or supervisors (agents class 2) and the businesses they work in (agents class 3). That's right – agents can also represent collectives such as schools, workplaces, countries, or other collectives. Any number and scale of agents you can imagine is theoretically possible.

Now imagine each of your police, criminals, and potential victim agents in separate jars are now combined in a single environment, forcing them to interact. How do they interact? Who do they interact with? What is the nature of their interaction? It could be spatial (e.g. a set of police agents might move towards a set of suspected criminal agents who then attempt to escape), it could be educational (e.g. a set of teacher agents might pass information to a set of student agents who then pass information on to parent agents), or it could be behavioural (e.g. individuals from a set of job recruiters could compete to contact job hunters from a set of workers and then connect them to a set of employers to collect a financial reward). In fact, this is the type of behaviour that ABMs are especially good at; modelling interactions, incentives, and exchanges at the level of individual agents to understand patterns of behaviour at the larger system and societal scale.

However, I hinted in the paragraphs earlier that the structure of the artificial society should come from the behaviour of the agents at the individual level – that is, the model should be bottom-up. This is the last, and most important point I want to emphasise when thinking about a problem you'd like to tackle using ABM. That is, the difference between 'interesting' and 'boring' ABMs. The difference is this:

ABM is best used in situations where you are presented with macro-system behaviour that you observe in the world but do not really understand (or perhaps no-one does), but you want to explain *how* it emerges from the interaction of individuals. The key here is that you should not really understand how the system works before attempting to model it. You might have a hunch or hypothesis, but it is thus far a mystery, and one that you (probably) suspect existing theorists or policymakers have not adequately addressed. Models that explore explanatory mechanisms like this are what I would consider 'interesting' problems because they can provide new insight. Examples of macro patterns where conjecture remains around how those patterns are created could include housing and financial markets, political movements and revolutions, working conditions and wages, bullying, gender or racial employment biases, disease spread and injury models, geographic voting patterns, bushfire behaviour models, or the development of professional, academic, and social networks on or off-line.

By contrast, ABM should be sparingly used by social scientists to investigate 'boring' problems that are simply scaled-up versions of micro behaviours operating at the macro scale (see complexity discussion in Chapter 1). They should also not be used in circumstances where there is limited interaction, capacity for variation, or feedback between agents within the model. Examples of boring problems might be: logistics models where packages of goods simply move between locations, pedestrian or traffic models where origin-destination matrices are static; educational achievement models where the average score of a class on a test is simply a function of how much time students spend in the classroom; or models of hospital flow where patients are transferred between treatment units based on presenting symptoms.

Unfortunately, ABMs are occasionally, needlessly, used in these kinds of applications – often accompanied by 'whizz-bang' 3D graphics interfaces that provide an impression the model is far more sophisticated than it is. While models of this type can still be useful for testing 'what if' scenarios (e.g. "What if I buy a larger capacity truck to move my goods?", or "What if I employ more front line triage staff?"), they often do not test existing nor generate new theory because they are built on an existing top-down understanding of the system's mechanics. Their results can also more easily, verifiably and efficiently be produced using alternative, simpler means. In summary, if you already know how a macro societal pattern is created in the real world and if this pattern is easily predicted from the behaviour of individuals (e.g. the number of people who like ice-cream at the population level is simply the sum of people who like ice-cream at the local level, and this remains stable over time) you can create a simulation of this using ABM – and it will work without error – but it will likely be very boring, and you have not really used ABM in its most powerful guise. To summarise, interesting ABMs are best used for exploration of the unknown rather than simulation of the familiar.

Procedure and Advice

The procedure set out below is directed towards the reader who has not previously tackled the development of an ABM. It is therefore focused on advice on how to think about the model under construction rather than how to 'code' or implement a model. This will be useful to those of you who have HFE domain knowledge and want to work with an agent-based modeller who does not have the domain knowledge that you do. You can, of course, learn to code and construct models, yourself, and for that, there are many useful texts to get you started (e.g. Gilbert, 2019; Kahn, 2015; Macal and North, 2005; Wilensky and Rand, 2015).

However, for the 'conceptual' modeller who wants to convert their ideas into action, taking the information below to your first few meetings with a modeller will supercharge the process of model development and answer many of the questions they are likely to have. Further, as the advice here is meant for the relative novice, it excludes information related to participatory modelling workshops or other methods that would normally be considered important when developing models alongside third parties. More advanced readers are encouraged to consider texts and frameworks for model building and explanation such as van Dam et al. (2012), and Grimm et al. (2020).

A flowchart depicting the ABM procedure is presented in Figure 12.2. Step-by-step guidance is presented below.

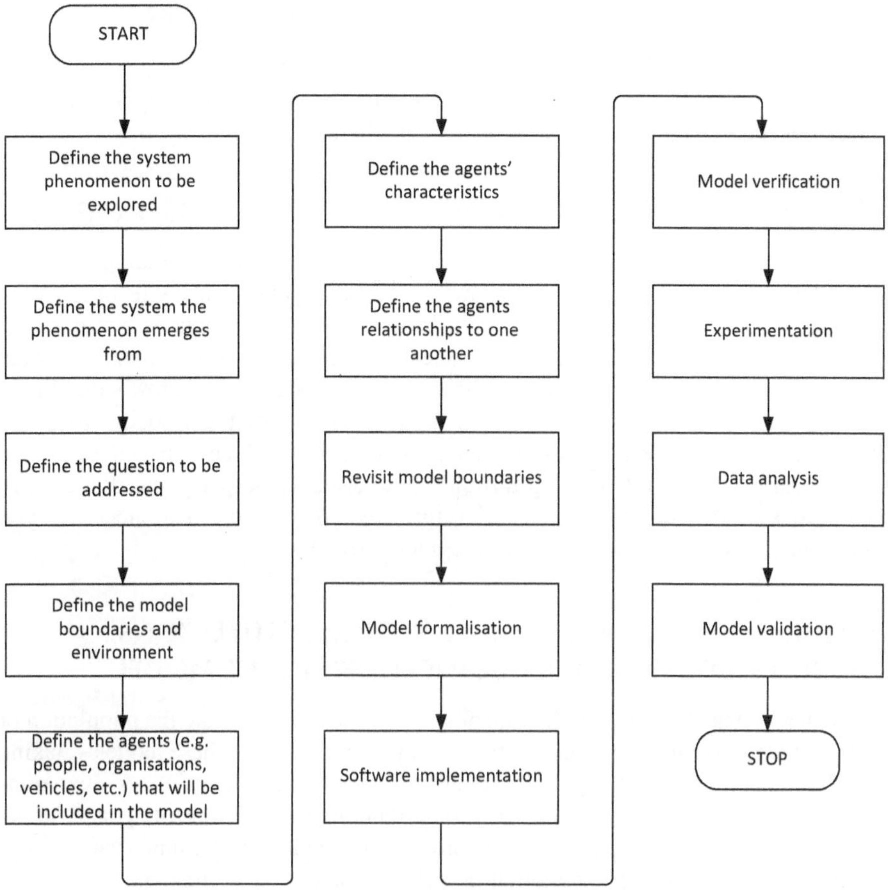

FIGURE 12.2 A 14-step flowchart for defining, building, and analysing agent-based models.

STEP 1: DEFINE THE SYSTEMS PHENOMENON THAT YOU WANT TO EXPLORE

Remember, this should ideally be a contested or intriguing social phenomenon that is currently poorly understood.

STEP 2: DEFINE THE SYSTEM THE PHENOMENON EMERGES FROM

The system does not need to be drawn from reality. In fact, if you are thinking about tackling an ABM from start to finish by yourself, it is probably best to model an abstract or simple 'toy' system that interests you in the first instance.

STEP 3: DEFINE THE QUESTION YOU WANT TO ADDRESS

For example, "No-one instructs society to act like this from the top-down, so what are the micro-level interactions between individual agents in this system that create macro-phenomena of (insert macro-observation here)?"

STEP 4: DEFINE THE MODEL BOUNDARIES AND ENVIRONMENT

Given that ultimately, 'everything is connected to everything', deciding what the boundaries of the system are and the environment that your agents will 'live' within is very important. However, your model should also be as simple as possible. So, what features of the system absolutely need to be included in the model and what should be left out? How long should the system be modelled for? How should time in the model be counted? Is 'space' in the mode real or abstract (e.g. social distance)? What resources are available in the model environment? When will you know that the model has finished? Working through these issues is not easy, but don't worry about making the boundaries 'perfect' at this stage. It is almost guaranteed that you will revisit this question as model development progresses.

STEP 5: DEFINE THE AGENTS (E.G. PEOPLE, INSTITUTIONS, VEHICLES, ETC.) THAT WILL BE INCLUDED IN THE MODEL

Now that you have a sense of the model's boundaries, clearly state the population or sets of agents you are interested in including. Define classes of these models – taking our bug analogy from above, is it enough to have a single set of 'bugs' who simply have different characteristics? Or should there be three classes of agents: 'ants', 'caterpillars', and 'flies'? The architecture of your model will be important when it comes to its ultimate implementation in code by you or your technician.

STEP 6: DEFINE THE AGENTS' CHARACTERISTICS

What abilities do your agents have? On what qualities do each of the agents in your model vary? How much do these characteristics vary between agents? How are these characteristics defined? Do they use nominal, ordinal, or scale values? How will these characteristics be assigned to each of your agents? Do these values change over time? Do your agents have goals? What behaviour do they exhibit in order to achieve those goals? Do they receive punishments? What decisions do they make? Under what circumstances do they make these decisions? What do agents perceive in the environment? Do they move? Why do they move? Do they ever die? If so, are they replaced? Any of these questions may or not be relevant to the agents in your model. There is an infinite range of characteristics and behaviours that you can prescribe.

STEP 7: DEFINE THE AGENTS' RELATIONSHIPS TO ONE ANOTHER

How do each of the agents relate to one other? What information do they exchange? Under what circumstances do they exchange it? How does the behaviour of some agents affect the behaviour of others? How does the behaviour of the agent affect its characteristics? How does the behaviour of other agents the agent interacts with affect its characteristics? For example, when an ill patient with a health 'score' of 3/10 visits a doctor, what happens to their level of health? Causal loop diagrams (see Chapter 8) and state charts can be useful thinking tools help at this stage.

STEP 8: REVISIT YOUR MODEL BOUNDARIES

Is the scope of your model still correct? Are all the agents that need to be included now included? Are redundant agents (those that do not interact with others) removed?

STEP 9: MODEL FORMALISATION

At this stage, the model transfers from the domain knowledge expert to the computational modeller. The model formalisation phase transcribes the model 'story' into pseudo-code, which is a generic language enabling integration into your software platform of choice. There are a range of ABM platforms available, however a platform that is free, very user-friendly and has an extensive examples and support community is NetLogo (Wilensky, 2019). Search for the download in your browser.

STEP 10: SOFTWARE IMPLEMENTATION

At this stage, the model will be coded into agent-based modelling platform by the model developer – that might be you.

STEP 11: MODEL VERIFICATION (THIS IS A LONG PROCESS)

Here, the model is tested under a variety of circumstances and settings to confirm that the code is correct, and the model contains no technical errors that need to be fixed before observing the model results. This is essentially a relationship confirmation and 'bug fixing' phase and might take you all the way back to Step 4 on occasion.

STEP 12: EXPERIMENTATION

Once you are confident that the model structure and code is operating correctly, this stage is where you manipulate the model and test its results.

STEP 13: DATA ANALYSIS

Depending on how you have built the model, data produced by it will be output either in the model interface enabling you to simply 'see' what is going on, or in.csv format enabling you to analyse results using your chosen standard analysis program. Results from 'toy' models will most likely be on-screen. For researchers familiar with programs such as R, or programming libraries like Numpy, various plugins exist for NetLogo that enable the researcher to move between these applications. This is particularly useful for models that produce large datasets. Even 'simple' ABMs can produce datasets that are many millions of rows long. Deciding how long you need to run the model for and what variables you need to record (see Step 4) is also important for this stage.

STEP 14: MODEL VALIDATION

This is your point of reflection. How did the model perform? Did it reproduce the effects that you expected? Did it reproduce patterns observed in the real world? Do you think the mechanisms that created those patterns are realistic or constitute a 'candidate mechanism' (Joshua M. Epstein and Axtell, 1996)? Did anything unexpected occur? Changes to which variables contributed to the patterns you observed? Your model is just an abstraction of reality, so it is probably wrong, but is it useful (Box, 1976)?

TRAINING AND APPLICATION TIMES

Agent-based modelling is a beguiling method that can take some time to become competent in. Not only do you have to start 'thinking' like an agent-based modeller (from the bottom-up), but you also need to learn how to translate the concepts and interactions that you have in your head, into code used by the modelling software. Typically, a simple model that is of either a publishable or practically applicable standard will take 3 to 6 months to develop, and a similar time to either embed in a process or write-up and have published.

RELIABILITY AND VALIDITY EVIDENCE

Because it is an unfamiliar technique to many, ABM suffers from a dose of scepticism within disciplines outside computer science. It is often challenging for reviewers or others in applied settings to 'trust' a model – especially if they have not been involved in its development and can only interact with it in a static environment. These and similar issues are described well by Sondoss et al. (2020) and entertainingly by Waldherr and Wijermans (2013) in their paper, 'Describing social simulation models to sceptical minds'. Other researchers have attempted valiantly to create standardised frameworks for the development and description of ABMs in order to improve readability, transparency, and reliability, the success of which has been mixed (e.g. Grimm et al., 2010; Grimm et al., 2017). Open data, open modelling (e.g. comses.net), community modelling (http://modelingcommons.org), and code hosting platforms such as GitHub (github.com) and Bitbucket (bitbucket.com) have also assisted in creating greater transparency, checks, and reproducibility for model developers.

While it is a relatively simple task for a modeller to seek historic data for inputs and to attempt data calibration for outputs, validation (and in many cases the want for prediction) is perhaps not the best test of whether an ABM is been useful or not. Indeed, Joshua Epstein's (2008) lecture published in the Journal of Artificial Societies and Social Simulation (jasss.soc.surrey.ac.uk), 'Why model' provides 16 reasons other than prediction to use ABM. The most powerful element of which, I think, is to *explain*. Funnily enough, I had the good fortune of listening to Professor Epstein give a lecture in 2016 on predicting outbreak and spread of the Zika virus in New York during a conference devoted to the 20th Anniversary of his and Robert Axtell's seminal text, *Growing Artificial Societies: Social Science from the Bottom Up* (1996). After the lecture, I asked him why, given he had probably heard of a lecture (given by himself) on 16 reasons *other* than prediction to do modelling, he presented a model of prediction. His response was, "I still need to get paid, and prediction is what people will pay you for." Sound familiar?

A perhaps more important point about reliability and validity he made in both this and the previous lecture I mentioned, however, was emphasis often placed on model validation. "But can you validate your model?" is a common question that ABM researchers are asked. Epstein's response is to ask the questioner if by 'validate' they mean he can write it down explicitly to calibrate against existing data or relationships. If that is the question, then the answer is "Yes". In his words, "It [the model] may be baloney, but at least it's replicable baloney." This is his major point. ABM enables and requires a model to be explicitly defined. Most people are working with models every day, but they are mental, implicit models. ABM demands a discipline that requires the implicit to be defined, refined, and tested.

And here, too, is where the issue of candidate mechanisms for explaining social phenomena arise. Even if a model can accurately track and predict growth and dynamics in the model's obscure and complex artificial society, it still does not prove that the model represents the true nature of dynamics within a real-world system. All it means is that the mechanism cannot be dismissed as a possible candidate.

TOOLS REQUIRED

A number of free and commercial ABM tools exist for systems and social science researchers. For social scientists, the most popular non-commercial application available, containing the broadest community support and widest range of example and shared models is NetLogo (Wilensky, 2019). Other applications (and associated languages) worthy of note that use general programming languages the reader may be already familiar with are Mesa (Python) (Kazil et al., 2020), Agents.jl (Julia) (Datseris et al., 2021), and RePast (Java) (North et al., 2013).

CASE STUDY EXAMPLE: AUTONOMOUS VEHICLE AND VULNERABLE ROAD USER CRASHES

The following example is a short paper submitted as an extended Abstract to the 2019 Agent-Based Modelling of Urban Systems workshop, which is part of the broader AAMAS conference. While it does not go into great detail, it provides the reader with an understanding of the structure and format that ABM papers and publications can potentially take, and the types of issues they can explore.

BACKGROUND

Around the world, road crashes kill 1.35 million people per year and injure up to 50 million more (World Health Organization, 2018). Road trauma is now the world's

5th leading cause of death and disability (Abajobir et al., 2017). Global road trauma continues to rise as developing countries embrace the motor vehicle, which is replacing active transport mode-share. The transition period that developing countries face, which sees mixed mode-share between vulnerable road users (e.g. pedestrians and cyclists) alongside motor vehicles delivers very high rates of injury and trauma among these populations (Stevenson et al., 2016).

While developing countries move inexorably towards car-based transport systems, advanced economies are becoming increasingly aware of the economic and health benefits associated with urban design that discourages car-use and encourages more cycling and walking. However, in cities where poor separation between motorised vehicles and vulnerable road users exists (e.g. traffic modes mix) injury rates per km travelled are higher. It is therefore difficult for cities to at once encourage increased rates of cycling without experiencing concomitant (though not proportional) increases in cycling crashes (Stevenson et al., 2016). While investment in separated cycling infrastructure is one solution to this problem, high financial and political costs are often associated with such policies when attempted in already car-dominated countries such as the United States, Canada, Australia, and New Zealand (Pucher et al., 2010; Reynolds et al., 2009).

The deployment of Autonomous vehicles (AVs) has been promoted as an alternative solution to the issue of mixed mode interaction. If AVs can learn to accurately identify other vehicles and vulnerable road users and either evade or avoid situations leading to crashes, the potential benefits in terms of reduced injuries and trauma could be significant. However, while the ideal of flawlessly performing AVs exists, the technology required to ensure AVs can operate safely in the extreme variety of unpredictable and novel situations vehicles face, to learn from these, and to then to make appropriately 'moral' decisions in in the case of inevitable collisions, is far from ready (Hancock, 2018). Further, once deployed, AVs will be held to a higher safety standard than human drivers. Perhaps driven by the marketing 'hype' surrounding AV technology, community expectations hold that AVs will be *safer* than human drivers (Anderson et al., 2018). In this context, the financial and legal consequences for AV manufacturers' vehicles contributing to crashes may also be higher than for fallible human drivers.

However, technological barriers aside, if AV manufacturers achieve 'flawlessly performing' vehicles that never collide with humans, humans' responses to the introduction of AVs are still uncertain. One of humans' great strengths is that we are adaptable; we learn to manipulate and exploit our environment. Consistent with other learned behaviours across multiple domains, if AVs behave differently around humans (i.e. to drive conservatively so as to avoid collisions at all costs), humans may begin to behave differently around AVs than they had previously around 'manual' cars that bear the threat of injury. This dynamic adaptation could lead to new situations and crash risks not previously present or imagined within transport systems.

The goal of this case study was to explore changes in car-cyclist crash frequency associated with the introduction of AVs into an existing transport system where AVs perform 'flawlessly', human drivers adapt to the presence of cyclists, and cyclists adapt to the presence of AVs.

METHOD

The model was an adaptation of a previously published (Thompson et al., 2016) NetLogo (Wilensky, 2019) agent-based model of behavioural adaptation by car drivers to the increased presence of cyclists represented within a simple, 'Manhattan grid' configuration. In the prior model, drivers became increasingly aware of cyclists as more cyclists entered the system, and through an individual learning process represented by Rescorla–Wagner (R–W) models (Rescorla and Wagner, 1972), adapted their driving behaviour and likelihood of driving safely around cyclists in response. In this paper, we extended the representation to include an additional class of vehicle (AVs) and also provided cyclists with an ability to learn to recognise and adapt behaviour towards both human-driven cars (cars) and AVs. The means of adaptation for cyclists also followed an R–W configuration, with each cyclist possessing an accuracy of perception, a desire to act well, a memory for previous events, a maximum level of association, a timescale for building up knowledge, and a timescale for forgetting. The full model was published in the Journal Ergonomics (Thompson et al., 2020).

Assumptions made in the model were that AVs acted flawlessly by exhibiting extreme care, perfect identification of other vehicles, and always yielded to other cars and cyclists. AVs never collided with other vehicles. Human drivers acted identically to the manner they had done in the previous model; while desiring to drive as safely as they could around cyclists, their performance was randomly imperfect, and their skill-levels changed in response to increasing or decreasing exposure to cyclists, consistent with the behavioural adaptation hypothesis.

Cyclists' adaptation consisted of them learning that AVs would not crash into them. With increasing exposure, cyclists learned that they could safely progress through intersections without needing to yield to AVs because they operated flawlessly. Like human drivers, however, cyclists' perceptions were also not perfect; they did not always recognise the difference between cars and AVs.

To test the comparative effect on crashes that the introduction of AVs had, the model was run using two experiments. In the first, a comparison of crash frequency was made between conditions consisting of a population of 2,000 cars and 500 bicycles, with crash frequencies after 500 additional manually operated cars (MVs) were deployed in the system at time-step 250 of 750. In the second experiment, an additional 500 autonomous vehicles (AVs) were instead deployed. Cyclist numbers remained constant in each experiment.

RESULTS

Results of condition 1 (deployment of an additional 500 MVs) demonstrated that the mean number of collisions in the system increased from 3.7 per time-step to 4.4 after MVs were introduced. In the second experiment – and despite AVs operating 'flawlessly' and never colliding with any other vehicle – mean crashes per time-step also increased from 3.7 to 4.1. The increased crash frequency in the AV deployment condition was not caused by AV's behaviour, but occurred between human-operated cars and cyclists due to cyclists' adjusted, imperfect expectations about the behaviour and capability of human-operated and autonomous cars they were exposed to in the system (see Figure 12.3). Cyclists who eventually became used to AVs occasionally expected cars to operate as AVs did. They then failed to yield and were more likely to be crashed into by human-driven vehicles. For identical reasons, subsequent experiments also showed that when human-operated cars in the system were exchanged over time by AVs, cyclist crashes did not follow a parallel reduction in frequency over time, but reduced more slowly in the early stages of replacement before accelerating in the latter stages as full replacement neared (Thompson et al., 2020, Figure 12.4).

FIGURE 12.3 View of the NetLogo model indicating the system's Manhattan grid configuration and showing the position of vehicles at one moment in the system.

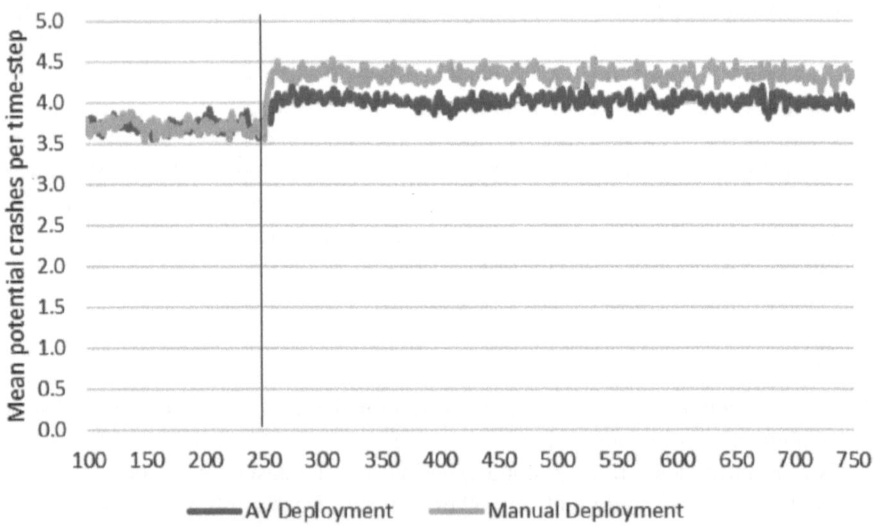

FIGURE 12.4 Count of crashes per time-step with addition of either autonomous or manual vehicles at time-step 250, indicated by the dark vertical line.

CONCLUSIONS

Humans are highly proficient at learning and adapting quickly to new environments, resulting in novel patterns of interaction. Regardless of whether AVs can learn to operate flawlessly in a transport system or not, humans will display adaptive behaviours in response to AVs' presence, especially when there is incentive (such as reducing travel time) to manipulate interactions. Such adaptation potentially creates novel crash situations and risks that are not typically in-scope of discussion when considering the safety benefits of AVs, which tend to focus on the role of vehicle operators (Hancock, 2018). Dynamic modelling perspectives that include surrounding humans' potential responses to AVs behaviour should be included in discourse and modelling of human–AV interaction.

RECOMMENDED READING

The following texts are offered as starting points for assisting to further understand ABM, and to jump into some practical examples.

- Agent-based modelling (Gilbert, 2008);
- An introduction to agent-based modelling: modelling natural, social, and engineered complex systems with NetLogo (Wilensky and Rand, 2015);
- An introduction to agent-based modelling (Kahn, 2015); and
- Tutorial on agent-based modelling and simulation (Macal and North, 2005).

More than anything, though, I would recommend readers jump into NetLogo (Wilensky, 2021) and open examples from the myriad applications available in the Models Library – especially those from the social science category. Read the 'info' tabs on each of the models and follow the prompts they provide. Have fun experimenting with these models by testing their extremes. They are artificial societies – you can't really break them.

NOTES

1 The consequences of this would make for an interesting modelling question.
2 Also relevant here is the quote from Richard Feynman, "What I cannot create, I do not understand."

REFERENCES

Abajobir, A. A., Abate, K. H., Abbafati, C., Abbas, K. M., Abd-Allah, F., Abdulkader, R. S., … Aboyans, V. (2017). Global, regional, and national disability-adjusted life-years (DALYs) for 333 diseases and injuries and healthy life expectancy (HALE) for 195 countries and territories, 1990–2016: a systematic analysis for the Global Burden of Disease Study 2016. *The Lancet*, 390(10100), 1260–1344.

Anderson, J. M., Kalra, N., Stanley, K. D., & Morikawa, J. (2018). *Rethinking Insurance and Liability in the Transformative Age of Autonomous Vehicles*: RAND.

Box, G. E. P. (1976). Science and Statistics. *Journal of the American Statistical Association*, 71(356), 791–799. doi:10.2307/2286841

Datseris, G., Vahdati, A. R., & DuBois, T. C. (2021). Agents. jl: A performant and feature-full agent-based modelling software of minimal code complexity. *arXiv preprint arXiv:2101.10072*

Epstein, J. M. (2008). Why model? *Journal of Artificial Societies and Social Simulation*, 11(4), 12. Retrieved from http://jasss.soc.surrey.ac.uk/11/4/12.html

Epstein, J. M. & Axtell, R. (1996). *Growing Artificial Societies: Social Science from the Bottom Up.*

Foster, P. C., Mlot, N. J., Lin, A., & Hu, D. L. (2014). Fire ants actively control spacing and orientation within self-assemblages. *The Journal of Experimental Biology*, 217(12), 2089. doi:10.1242/jeb.093021

Gilbert, N. (2008). *Agent-Based Models*: Sage.

Gilbert, N. (2019). *Agent-Based Models* (Vol. 153): Sage Publications, Incorporated.

Grimm, V., Berger, U., DeAngelis, D. L., Polhill, J. G., Giske, J., & Railsback, S. F. (2010). The ODD protocol: A review and first update. *Ecological Modelling*, 221(23), 2760–2768.

Grimm, V., Polhill, G., & Touza, J. (2017). Documenting social simulation models: The ODD protocol as a standard. In *Simulating Social Complexity* (pp. 349–365): Springer.

Grimm, V., Railsback, S. F., Vincenot, C. E., Berger, U., Gallagher, C., DeAngelis, D. L., ... Ayllón, D. (2020). The ODD protocol for describing agent-based and other simulation models: a second update to improve clarity, replication, and structural realism. *Journal of Artificial Societies and Social Simulation*, 23(2), 7. doi:10.18564/jasss.4259

Hancock, P. A. (2018). Some pitfalls in the promises of automated and autonomous vehicles. *Ergonomics*, 1–17. doi:10.1080/00140139.2018.1498136

Kahn, K. (2015). An introduction to agent-based modeling. *Physics Today*, 68, 55.

Kazil, J., Masad, D., & Crooks, A. (2020). *Utilizing python for agent-based modelling: the mesa framework. Paper presented at the International Conference on Social Computing, Behavioral-Cultural Modeling and Prediction and Behavior Representation in Modeling and Simulation.*

Macal, C. M. & North, M. J. (2005). *Tutorial on Agent-Based Modeling and Simulation*. New York: IEEE.

North, M. J., Collier, N. T., Ozik, J., Tatara, E. R., Macal, C. M., Bragen, M., & Sydelko, P. (2013). Complex adaptive systems modeling with Repast Simphony. *Complex Adaptive Systems Modeling*, 1(1), 1–26.

Pucher, J., Dill, J., & Handy, S. (2010). Infrastructure, programs, and policies to increase bicycling: An international review. *Preventive Medicine*, 50, *Suppl 1*, S106–125. doi:10.1016/j.ypmed.2009.07.028

Rescorla, R. A. & Wagner, A. R. (1972). A theory of Pavlovian conditioning: Variations in the effectiveness of reinforcement and nonreinforcement. *Classical Conditioning II: Current Research and Theory*, 2, 64–99.

Reynolds, C. C., Harris, M. A., Teschke, K., Cripton, P. A., & Winters, M. (2009). The impact of transportation infrastructure on bicycling injuries and crashes: A review of the literature. *Environ Health*, 8, 47. doi:10.1186/1476-069X-8-47

Sondoss, E., Tatiana, F., Anthony, J. J., Albert, J. K., Moira, L. Z., Ioannis, N. A., ... Steve, J. L. (2020). Eight grand challenges in socio-environmental systems modeling. *Socio-Environmental Systems Modelling*, *2*(0). doi:10.18174/sesmo.2020a16226

Stevenson, M., Thompson, J., de Sa, T. H., Ewing, R., Mohan, D., McClure, R., ... Woodcock, J. (2016). Land use, transport, and population health: Estimating the health benefits of compact cities. *Lancet*, 388(10062), 2925–2935. doi:10.1016/S0140-6736(16)30067-8

Thompson, J., Read, G. J. M., Wijnands, J. S., & Salmon, P. M. (2020). The perils of perfect performance; considering the effects of introducing autonomous vehicles on rates of car vs cyclist conflict. *Ergonomics*, 1–22. doi:10.1080/00140139.2020.1739326

Thompson, J., Savino, G., & Stevenson, M. (2016). A model of behavioural adaptation as a contributor to the safety-in-numbers effect for cyclists. *Transportation Research Part A: Policy and Practice*, 85, 65–75. doi:10.1016/j.tra.2015.12.004

van Dam, K. H., Nikolic, I., & Lukszo, Z. (2012). *Agent-Based Modelling of Socio-Technical Systems* (Vol. 9): Springer Science & Business Media.

Waldherr, A. & Wijermans, N. (2013). Communicating social simulation models to sceptical minds. *Jasss-the Journal of Artificial Societies and Social Simulation*, 16(4), 13. Retrieved from Go to ISI://WOS:000330124700013

Wilensky, U. (2019). NetLogo Version 6.1.0 (Version 6.1.0). United States of America: Centre for Connected Learning and Computer-Based Modeling, Northwestern University. Retrieved from ccl.northwestern.edu/NetLogo

Wilensky, U. (2021). NetLogo Version 6.2.0 (Version 6.1.0). United States of America: Centre for Connected Learning and Computer-Based Modeling, Northwestern University. Retrieved from ccl.northwestern.edu/NetLogo

Wilensky, U. & Rand, W. (2015). *An Introduction to Agent-Based Modeling: Modeling Natural, Social, and Engineered Complex Systems with NetLogo*: Mit Press.

World Health Organization. (2018). *Global Status Report on Road Safety 2018*. World Health Organization.

13 System Dynamics

BACKGROUND

Stakeholders within complex systems are required, more than ever, to integrate different forms of scientific and expert-based evidence to inform decision-making (Currie et al., 2018). However, making an appropriate decision and identifying novel solutions, whether it be to advance a political, economic, or health-related agenda, is becoming increasingly challenging. Traditional analytical methods and statistical techniques are no longer appropriate to inform evidence-based decision-making due to the presence of non-linear cause-effect relationships and feedback loops within complex systems. Fortunately, sophisticated computational methods are available and can be applied to better understand the elusive dynamics of complex systems, of which one approach is system dynamics modelling.

System dynamics is a computational (i.e. computer-based) modelling method which is used to analyse the behaviour of complex systems and problems over time (Sterman, 2000). Originally created in the 1950s at the Massachusetts Institute of Technology (Forrester, 1958), applying system dynamics involves the development of quantitative models which incorporate stocks and flows, feedback loops, and information (auxiliary) variables. Equations are built into system dynamics models to enable simulation of behaviour in terms of growth and decay and the testing of different policy interventions.

System dynamics models are typically built via a group modelling process based on an initial Causal Loop Diagram (CLD) of the system or problem in question (see Chapter 8). An overview of the system dynamics modelling process is presented in Figure 13.1.

DOMAIN OF APPLICATION

System dynamics is highly flexible and has been used extensively to study a wide range of complex problems in many domains. A recent systematic review identified 20 broad application areas including business performance, project management, resource management, supply chain management, transport and traffic safety, pollution and emissions, sustainability, water, agriculture, animals and fauna, land use, farming, disease, healthcare, research and development, and social and urban issues (Zanker et al., 2021). Other notable areas in which system dynamics has been applied include workplace safety (Shire et al., 2018), climate adaptation (Gohari et al., 2017), COVID-19 (Ibarra-Vega, 2020), terrorism (Lin et al., 2007), and defence (Coyle et al., 1999).

DOI: 10.1201/9780429281624-13

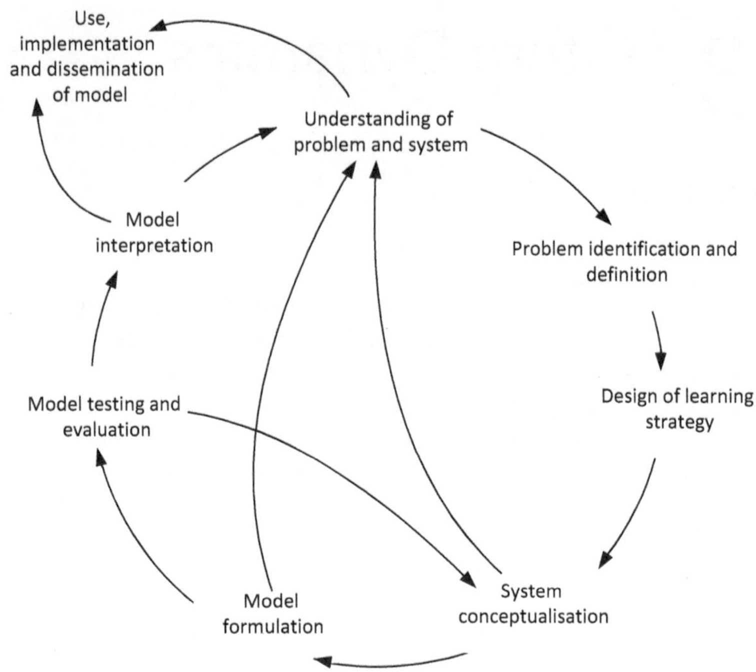

FIGURE 13.1 Overview of system dynamics modelling (adapted from Martinez and Richardson, 2002).

PROCEDURE AND ADVICE

A flowchart depicting the system dynamics procedure is presented in Figure 13.2. Step-by-step guidance is presented below. The system dynamics society website (https://systemdynamics.org/) is an extremely useful resource for guidance materials, including tutorials, guides, exemplar models, and articles describing system dynamics models and best practices.

STEP 1: DEFINE ANALYSIS AIMS AND ANALYSIS BOUNDARIES

The first step involves clearly defining the aims, focus, and boundaries of the analysis. The method is typically used to understand complex problems that are resilient to new interventions; hence it is important at this stage to define clearly the problem being addressed. Consideration of scope and the analysis boundaries is also important to avoid being too narrow and thus not taking an appropriate systems-thinking lens, but also not having such a broad scope that the analysis becomes unwieldly. This ensures that recommendations and the proposal of new interventions following the analysis will have a greater chance to make a positive difference within the system of interest.

The model presented in the case study example at the end of this chapter is derived from a study that focussed on fatal road crashes caused by drivers under the influence

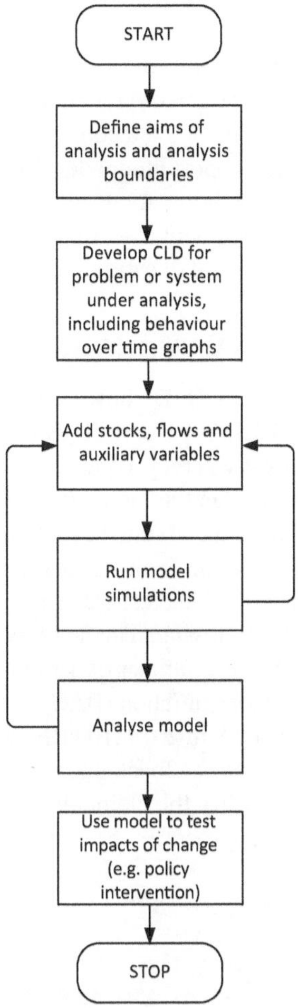

FIGURE 13.2 System dynamics procedure following definition of aims and collection of data.

of alcohol, with the aim being to test the impact of different policy interventions. The following problem, aims and analysis boundaries were defined during step 1:

- **Problem definition**: road trauma caused by drink driving-related crashes;
- **Aim(s) of the analysis**: simulate the impact of different policy interventions designed to prevent drink driving-related crashes and trauma; and
- **Analysis boundaries**: scope was limited to the problem of drink driving within Australia, and did not include other modes of transport such as bicycles, motorcycles and heavy goods vehicles, or other road transport jurisdictions.

STEP 2: DATA COLLECTION

Once the problem, aims of the analysis, and analysis boundaries are clearly defined, the next step involves collecting targeted data about the problem and the system in which the problem occurs. The specific data collected is dependent on the study aims and resources available. Data to support model development can be derived from various sources including SME workshops, focus groups, structured or semi-structured interviews, literature reviews, documentation review (e.g. incident reports, accident investigations, coronial inquests), and questionnaires and surveys.

STEP 3: DEVELOP CAUSAL LOOP DIAGRAM

Once sufficient data are gathered, a CLD showing key variables and feedback loops should be developed. The full CLD procedure is presented in Chapter 8, but typically involves a group modelling process whereby a draft model is developed and subsequently refined through a process of review and iteration. It is important to note that part of the CLD development process involves specifying the reference mode or dynamic behaviour over time that is of interest. For example, in the drink driving model presented in the case study in this chapter, the behaviour of interest was annual fluctuations in the number of fatal road crashes caused by drink driving. This is shown in Figure 13.3, where in the years prior to model development, the number of drivers and motorcycle riders killed in Australia with a Blood Alcohol Concentration (BAC) above the legal limit rose and fell annually between 136 in 2013 and 98 in 2019 (BITRE, 2020).

Whilst CLDs are useful to show feedback and non-linear interactions within systems, they cannot be used to quantify the elements of a system. This is where stocks and flows, and mathematical approaches become useful when applying system dynamics.

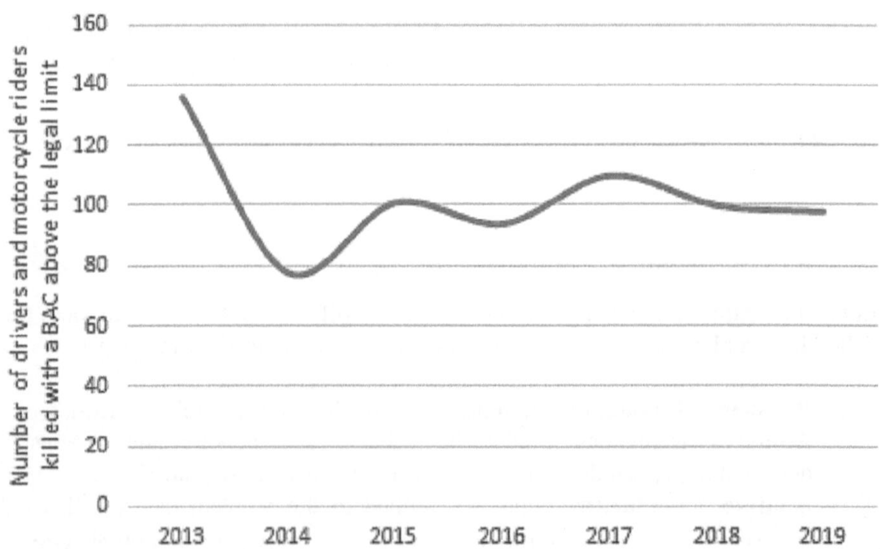

FIGURE 13.3 Fluctuating annual deaths in Australia involving drivers and motorcycle riders with a Blood Alcohol Concentration (BAC) above the legal limit.

STEP 4: BUILD STOCK AND FLOW DIAGRAM

Steps 4 to 7 should be undertaken using a system dynamics software tool such as VenSim (https://vensim.com) or Stella Architect (https://www.iseesystems.com/store/products/stella-architect.aspx); however, it is fine for novice analysts to undertake Steps 3 and 4 using pen and paper until they become proficient with the relevant software tool.

Development of the stock and flow diagram begins with a process of identifying the stocks, flows, and auxiliary variables to be included in the system dynamics model. Stocks represent entities which accumulate or dissipate over time, characterise the state of the system, and drive decision-making and action (Sterman, 2000). Example stocks include bank account balances, the number of customers that a business has, and the population of a given country. The amount contained within a stock increases and decreases through inflow and outflow rates (Sterman, 2000). For a bank balance stock, the inflow comprises money paid into the account (e.g. through work salary) and the outflow includes any expenditure which removes money from the account (e.g. payment of bills). For the customers stock, the inflow comprises newly acquired customers, and the outflow includes lost customers who have taken their business elsewhere. For the population example, the inflow is the country's birth rate, and the outflow is the country's death rate (Sterman, 2000). This is shown in Figure 13.4 which shows a simple example of a population stock and flow diagram.

Thus, inflows and outflows are the rates at which given quantity is being added to or subtracted from the stock. It is important at this stage to define and add initial values for each of the stocks included in the stock and flow diagram. Each stock should have an associated initial value on which the model will base its simulations.

Once all stocks, flows, and auxiliary variables have been identified, the analyst should use the software tool to build an initial stock and flow diagram.

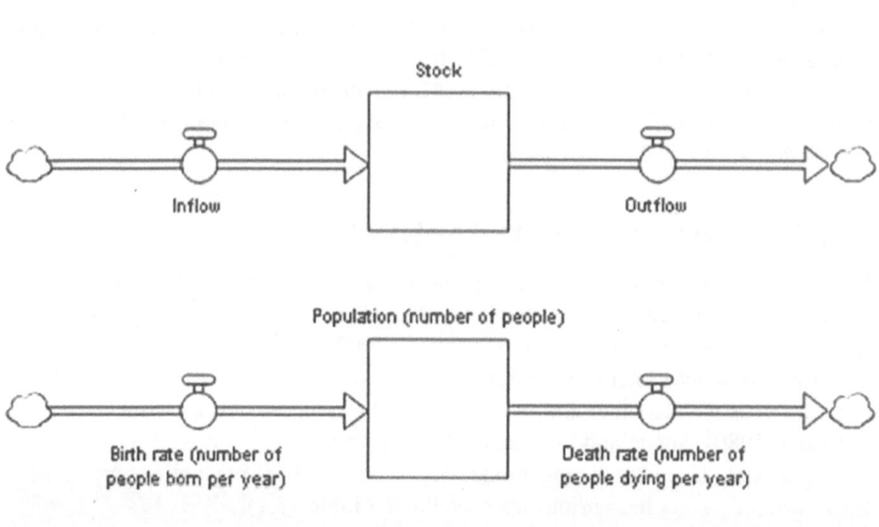

FIGURE 13.4 Example stock and flows diagram for population growth (Adapted from Sterman, 2000).

STEP 5: ADD MODEL EQUATIONS

Equations are built into the system dynamics model to enable simulation of growth and decay over time. Whilst many different types of equations can be used, the most common type in system dynamics are differential equations which are used to study and model rates of change. Thus, the next step involves adding the model equations via the software tool's equation editor or equation tool function. This involves defining units and adding initial values and equations for each of the stocks and connected variables in the stock and flow diagram. Once these have been included it is important to use the software tool's units check function to ensure that the logic of the model is correct. Once any errors are corrected and the units check is successfully completed the model is ready for simulation.

STEP 6: RUN MODEL SIMULATIONS

The next step involves finalising the model settings and running a series of initial model simulations. This involves specifying the units of time for the model (e.g. months or years) and the initial and final time points based on the time frame over which you want to simulate behaviour. For example, when using months as the unit of time and simulating the model over a 5-year period, the initial time point would be set at 0 and the final time point would be set at 60. The time step should also be set at this point. This reflects the amount of time the software runs the model per unit of time, and it is generally recommended that the smallest time step value be selected for greatest model precision (Zimmerman, 2021).

Once the model settings are finalised simulations of the model should be used until the analyst is satisfied that the model is running appropriately, and the outputs have face validity. This involves using the run feature in the software tool and then viewing the simulation results. The results can be viewed using the software tool's graph function which produces graphs that depict the growth and decline of the model's stocks over the simulation period (these can be viewed individually in separate graphs or together on the same graph). If the model does not run, then something may be wrong with the model settings. Moreover, if the model produces spurious results, it may be that the equations need work. After several iterations of simulation, model validation can be undertaken.

STEP 7: MODEL ANALYSIS AND VALIDATION

Model validation involves comparing the behaviour of the model with observed behaviour in the real-world system on which the model is based. Unfortunately, model testing is often inadequate (Sterman, 2002). There are various tests available for validating system dynamics models and testing can involve the analyst(s) themselves as well as other analysts not involved in model development (Senge and Forrester, 1980). Senge and Forrester (1980) describe three forms of model testing: testing model structure, testing model behaviour, and testing policy implications. Each form of testing has various tests available (Table 13.1).

TABLE 13.1
System Dynamics Model Validity Tests

Test	Description
	Tests of model structure
Structure verification test	Involves comparing the structure of the model with the structure of the real-world system on which the model is based
Parameter verification test	Involves comparing model parameters with observed parameters in the real-world system on which the model is based
Extreme conditions test	Involves testing that the model can permit combinations of extreme conditions for its variables
Boundary adequacy test	Involves testing that the model includes all relevant structure from within the real-world system on which the model is based
Dimensional consistency test	Involves running dimensional analyses of the model's equations
	Tests of model behaviour
Behaviour reproduction test	Involves testing how well model generated behaviour matches observed behaviour in the real-world system on which the model is based
Behaviour prediction test	Involves testing how well model generated behaviour predicts future behaviour in the real-world system on which the model is based
Behaviour anomaly test	Involves identifying anomalies whereby model behaviour contradicts real-world behaviour and determining and correcting responsible model elements
Family member test	
Surprise behaviour test	
Extreme policy test	Involves making extreme modifications to model elements and running simulations to determine if the expected effect on behaviour arises
Boundary adequacy test	Involves testing whether the model is equipped with the structure required to address the problem it was designed for
Behaviour sensitivity test	Involves testing the model's sensitivity to changes in parameter values
	Policy implications
System improvement test	Involves testing whether policies identified as improving system behaviour in the model improve behaviour in the real-world
Changed behaviour prediction test	Involves testing whether the model can correctly predict changes brought about by policy change
Boundary adequacy test	Involves examining how changes to the model boundary would influence policy recommendations derived from the model
Policy sensitivity test	Involves testing whether policy recommendations are influenced following changes to parameter values

(Adapted from Senge and Forrester, 1980).

STEP 8: USE MODEL TO TEST IMPACTS OF CHANGE (E.G. POLICY INTERVENTIONS)

The final step involves testing the impact of different policy interventions by modifying model parameters to reflect different policy interventions and then running model simulations. For example, in the drink driving model presented in the case study in this chapter, this step involved testing the following two policy interventions when implemented individually and then together as part of an integrated policy (Salmon et al., 2020).

1. **Standard road safety target-based policy**. The first policy was a standard road safety policy which aimed to achieve a specific reduction in drink driving-related fatalities via enforcement and education. This included the target of a 50% reduction in drink driving-related fatalities to be achieved over the 30-year simulation period.
2. **Public health policy**. The second policy was a public health policy which was designed to reduce alcohol misuse in the community through a set of standard public health measures including taxation of alcohol, restriction of sales, improving responsible sales and serving, and increasing diagnoses and recovery programs (Salmon et al., 2020).

Each policy was assessed based on the reduction in drink driving-related fatalities achieved over the 30-year simulation period.

ADVANTAGES

- System dynamics models are used to dynamically simulate behaviour over time, providing a distinct advantage over other 'static' systems analysis methods;
- System dynamics includes qualitative (CLDs) and quantitative modelling approaches, the latter of which is useful for stakeholders;
- Model values and equations can be modified to show the effect of different interventions on the behaviour of the system. Accordingly, system dynamics is useful for testing the impacts of different policy interventions;
- The initial CLD ensures that the system dynamics model is developed based on a good understanding of the problem or system in question;
- System dynamics is a generic method and can be applied to any issue in any domain;
- Various on-line courses, workshops and training materials are available to support model building, analyses, and interpretation;
- There is a long history of applications in many different areas;
- System dynamics is a highly flexible method and can be used with varying degrees of granularity and scope; and
- Various software programs are available to support the development of system dynamics models, including VenSim and Stella Architect.

DISADVANTAGES

- The method can be difficult to grasp for some analysts and requires a good understanding of mathematics and equations;
- Model development and refinement can be time consuming, particularly for complex problems. This is exacerbated by the need for an initial CLD;

- The quality and validity of system dynamics models and outputs is dependent on the involvement of multiple SMEs during development;
- Model testing and validation can be time consuming compared to other methods;
- System dynamics models can become overly complex, unwieldy, and difficult to interpret; and
- Some of the software tools require expensive licences, though tools such as VenSim are available to download for free.

RELATED METHODS

A qualitative CLD is normally developed and used to inform development of the quantitative system dynamics model.

APPROXIMATE TRAINING AND APPLICATION TIMES

System dynamics can be complex for some and requires extensive training and practice. Various courses are available on-line. Application time depends on the problem being tackled; however, development of system dynamics models typically requires significant time as the process involves initial data collection activities, development of a CLD, development of the system dynamics model, and multiple simulations and iterations of the model. This may require multiple group modelling sessions with SMEs to develop and refine both the CLD and the system dynamics model. System dynamics models can, however, also be entirely conceptual, requiring little to no 'real-world' data to either create or qualitatively validate among model builders and stakeholders.

RELIABILITY AND VALIDITY

Model validation is a key step in the process and various tests have been discussed in the literature (e.g. Senge and Forrester, 1980). Testing the validity of system dynamics models involves comparing model behaviour with empirical reality (ibid.). According to Senge and Forrester (1980), there is no single best test for model validation and various tests exist for validating model structure and behaviour (See Table 13.1). Confidence in the model builds over time as more tests are applied and successfully passed (ibid.).

TOOLS NEEDED

CLDs and the initial system dynamics model can be created using pen and paper or a whiteboard and marker. Once a draft system dynamics model is developed a software program such as VenSim or Stella Architect should be used to create the model and run simulations.

CASE STUDY EXAMPLE: DRINK DRIVING-RELATED ROAD TRAUMA

In Salmon et al. (2020) we developed a system dynamics model to simulate the impacts of different road safety policy interventions on drink driving-related road trauma. It goes without saying that drink driving continues to represent a significant road safety issue worldwide. In Australia and the United States, for example, drink driving is the primary cause of around a third of all fatal road crashes, (National Road Safety Partnership Program, 2018; NHTSA, 2016). Preventative measures for drink driving include increased detection, screening, enforcement, and education (Salmon et al., 2020). Whilst these approaches have been shown to be effective (Tay, 2005), a lack of recent progress in reducing the prevalence of drink driving indicates a need to explore new solutions (Salmon et al., 2020).

As part of a research program which aimed to produce novel interventions designed to prevent drink driving-related road trauma, we developed a system dynamics model that could be used to simulate the reduction in road trauma brought about by different policy interventions (Salmon et al., 2020). We were particularly interested in the benefits of integrating standard road safety interventions with public health interventions aiming to tackle alcohol addiction.

DRINK DRIVING-RELATED ROAD TRAUMA CLD

Initially, we developed a CLD for drink driving-related road trauma via a group modelling process (see Chapter 8). The final CLD is presented in Figure 13.5. As described in Chapter 8, the model shows that a key issue influencing the level of drink driving-related road trauma is the societal issue of alcohol addiction and consumption. A proportion of problem drinkers and indeed occasional drinkers either intentionally or unintentionally drive under the influence of alcohol, which in turn leads to crashes and injuries and fatalities. As injuries and fatalities rise, the gap between the desired road safety goal in terms of an acceptable level of drink driving-related trauma and the actual level of drink driving-related trauma increases, which

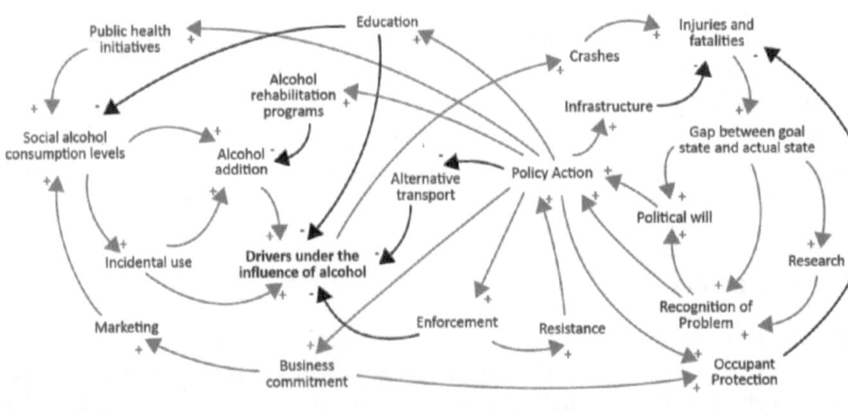

FIGURE 13.5 Causal loop diagram showing the variables which interact to create drink driving-related trauma (Adapted from Salmon et al., 2020).

facilitates recognition of the problem and increases the political will to address it. As these two variables reach a certain threshold, policy action is triggered, creating a greater impetus for road safety interventions that aim to prevent drivers from drink driving (Salmon et al., 2020).

The primary road safety interventions included in the CLD are enforcement and education. Enforcement includes standard deterrents such as random breath testing, stricter laws and penalties, and in-vehicle alcohol interlock devices (Salmon et al., 2020). Education includes targeted activities designed to prevent drivers from driving under the influence, such as television and social media campaigns and roadside advertising (Salmon et al., 2020). In addition, remedial education is used to educate drivers who have been caught drink driving. Injury prevention measures designed to limit trauma once crashes have occurred include improved occupant protection in vehicles and safer road infrastructure (e.g. wire rope road safety barriers). Finally, increasing the availability of alternative transport is designed to reduce the need to drive for people under the influence who need to travel (Salmon et al., 2020).

The CLD also includes public health interventions which attempt to address alcohol misuse and addiction, including public health initiatives (e.g. screening for alcohol misuse, health clinics, outreach, education campaigns) and alcohol rehabilitation programs (Salmon et al., 2020). Alcohol marketing and advertising is included in the CLD as a variable that can influence societal alcohol consumption levels, with aggressive advertising facilitating increased sales and consumption (Saffer and Dave, 2002).

As the road safety and public health interventions take effect, there is a decrease in drink driving-related crashes and fatalities based on a reduced number of drivers driving under the influence of alcohol. As described in Salmon et al. (2020), as the gap between the goal road safety state and actual road safety state decreases, diminished political will then acts to dampen policy action, which reduces 'road safety effort' and public health initiatives. This dynamic creates the annual fluctuations in drink driving-related trauma rates seen in Figure 13.3.

A system dynamics model was subsequently developed based on the following dynamic hypotheses relating to drink driving-related trauma (Salmon et al., 2020):

1. The number of people in the community at a particular point in time who routinely consume alcohol in a harmful way (problem drinkers) is a balance between the number of people becoming problem drinkers since the previous point in time, and the number ceasing that behaviour over that time period. (Salmon et al., 2020).
2. The number of drink drivers at any point in time can be assumed to be a fraction of the total number of problem drinkers in the community. The number of drink drivers increases as the number of problem drinkers increases or the fraction of problem drinkers who drink drive increases (Salmon et al., 2020). Increases in the number of fatalities increase the gap between the actual number of road fatalities and the number considered by the community to be acceptable, creating community concern and recognition of a problem. In turn this leads to an increase in road safety policy action which aims to reduce fatalities through increased levels of education and enforcement. The resulting reduction in fatalities

decreases the gap between actual and acceptable deaths and reduces community concern, which results in a reduction in road safety policy action, road safety expenditure, and level of education and enforcement (Salmon et al., 2020).

3. Increased government expenditure on road safety and the resulting increases in education and enforcement leads to a community compliance burden and resistance to government intervention. This acts to limit policy action, which leads to a reduction in interventions, and eventually results in the number of drink drivers and drink driving-related fatalities rising again. (Salmon et al., 2020).

PROBLEM STATEMENT AND REFERENCE MODE

Next, we articulated a problem statement to be solved by the model (Saeed, 1992). Specifically, we defined the problem statement in terms of a reference mode containing a trend and a cycle (Salmon et al., 2020):

1. Road safety interventions aimed at addressing drink driving-related deaths are achieving diminishing returns. Reductions in drink driving-related road crash deaths are levelling out above intended targets in an exponential decay trend.
2. Additional to the trend there is an overlaying cycle with short-term fluctuations in drink driving deaths that oscillates around the long-term trend.

MODEL DEVELOPMENT

A draft system dynamics model was developed using the Stella Architect dynamic modelling software program. The key outcome variable was serious injuries and fatalities arising from road crashes caused by drink driving, with a time horizon of 30 years and simulation period of one year. The model used a high level of aggregation and abstraction to reflect archetypal behaviour and relationships rather than a particular data set or context. Only components essential to the model's performance were included to maintain parsimony (Daganzo et al., 2012). Numerical values in the model were generic and where relevant, values were normalised with the multiplier of 1 representing the status quo. Effects were expressed in terms of variance from status quo (Salmon et al., 2020).

The model comprised one module for each of hypotheses described above: 1) A prevalence of problem drinkers module; 2) A road safety module; and 3) A compliance resistance module. The problem drinkers module is presented in Figure 13.6. This includes a single stock comprising the number of people habitually misusing alcohol with two minor loops that determine the actual number of people in the community during the simulation period (one year) habitually consuming alcohol in a harmful way. This number grows if the new cases rate exceeds the cases recovering rate (Salmon et al., 2020).

The number of people habitually abusing alcohol (Figure 13.7) influences the number of people who habitually drink drive according to the nominated fraction.

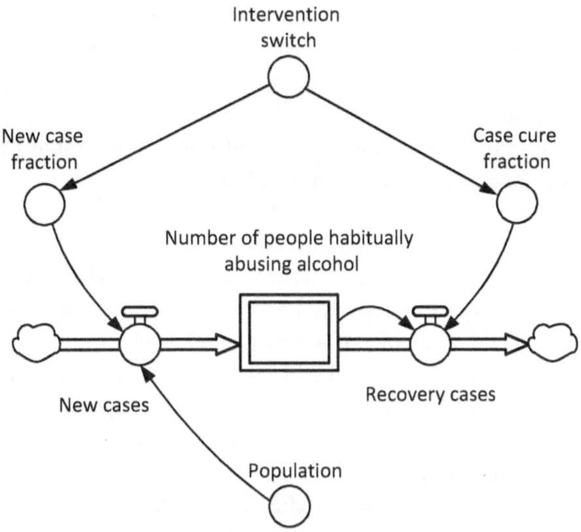

FIGURE 13.6 Prevalence of problem drinkers module (QLD = Queensland) (Adapted from Salmon et al., 2020).

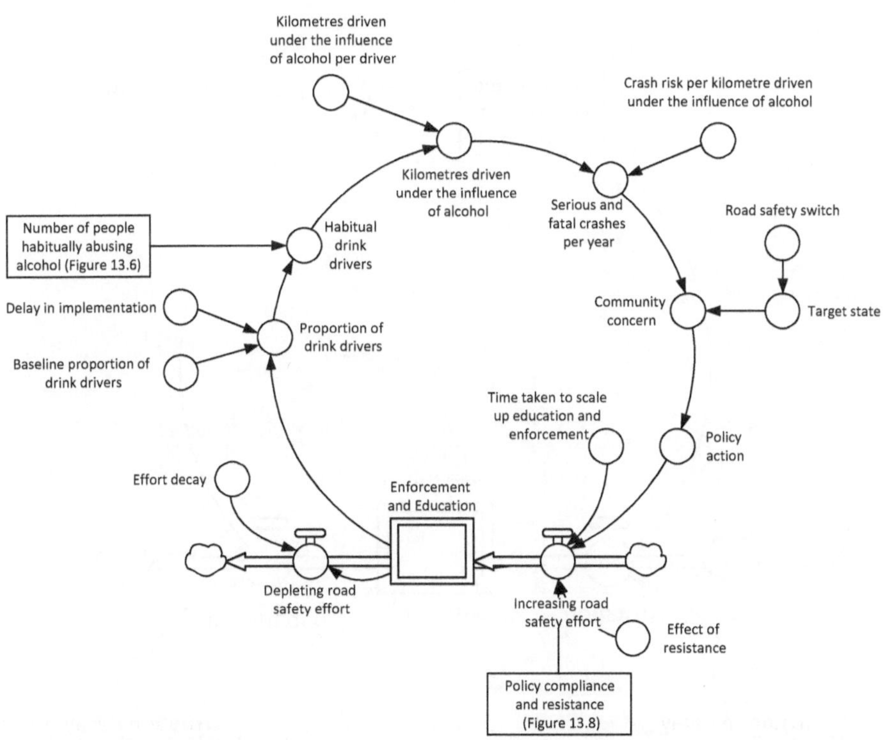

FIGURE 13.7 Road safety module (Adapted from Salmon et al., 2020).

The average kilometres driven by habitual drink drivers per year, multiplied by the risk of serious drink driving-related crashes per kilometre results in a number of drink driving-related crashes per year (Salmon et al., 2020). The level of community concern equals the number of crashes divided by the acceptable (or target) number of deaths. Community concern operates as a multiplier on the input to the road safety intervention enforcement and education stock which is normalised to 1 to represent the amount of effort involved to produce the current baseline proportion who drink drive, as the loop is closed and maintained in equilibrium to the number of deaths (Salmon et al., 2020). The enforcement and education stock is based on an annual road safety budget that is fully expended and replenished the next year at a multiple of the business-as-usual level of enforcement and education as determined by the policy action and compliance resistance inputs (Salmon et al., 2020). The road safety loop also contains two information delays to cover the gap between recognising the problem and the resource distribution logistics involved in scaling up societal level solutions (Salmon et al., 2020).

The policy compliance resistance module is presented in Figure 13.8. This represents dynamic and changeable levels of policy resistance and compliance. Increasing levels of enforcement and education create a public compliance issue which provides resistance to road safety policies and acts to eventually diminish road safety effort (Salmon et al., 2020).

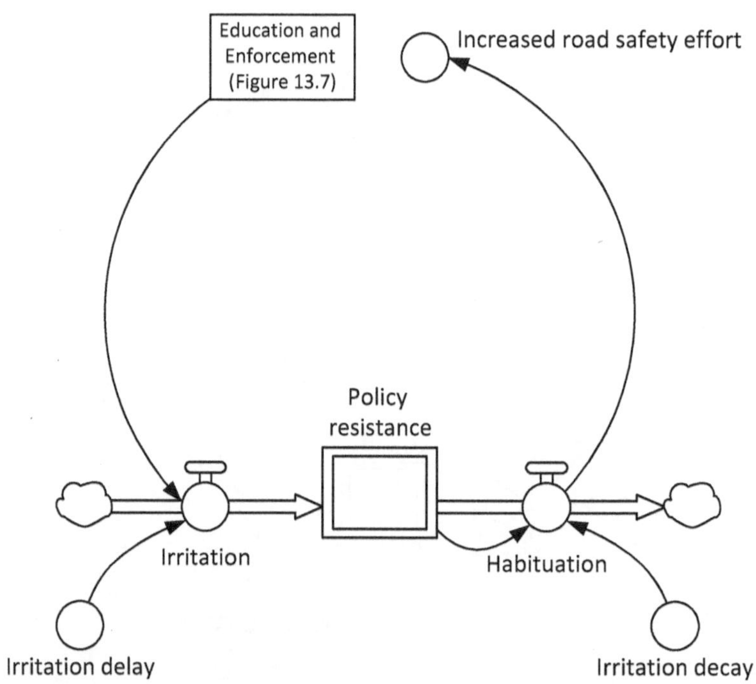

FIGURE 13.8 Policy compliance resistance module (Adapted from Salmon et al., 2020).

POLICY DESIGN AND SIMULATION

We used the model to test two policies when implemented individually and then together as part of an integrated policy (Salmon et al., 2020):

1. **Standard road safety target-based policy**. The first policy was a standard road safety policy which aimed to achieve a specific reduction in drink driving-related fatalities via enforcement and education. This included the target of a 50% reduction in drink driving-related fatalities to be achieved over the 30-year simulation period.
2. **Public health policy**. The second policy was a public health policy which was designed to reduce alcohol misuse in the community through a set of standard public health measures including taxation of alcohol, restriction of sales, improving responsible sales and serving, and increasing diagnoses and recovery programs.

RESULTS

OBSERVED BEHAVIOUR OF THE MODEL

Road Safety policy: When a 50% reduction in road toll target was set, the model was pushed off equilibrium three years into the simulation (see the left graph in Figure 13.9). Following this there was a large reduction in the number of deaths, and then an oscillation with damping to reach a new equilibrium at about 30 years. This new equilibrium represented a 30% reduction in serious and fatal crashes, meaning that the target 50% reduction was not achieved. This was a close replication of the expected reference mode. That is, the reduction in drink driving related deaths demonstrated diminishing returns with reductions levelling out above intended targets in an exponential decay trend. Additional to this trend there is an overlaying observed cycle with short period fluctuations in drink driving deaths that oscillates around a

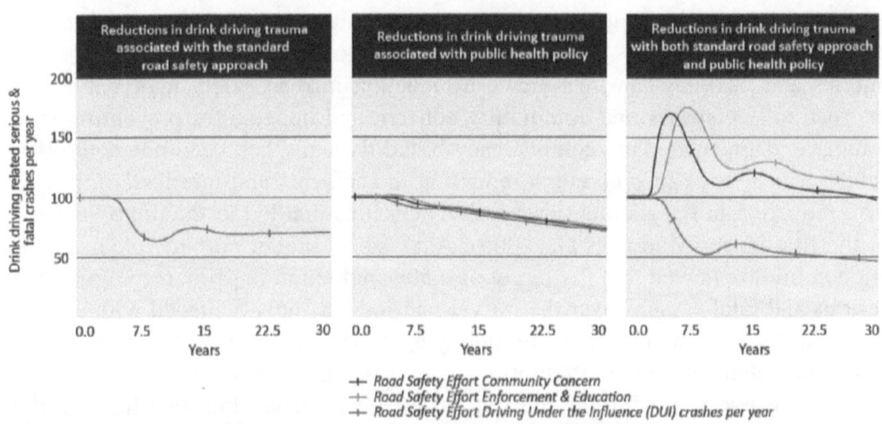

FIGURE 13.9 Model outputs under different policy settings (Salmon et al., 2020).

long-term trend pattern with attenuation in this oscillation over time (Salmon et al., 2020). This can be explained by a balance between two dominant loops. As fatalities decrease, community concern around road safety diminishes and effort to prevent drink driving-related trauma declines as a result. As the current system (apart from education and enforcement efforts) is set to maintain the baseline fatality rate (which is higher than the target) the result is a compromise between the baseline rate and the target rate (Salmon et al., 2020).

Public health policy intervention: The graph in the middle of Figure 13.9 shows the observed behaviour when the public health policy solution was included. This resulted in a steady reduction in deaths, but also a corresponding reduction in community concern and road safety effort. This suggests that the benefit of public health interventions was not being realised because the road safety effort rested on its laurels (Salmon et al., 2020).

Combined interventions: The graph to the bottom right-hand side of Figure 13.9 shows the effect when both policies are implemented together in an integrated manner. The simulation suggests that the combined policy is able to achieve the target of a 50% reduction in drink driving-related trauma within the 30-year period. The graph also shows a close correlation between community concern and the ensuing increase in enforcement and education, with a substantially reducing effort after the initial drop is reached to maintain ongoing equilibrium at the target level. When both interventions work together it removes much of the feedback effect trying to maintain the higher rate of a system at status quo (Salmon et al., 2020). The public health policy effect changes key parts of the system that allows the road safety targets to drive the changes that road safety practitioners are striving for.

CONCLUSIONS

The drink driving-related road trauma system dynamics model provides an explanation of why drink driving-related trauma rates tend to fluctuate annually. As described in Salmon et al. (2020), the model suggests that increased road safety policy effort leads to an increased level of intervention through avenues such as enforcement and education, which over time acts to reduce the number of drink drivers and related crashes and fatalities. This reduction in the number of drink driving-related crashes, injuries and fatalities can then create a reduction in road safety interventions as the road toll decreases and community concern and impetus for preventing road trauma is dampened. This complacency-based dynamic has been noted in other safety critical systems whereby a reduction in accidents and injuries is followed by a reduction in the accident prevention activities that led to the improvements in the first place (Marais et al., 2006). Analysis of model outputs under different conditions showed that the target of a 50% reduction in drink driving-related serious and fatal crashes over the 30-year period was only achieved with a combined road safety and public health policy approach (Salmon et al., 2020). It was thus concluded that when attempting to prevent drink driving-related crashes, more expansive interventions that target not only drivers but also the societal issue of alcohol addiction and increasing alcohol consumption levels are required (Salmon et al., 2020).

RECOMMENDED READING

Senge, P. M., & Forrester, J. W. (1980). Tests for building confidence in system dynamics models. *System Dynamics, TIMS studies in management sciences*, 14, 209–228.

Sterman, J.D. (2000). *Business Dynamics: Systems Thinking and Modeling for a Complex World.* Boston: Irwin McGraw-Hill.

REFERENCES

BITRE (2020). Road trauma Australia: Annual Summaries, https://www.bitre.gov.au/publications/ongoing/road_deaths_australia_annual_summaries, accessed 1st September 2021.

Coyle, J. M., Exelby, D., & Holt, J. (1999). System dynamics in defence analysis: Some case studies. *Journal of the Operational Research Society*, 50(4), 372–382.

Currie, D. J., Smith, C., & Jagals, P. (2018). The application of system dynamics modelling to environmental health decision-making and policy – a scoping review. *BMC Public Health*, 18, 402.

Daganzo, C. F., Gayah, V. V., & Gonzales, E. J. (2012). The potential of parsimonious models for understanding large scale transportation systems and answering big picture questions. *EURO Journal on Transportation and Logistics*, 1(1–2), 47–65.

Forrester, J. W. (1958). Industrial dynamics – a major breakthrough for decision makers. *Harvard Business Review*, 36(4), 37–66.

Gohari, A., Mirchi, A., & Madani, K. (2017). System dynamics evaluation of climate change adaptation strategies for water resources management in central Iran. *Water Resources Management*, 31(5), 1413–1434.

Ibarra-Vega, D. (2020). Lockdown, one, two, none, or smart: Modeling containing covid-19 infection. A conceptual model. *Science of the Total Environment*, 730, 138917.

Lin, C. H., Liou, D. Y., & Wu, K. W. (2007). Opportunities and challenges created by terrorism. *Technological Forecasting and Social Change*, 74(2), 148–164.

Marais, K., Saleh, J. H., & Leveson, N. G. 2006. Archetypes for organisational safety. *Safety Science*, 44(7), 565–582.

Martinez, I. & Richardson, G. P. (2002, July). An expert view of the system dynamics modeling process: Concurrences and divergences searching for best practices in system dynamics modeling. In *Proceedings of the 20th International Conference of the System Dynamics Society* (Vol. 28).

National Highway Traffic Safety Administration. (2016). Distracted driving 2014. Traffic safety facts research note (report no. DOT HS 812 260).

National Road Safety Partnership Program. (2018). Fact sheet: drink driving, https://www.nrspp.org.au/resources/fact-sheet-drink-driving/, accessed 20 May 2019.

Saffer, H. & Dave, D. 2002. Alcohol consumption and alcohol advertising bans. *Journal of Applied Economics*, 34(11), 1325–1334.

Salmon, P. M., Read, G. J. M., Thompson, J., McLean, S., & McClure, R. (2020). Computational modelling and systems ergonomics: A system dynamics model of drink driving-related trauma prevention. *Ergonomics*, 63(8), 956–980.

Senge, P. M. & Forrester, J. W. (1980). Tests for building confidence in system dynamics models. *System Dynamics, TIMS Studies in Management Sciences*, 14, 209–228.

Shire, M. I., Jun, G. T., & Robinson, S. (2018). The application of system dynamics modelling to system safety improvement: Present use and future potential. *Safety Science*, 106, 104–120.

Sterman, J. D. (2000). *Business Dynamics: Systems Thinking and Modeling for a Complex World.* Boston: Irwin McGraw-Hill.

Sterman, J. D. (2002). All models are wrong: Reflections on becoming a systems scientist. *System Dynamics Review: The Journal of the System Dynamics Society*, 18(4), 501–531.

Tay, R. (2005). The effectiveness of enforcement and publicity campaigns on serious crashes involving young male drivers: Are drink driving and speeding similar? *Accident Analysis & Prevention*, 37(5), 922–929.

Zanker, M., Bureš, V., & Tučník, P. (2021). Environment, business, and healthcare prevail: A comprehensive, systematic review of system dynamics application domains. *Systems*, 9(2), 28.

Zimmerman, N. (2021). Romeo and Juliet: introduction to the modelling process. System Dynamics Society Seminar, 18th August 2021.

Section VI

Many Model Systems Thinking

14 A Many-Model Systems Thinking Approach

INTRODUCTION

This book has introduced 12 systems thinking methods that researchers and practitioners can apply to help understand and respond to complex problems. In this final chapter we consider how these methods can usefully be applied to perhaps the most complex of all problems – global-scale risks and existential threats. Alarmingly, society faces an increasing number of such problems. In their annual global risks report for 2021, the World Economic Forum identified the current top ten global risks based on impact: 1) climate action failure, 2) weapons of mass destruction, 3) biodiversity loss, 4) extreme weather, 5) water crises, 6) information infrastructure breakdown, 7) natural disasters, 8) cyberattacks, 9) human-made environmental diseases, and 10) infectious diseases (World Economic Forum, 2021). Beyond this, there are various emerging global risks that threaten global health and wellbeing, including artificial intelligence, automation replacing human workers, the genetic modification of humans, an ageing population, and otherworld settling (Salmon et al., 2021).

The ongoing COVID-19 pandemic provides a sobering example of how inadequate preparation and response to global risks can result in millions of deaths and enormous economic and societal costs. At the time of writing, the pandemic has killed over 6 million and the total economic cost is estimated to be in the region of $28 trillion dollars (International Monetary Fund, 2020). Whilst there are many factors which led to failures around the management of COVID-19, one critical aspect when managing global risks and existential threats is the methods that stakeholders use to describe, understand, and help respond to the issue. Arguably, there are few appropriate scientific methods to support an effective response to global risks and existential threats (Aven and Zio, 2021; Salmon et al., 2021).

As global issues and existential threats are complex, multifactorial, dynamic, and heavily interrelated, they are extremely difficult to describe, understand, and respond to successfully. It is our contention that the methods described in this book have a critical role to play in understanding and responding to complex global risks and existential threats (Salmon et al., 2019b, 2021; Thatcher et al., 2018; 2020). This is based on the fact that the critical factors underpinning such issues are human-centred, and the capacity for systems thinking methods to model the structure and behaviour of complex systems and identify 'leverage points' where interventions can have significant effects on behaviour (Salmon et al., 2019b, 2021). It is also our view that the analytical and explanatory power of the methods can be enhanced by applying them together in an integrated manner. In complexity science, the so-called 'many-model' philosophy argues that complex multifactorial problems are better understood and solved through the application of multiple modelling approaches, rather than the

DOI: 10.1201/9780429281624-14

application of only one (Page, 2016; 2017). It is our view that this form of thinking is required when attempting to understand and respond to global risks and existential threats. Further, we believe the utility of the methods described in this book can be enhanced by applying them together as part of a many-model systems thinking approach. To demonstrate, in this chapter we outline a many-model systems thinking approach which offers a way of integrating the methods described in this book to use as part of a powerful framework for responding to large-scale and highly complex issues. We then demonstrate the approach via a case study in the area of road safety (Salmon and Read, 2019).

A MANY-MODEL SYSTEMS THINKING APPROACH

Within complexity science a many-model thinking approach to complex problems is advocated when faced with complex and seemingly intractable problems (Page, 2016; 2017). This involves using a number of different modelling methods to create multiple analyses, or 'models' of the problem being investigated, with the aim being to gather insight regarding causal mechanisms, potential interventions, and to simulate and assess their likely impact. Page (2016; 2017) argues that, by developing a diverse set of models rather than one in isolation, analysts can reach a deeper understanding, make better predictions, and ultimately design and implement better interventions. Hence, in the case of the methods presented in this book, it is suggested that more utility will be achieved by applying a combination of the methods rather than only one in isolation.

As can be seen by the different book sections the 12 methods presented can be categorised into four categories of method: 1) risk assessment, 2) systems analysis and design, 3) accident analysis, and 4) computational modelling. The methods in each category can therefore be used to provide a different perspective on the particular issue being addressed. Risk assessment methods provide a proactive description of the hazards and adverse outcomes that could arise, whilst accident analysis methods are used to generate in-depth analyses of incidents after they have occurred. Systems analysis methods are used to develop models to help understand the structure, composition, and behaviour of a particular system, and system design methods are used to help develop design interventions which aim to optimise system behaviour and performance. Finally, computational modelling methods are used to simulate the behaviour of systems over time, often using the outputs of systems analysis methods as their basis on which to simulate behaviour. Putting these different forms of analysis together can thus be extremely useful, as multiple perspectives on the issue are provided (Salmon and Read, 2019).

We therefore propose that the methods described in this book can be used as part of a many-model systems thinking approach comprising the five steps outlined in Figure 14.1 (adapted from Salmon and Read, 2019). The specific combination of methods to be used is flexible, with various suitable methods available at each step.

Salmon and Read (2019) tested the utility of the many-model systems thinking approach – specifically integrating six systems thinking methods to examine road trauma: AcciMap, WDA, STAMP, EAST, the Sociotechnical Systems Design Toolkit (STS-DT; Read et al., 2018) and System dynamics.

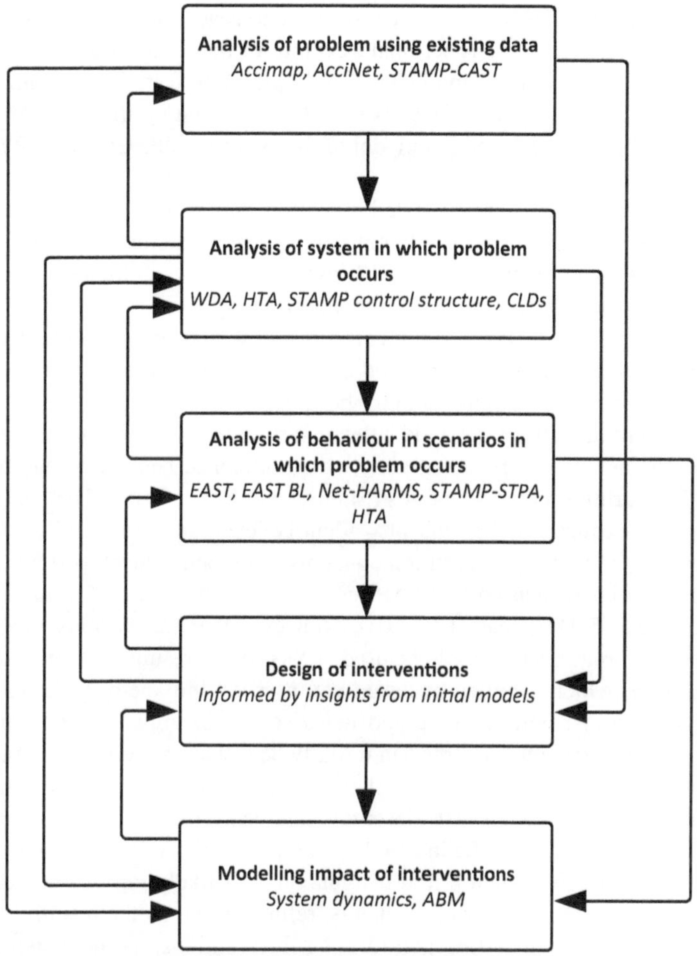

FIGURE 14.1 Many-model systems thinking approach with example methods for use during each step. (Adapted from Salmon and Read, 2019).

As shown in Figure 14.1, the first step involves the use of an accident analysis method to analyse in-depth the problem being tackled. In Salmon and Read's (2019) study, an initial ActorMap and AcciMap (Chapter 9) analysis provided an in-depth analysis of relevant stakeholders and the contributory factors which interact to create fatal road crashes; however, this step could also utilise AcciNet (Chapter 10) or STAMP-CAST (Chapter 11). For example, for a future study examining a government's inadequate response to COVID-19, either AcciMap, STAMP-CAST, or AcciNet could be used to analyse the systemic factors which led to the inadequate response.

The second step then involves modelling the sociotechnical systems in which the problem occurs. In Salmon and Read's (2019) study the STAMP control structure (Chapters 4, 11) was used to provide a detailed description of the control and feedback mechanisms currently in place to manage road safety. WDA (Chapter 6) was then used to describe the functional structure of the road transport system and identify the constraints influencing behaviour. For the future study examining a government's inadequate response to COVID-19, WDA and/or the STAMP control structure could be used to develop a model of the government's COVID-19 response system.

Once the system has been modelled, specific scenarios in which the problem arises should be analysed using the Event Analysis of Systemic Teamwork (EAST; Chapter 7). Here, Salmon and Read (2019) used EAST to examine specifically road user interactions at intersections. For the future study examining a government's inadequate response to COVID-19, specific response scenarios could be analysed using EAST, such as the introduction and easing of lockdown restrictions, vaccinating the population, and managing country borders.

These four initial analyses provide a detailed description of the issue, the system in which the issue occurs, the constraints which influence behaviour, and how systems may fail during specific scenarios (Salmon and Read, 2019). The analyses can be reviewed to extract insights and also identify leverage points and areas where interventions are required. Intervention concepts can then be developed by using the models and their insights as part of a participatory design process. Salmon and Read (2019) used the STS-DT (Read et al., 2018); however, there are various other participatory design approaches that could be used. This process results in a series of potential interventions which can then be tested via insertion into the original analyses to model likely impacts, both positive and negative; for example, Salmon and Read (2019) discuss how design concepts can initially be tested and refined through insertion into the WDA and assessment of their likely impact on the means-ends links (Read et al., 2017). For the COVID-19 study, interventions designed to optimise the vaccine delivery system could be inserted back into the WDA, STAMP control structure, and EAST models to gain initial insights on their likely impacts and effectiveness. In turn, the design concepts can then be refined based on the results.

Following a process of testing and refining interventions, the next step involves the use of computational modelling methods such as ABM (Chapter 12) and System dynamics (Chapter 13) to simulate the likely impacts of the proposed interventions (e.g. McClure et al., 2015; Salmon et al., 2020). Critically, development of the system dynamics model should be informed by the initial AcciMap, WDA, STAMP, and EAST analyses, ensuring validity of the model and its outputs (Salmon et al., 2020). With the COVID-19 example, this could involve developing a system dynamics model of the COVID-19 vaccine delivery system and testing its performance for a period of time with and without the proposed interventions. Again concepts can be refined based on the findings.

The output of the process is a set of novel interventions with initial evidence regarding their likely efficacy following implementation. The use of systems thinking methods ensures that the interventions target leverage points within the system whereby small interventions can have dramatic impacts on behaviour (Meadows, 2008). As a result, the interventions produced will likely be novel and target high impact areas of the system such as policy, system rules and regulations, feedback loops, system goals, and information flows.

CASE STUDY EXAMPLE: MANY-MODEL SYSTEMS THINKING IN ROAD SAFETY

As discussed above, Salmon and Read (2019) tested the many-model systems thinking approach by applying it to the problem of road trauma in Australia. This involved integrating a series of existing analyses to examine the insights to be gained from applying multiple methods together (see Table 14.1 for overview of analyses). A summary of the outputs and insights from each step in the many-model systems thinking approach is given below.

STEP 1. IN-DEPTH ANALYSIS OF PROBLEM

Salmon and Read (2019) utilised Newnam et al.'s (2017) AcciMap analysis of 21 Australian freight crashes from 2004 to 2014, which was developed based on a review of coronial road freight crash inquests (Figure 14.2).

According to Salmon and Read (2019), three key insights can be taken from the AcciMap analysis. First, that road freight crashes are caused by a network of interacting contributory factors spanning all levels of the road transport system. Of particular interest are the contributory factors relating to aspects of the road transport system beyond the driver, vehicle and road environment, including work scheduling, supervision, regulation, and policies and procedures. Second, the analysis demonstrates that the recommendations produced by Coroners in response to road freight crashes do not align with systems thinking, instead focusing mainly on the driver and vehicles (Newnam et al., 2017). Third and finally, the AcciMap raises questions around the suitability of road crash data collection and analysis systems. Existing systems do not align with systems thinking and thus not provide the capacity to gather data on crash contributory factors beyond the road users, vehicles, and road environment involved.

STEP 2: IN-DEPTH ANALYSIS OF THE SYSTEM IN WHICH THE PROBLEM OCCURS

In Salmon and Read's (2019) case study two methods were used to provide models of the road transport system: the STAMP control structure and WDA. Salmon and Read (2019) used Salmon et al.'s (2016) control structure model of the road transport system in Queensland, Australia, to show who currently shares the responsibility for road safety and how these responsibilities are enacted through control and feedback mechanisms (see Figure 14.3).

Salmon and Read (2019) took various insights from the control structure model. For example, although the control structure shows that there are many controls designed to prevent road trauma, some require strengthening (e.g. drink driving controls) and in some cases new controls may be required (e.g. controls to prevent driving under the influence of legal drugs). Salmon and Read (2019) also argued that the control structure reveals instances where feedback mechanisms could be improved so that stakeholders at the higher levels of the road transport system can be better informed when developing road safety policy, strategy, and interventions. Examples

TABLE 14.1
Summary of Models Used by Salmon and Read (2019) in Their Many-Model Systems Thinking Approach Road Safety Case Study

Model Type	Method	Theoretical Underpinnings	Analysis Boundaries	Scope and Aim of Analysis	Reference
Crash Contributory factor model	AcciMap	Systems theory	Road transport system	To examine the types of contributing factors and relationships that are considered in Australian road freight crash inquests	Newnam et al. (2017)
Network model	Event analysis of systemic teamwork	Network theory	Intersection system, drivers, cyclists, motorcyclists, pedestrians	To examine differences in behaviour across different forms of road user when negotiating intersections, identify factors that may play a role in conflicts between different road users at intersections, and gather information to inform the development of new intersection designs that consider all forms of road user	Salmon et al. (2014)
Control structure model	The Systems Theoretic Accident Model and Processes	Control theory	Road transport system	To identify what controls and feedback loops are enacted in pursuit of road safety To clarify who currently shares the responsibility for road trauma and to examine how the responsibilities are currently enacted	Salmon et al. (2016)
Constraints based model	Work domain analysis	Ecological psychology	Road transport system	To understand the functional structure of the road transport system in Queensland, Australia, and to generate insights for improving system design to promote safety	Salmon et al. (2019a)
Computational model	Systems dynamics	Systems theory Complexity theory	Road transport system, Land use planning, Public Health & Wider societal system	To model the relationships among land use, transport, economic development, and population health, and to simulate the impacts of different road safety, land use, and public health policies	McClure et al. (2015)

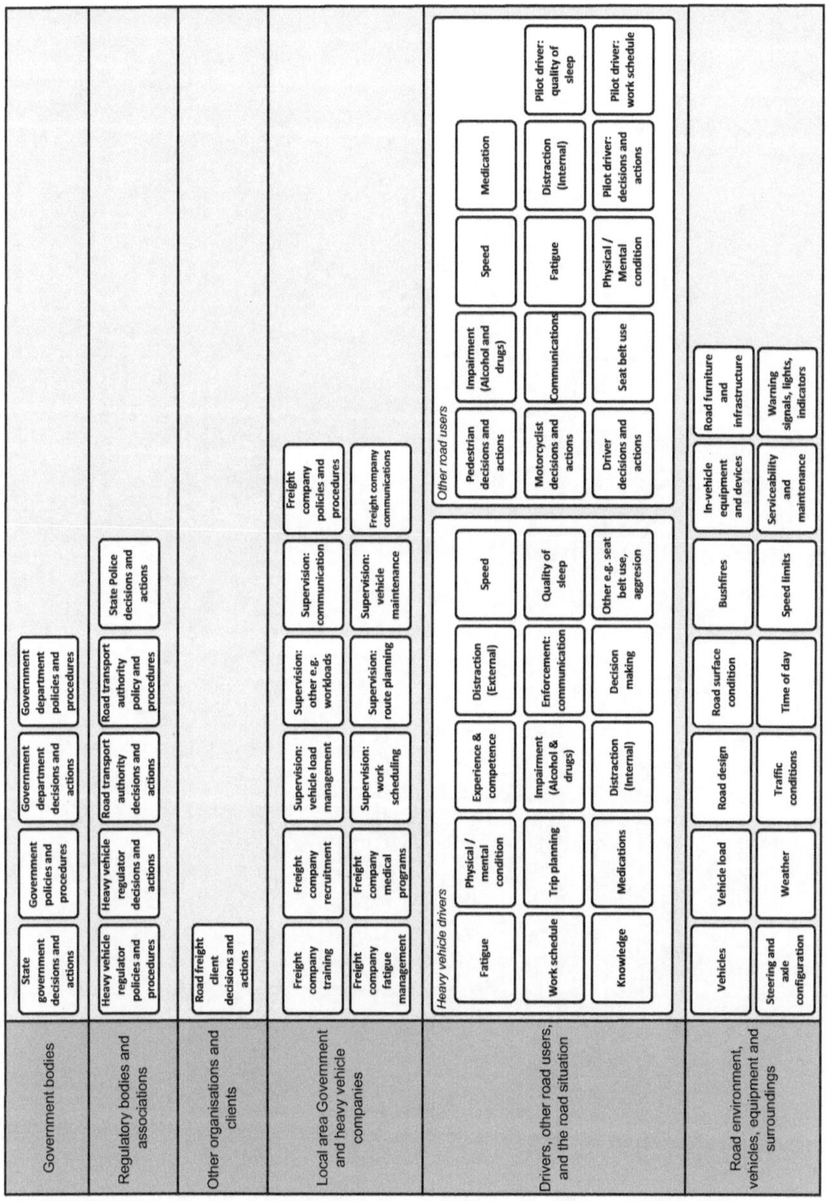

FIGURE 14.2 AcciMap showing road freight crash contributory factors. (Adapted from Newnam et al., 2017).

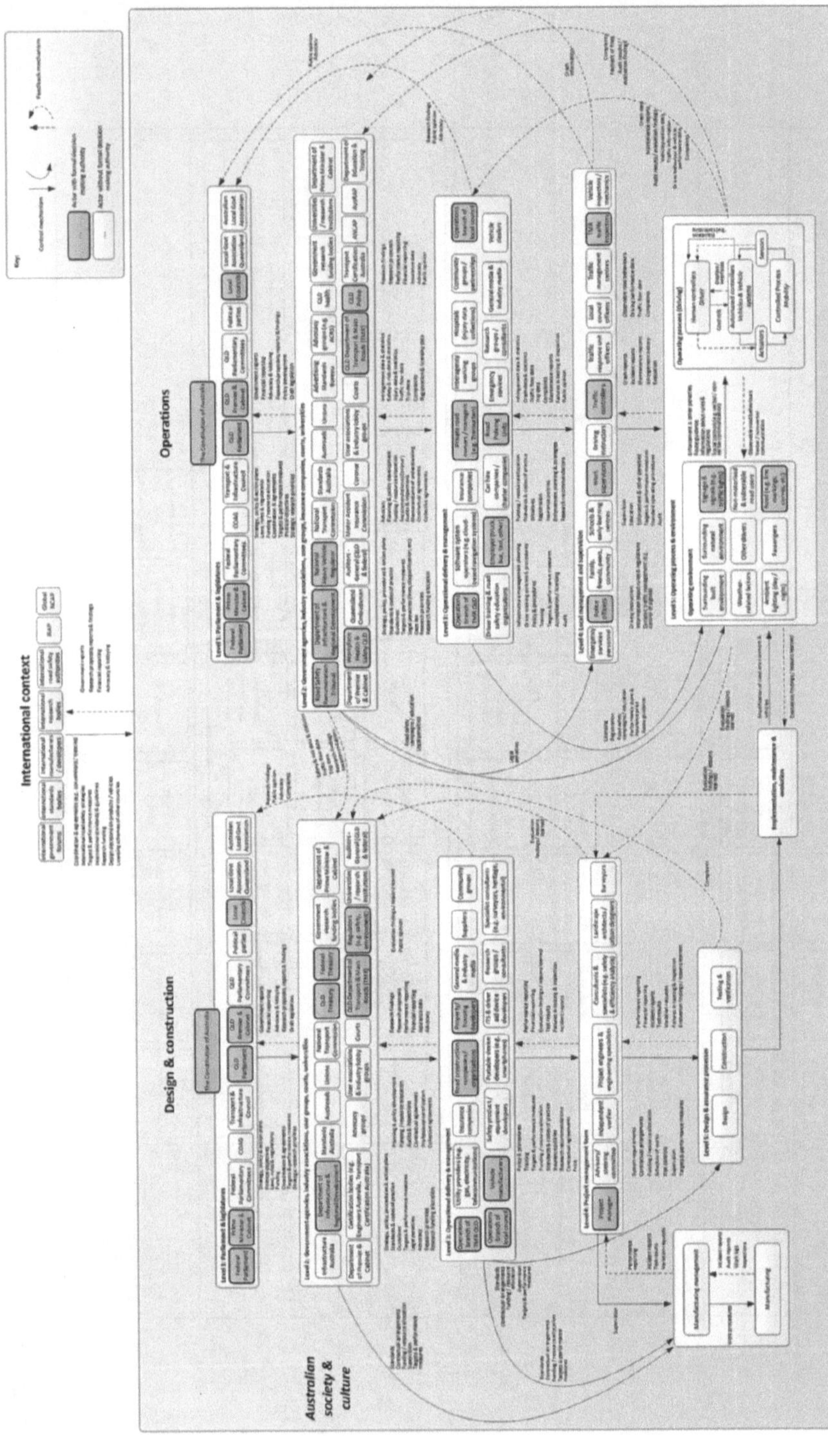

FIGURE 14.3 Qld road transport system design operations control structure. (Salmon and Read, 2019).

of feedback mechanisms that could be improved were crash reporting and analysis systems, information on road user behaviours, and feedback on the effectiveness of different road safety interventions (Salmon and Read, 2019).

The WDA (Figure 14.4) was developed to help understand the functional structure of the road transport system in Queensland, Australia, and to support the development of insights for improving road safety. According to Salmon and Read (2019) the WDA provides three key insights. First, based on an examination of the means–ends links connecting the nodes in the abstraction hierarchy, it was concluded that the structure of the road transport system may be more supportive of the functional purpose of 'economic growth' than it is of the functional purpose 'provide access'. As Salmon and Read (2019) point out, this suggests that economic imperatives have a strong influence on decision-making around road transport system design and operation. Second, the WDA indicated that potential interventions designed to prevent road trauma could target various aspects of the road transport system, including its objects, affordances, functions, values and priority measures, and functional purposes, which in turn suggests that there is scope to develop new and innovative road safety interventions that target aspects of the road transport system's functional structure (Salmon and Read, 2019). The third insight related to the importance of road safety policy and strategy, with the model showing clearly the critical role it plays in achieving road transport system purposes, values, and priorities. Based on this, Salmon and Read (2019) argued that new road safety strategy and policy may be required to facilitate additional reductions in road trauma.

STEP 3: ANALYSIS OF BEHAVIOUR IN SCENARIOS IN WHICH PROBLEM OCCURS

Salmon and Read (2019) used an EAST analysis of road user behaviour at intersections to demonstrate how the many-model thinking approach can move from the macro level of analysis to the micro. The aim of the EAST analysis was to examine specific scenarios where road trauma occurs, in this case intersections and specifically the interactions between drivers, cyclists, motorcyclists, and pedestrians when negotiating intersections. The EAST analysis was developed based on an on-road study incorporating all four types of road user as participants (Salmon et al., 2014). Example task, social and information networks from the EAST analysis of road user behaviour at intersections are presented in Figure 14.5.

Two key insights were derived from the EAST analysis. First, the outputs show that, even though pursuing the same goal of negotiating intersections safely, the tasks undertaken, interactions, and information used by each of the road users was different. A failure to consider these differences when designing road environments, road rules, and road user training programs may increase the likelihood of intersection collisions (Salmon and Read, 2019). Second, the EAST analysis revealed that current intersection designs do not support safe interactions between different types of road user. In particular, the analysis showed where key information was not provided through a lack of signage (e.g. no warnings for drivers to look for cyclists), where conditions in the intersection were not suited to all road users (e.g. a speed limit of

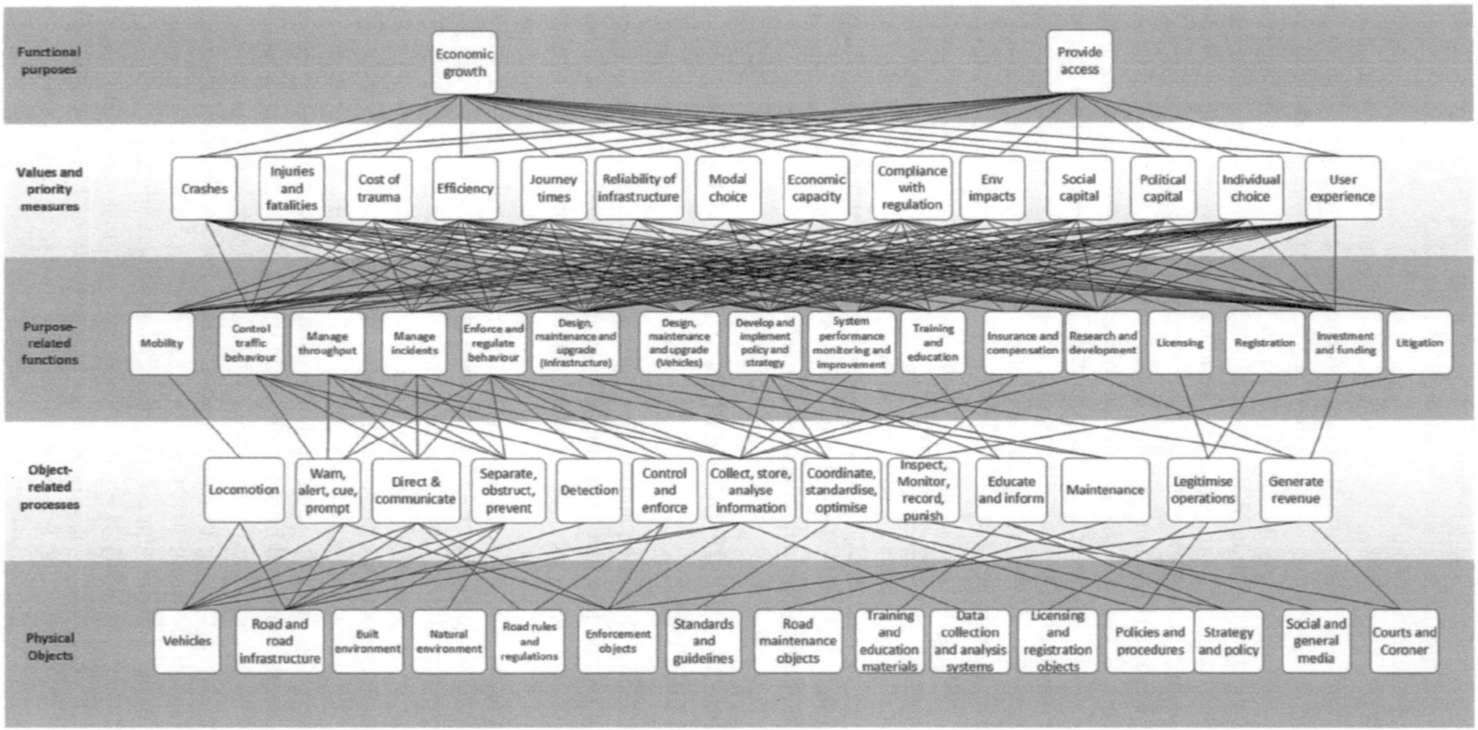

FIGURE 14.4 Work Domain Analysis of the QLD road transport system.

Source: Salmon et al., 2019a.

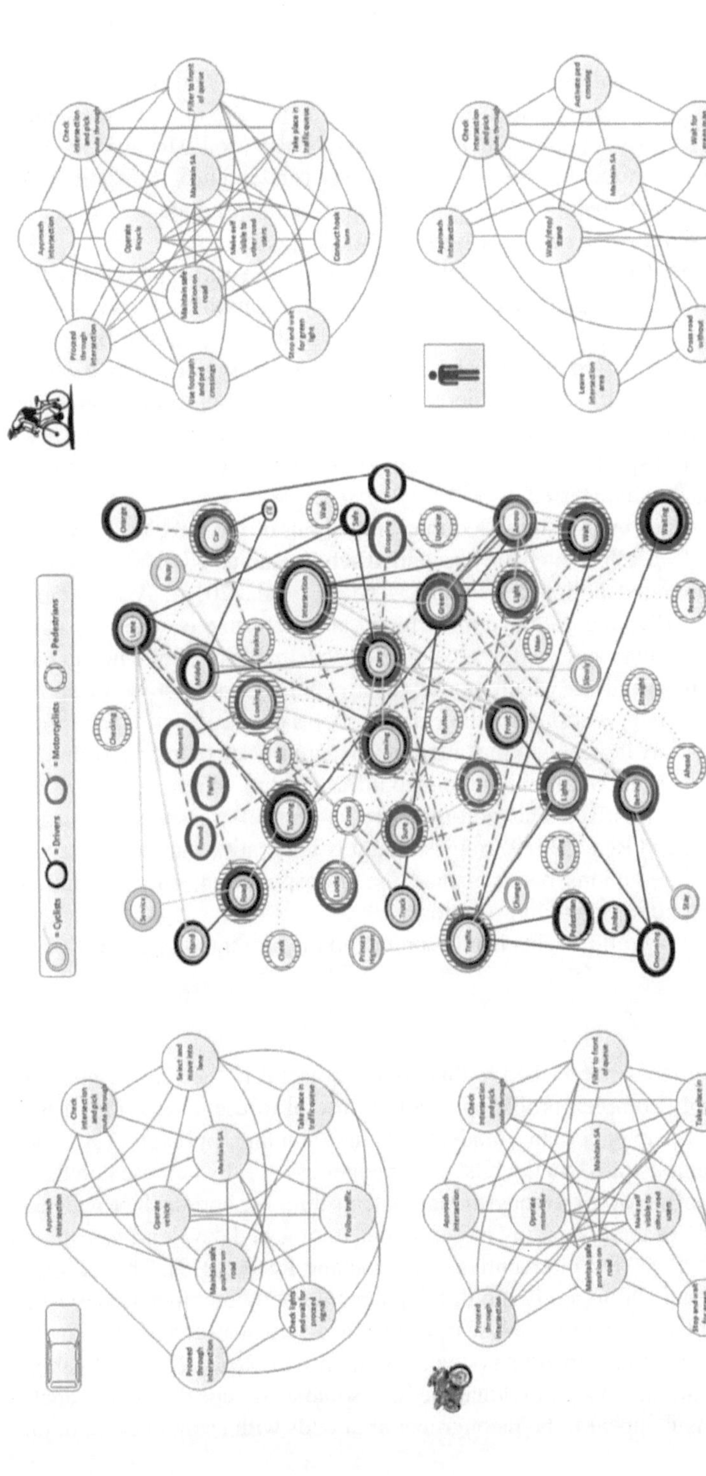

FIGURE 14.5 EAST intersection task and information networks. The information network in the middle of the figure shows the information used by each road user when negotiating intersections with the task networks on the left- and right-hand side of the figure showing the tasks undertaken by each road user when negotiating intersections. (Adapted from Salmon and Read, 2019).

80km/h) and where design features favoured certain types of road users over others (e.g. an absence of cycling lanes and motorcycle boxes).

A final component of Salmon's and Read (2019) study involved demonstrating how computational modelling (System dynamics) can be used to model road safety problems and the efficacy of different interventions. They used McClure et al.'s (2015) system dynamics model of the relationship between land use, transport, economic development, and population health which was used to simulate the impact of a series of land use-transport policies over a 10-year period in six major cities (Beijing, Copenhagen, Delhi, London, Melbourne, and New York). Specifically the simulations examined the impact of different policy interventions on road vehicle crash-related deaths and disability adjusted life years lost. The policies examined included one which aimed to reduce travel mode risk (road safety policy), one which aimed to change the distribution of transport modes (active transport policy), and one which aimed to improve transport infrastructure to change the distribution of transport modes (active transport infrastructure policy). A fourth policy combining all three initial policies was also tested.

The system dynamics model suggested that, in five out of the six cities, optimal reductions in vehicle crash-related deaths and disability adjusted life years lost occurred in the integrated policy condition (McClure et al., 2015). Accordingly, Salmon and Read (2019) suggested that the primary insight from the system dynamics model is that reductions and road trauma and improvements in public health can be achieved using interventions which reduce car use and increase active transport, so long as road safety interventions continue and appropriate modifications to infrastructure are made. A key insight was that attempts to increase active transport should be accompanied by appropriate modifications to infrastructure which reduce unsafe interactions between motorised and non-motorised forms of transport (McClure et al., 2015). Salmon and Read (2019) suggest that it is imperative that policy developers consider the interactions between land use, transportation, and population health when creating policies and interventions.

A summary of the overall insights from each method is presented in Table 14.2.

SUMMARY

Whilst each of the methods described in this book are useful in their own right, for large-scale and highly complex problems it will be useful to use a selection of the methods together as part of a many models systems thinking approach. The road safety case study demonstrated that, whilst each method is useful when used on their own, integrating them as part of a many-model systems thinking approach provided a powerful framework for understanding in-depth road trauma and developing, testing, and refining first-of-their-kind interventions (Salmon and Read, 2019). The findings indicate that it is beneficial to apply multiple systems thinking methods to describe the same problem, with the five analyses producing a series of coherent insights spanning all levels of road transport systems (Salmon and Read, 2019). Notably, some insights were unique to an individual method, some were repeated across analyses, and none of them appear to be incongruent or at odds with each other (Salmon and Read, 2019).

TABLE 14.2

Key Road Safety Insights Derived from Salmon and Read's (2019) Many Models Systems Thinking Approach Road Safety Case Study

Insight	Source
A systemic causal web. Road crashes are created by a network of multiple interacting contributory factors spanning all levels of the road transport system	AcciMap STAMP
Reductionist crash prevention strategies. Crash prevention strategies typically focus on the driver or environment and do not align with the key tenets of systems thinking	AcciMap
Inadequate data systems. Existing road crash data collection and analysis systems are inadequate as they do not support the identification of crash contributory factors at levels of the road transport system	AcciMap STAMP EAST
Non-homogeneous road users. Different road users experience the same road situations differently and exhibit a range of different behaviours as a result	EAST
Driver centric-road design. Road transport system design (e.g. infrastructure, policy, training) does not currently consider the different needs of road users (e.g. drivers versus cyclists). As a result, road environments, training programs, and road rules and regulations create conflicts between different forms of road user	EAST System dynamics
Sharing the load. There is a shared responsibility for road safety that spans a diverse set of actors and organisations that includes government and international organisations	AcciMap STAMP
Control problem or problem controls? Multiple road safety controls are enacted across the road transport system. As road transport systems evolve, new controls will be required, and existing controls require modification or strengthening	STAMP
Societal influences. Societal issues such as drug and alcohol addiction, an increasing desire to remain connected, and high demands on time play a key role in the behaviours that lead to road trauma	STAMP System dynamics Causal loop diagram
Lack of feedback mechanisms. Road transport systems currently do not have sufficient feedback mechanisms to ensure that policy makers are sufficiently informed regarding road transport system behaviour	STAMP
The need for fundamental change. New reductions in road trauma may only be achieved through fundamental changes to the functional structure of road transport systems	WDA System dynamics
The power of policy. Road safety policy and strategy is one leverage point that can be pulled to enact fundamental change within road transport systems	WDA Causal loop diagram System dynamics
Competing goals. Road transport systems have the competing goals of economic growth and providing access	WDA
An integrated policy approach. An integrated policy and strategy approach is required to achieve new reductions in road trauma	STAMP Causal loop diagram System dynamics
Smashing silos. Breaking down the silos that exist across transport, land use and public health systems will enable decision makers to develop integrated and holistic policies that have far greater road safety and public health impacts	Causal loop diagram System dynamics
Removing the motor vehicle. Reducing car use through improved land use planning and active transport infrastructure will lead to significant reductions in road trauma along with significant public health benefits	Causal loop diagram Systems dynamics
Active transport infrastructure. The potential health benefits of active transport will only be realised through significant modifications to land use and transport infrastructure	Causal loop diagram Systems dynamics

(Adapted from Salmon and Read, 2019.)

A useful feature of the many-model systems thinking approach is that it permits a progression of work from analysis of the problem and system through to the design of potential solutions and then simulation of the likely impact of proposed interventions (Salmon and Read, 2019). The specific combination of methods to use will depend on the project constraints and analysis boundaries; however, if the project scope allows it, it is recommended that all five steps from Figure 14.1 be employed. Whilst applying the many models systems thinking approach will incur high levels of resource usage, the exhaustive outputs represent a good return on investment. If project constraints do not permit application of all five steps, then a many models 'Lite' approach incorporating a selection of the steps and methods will still prove useful. For example, one less resource intensive approach might involve using AcciMap to analyse a particular incident, followed by the use of CWA to analyse the system, specific behaviours, and then to design and test appropriate solutions (Salmon and Read, 2019).

RECOMMENDED READING

Salmon, P. M. & Read, G. J. M. (2019). Many-model thinking in systems ergonomics: A case study in road safety. *Ergonomics*, 62(5), 612–628.

REFERENCES

Aven, T. & Zio, E. (2021). Globalization and global risk: How risk analysis needs to be enhanced to be effective in confronting current threats. *Reliability Engineering & System Safety*, 205, 107270.

International Monetary Fund. (2020). *World Economic Outlook Update*, June. https://www.imf.org/en/Publications/WEO/Issues/2020/06/24/WEOUpdateJune2020

McClure, R. J., Adriazola-Steil, C., Mulvihill, C., Fitzharris, M., Bonnington, P., Salmon, P. M., & Stevenson, M. (2015). Simulating the dynamic effect of land use and transport policies on the development and health of populations. *American Journal of Public Health* 105(S2), 223–229.

Meadows, D. H. (2008). *Thinking in Systems: A Primer*. Chelsea Green Publishing.

Newnam, S., Goode, N., Salmon, P. M., & Stevenson, M. (2017). Reforming the road freight transportation system? An investigation of Coronial inquests in Australia. *Accident Analysis and Prevention*, 101, 28–36.

Page, S. (2016). Many model thinking. *IEE, Winter Simulation Conference (WSC)*, 11–14 December, Washington, DC, USA.

Page, S. (2017). Many model thinking. Keynote address, *Computational Social Sciences of the Americas Annual Conference*, 19–22 October, Santa Fe, USA.

Read, G. J. M., Salmon, P. M., Lenne, M. G., & Goode, N. A. (2018). A sociotechnical design toolkit for bridging the gap between systems-based analyses and system design. *Human Factors and Ergonomics in Manufacturing and Service Industries*, 28(6), 327–341.

Read, G. J. M., Beanland, V., Lenne, M. G., Stanton, N. A., & Salmon, P. M. (2017). *Integrating Human Factors Methods and Systems Thinking for Transport Analysis and Design*. Boca Raton, FL: CRC Press.

Salmon, P. M., Lenne, M. G., Walker, G. H., Stanton, N. A., & Filtness, A. (2014). Using the Event Analysis of Systemic Teamwork (EAST) to explore conflicts between different road user groups when making right hand turns at urban intersections. *Ergonomics*, 57(11), 1628–1642.

Salmon, P. M. & Read, G. J. M. (2019). Many-model thinking in systems ergonomics: A case study in road safety. *Ergonomics*, 62(5), 612–628.

Salmon, P. M., Read, G. J. M., & Stevens, N. (2016). Who is in control of road safety? A STAMP control structure analysis of the road transport system in Queensland, Australia. *Accident Analysis and Prevention*, 96, 140–151.

Salmon, P. M., Read, G. J. M., Stevens, N. A., Walker, G. H., Beanland, V., McClure, R., Hughes, B.., Johnston, I., & Stanton, N. A. (2019a). Using the abstraction hierarchy to identify how the purpose and structure of road transport systems contributes to road trauma. *Transportation Research Interdisciplinary Perspectives*, 3, 100067.

Salmon, P. M., Read, G. J. M., Thompson, J., McLean, S., & McClure, R. (2020). Computational modelling and systems ergonomics: A system dynamics model of drink driving-related trauma prevention. *Ergonomics*, 63(8), 956–980.

Salmon, P. M., Stanton, N. A., Read, G. J. M., Stevens, N., Walker, G. H., & Hancock, P. (2019b). From systems ergonomics to global ergonomics: the world as a socio-ecological-technical system. *Chartered Institute of Ergonomics and Human Factors Annual Meeting*, April, UK.

Salmon, P. M., Stevens, N. J., McLean, S., Hulme, A., & Read, G. J. M. (2021). Human Factors and Ergonomics and the management of existential threats: a work domain analysis of a COVID-19 return from lockdown restrictions system. *Human Factors and Ergonomics in the Manufacturing and Service Industries*, 31(4), 412–424.

Thatcher, A., Nayak, R., & Waterson, P. (2020). Human factors and ergonomics systems-based tools for understanding and addressing global problems of the 21st century. *Ergonomics*, 63, 367–387.

Thatcher, T., Waterson, P., Todd, A., & Moray, N. (2018). State of Science: Ergonomics and global issues, *Ergonomics*, 61(2), 197–213.

World Economic Forum. (2021). The global risks report 2021. https://www.weforum.org/reports/the-global-risks-report-2020

Index

Note: Page numbers in *Italics* refer to figures; **bold** refer to tables.

D

E

F

G

H